GUIDED EXPLORATIONS OF THE MECHANICS OF SOLIDS AND STRUCTURES

Strategies for Solving Unfamiliar Problems

This book tackles the question: How can an engineer with a powerful finite element program but modest background knowledge of mechanics solve unfamiliar problems?

Engineering educators will find this book to be a new and exciting approach to helping students engage with complex ideas. Practicing engineers who use finite element methods to solve problems in solids and structures will extend the range of problems they can solve as well as accelerate their learning on new problems.

This book's special strengths include

- A thoroughly modern approach to learning and understanding mechanics problems
- Comprehensive coverage of a large collection of problems ranging from static to dynamic and from linear to nonlinear, applied to a variety of structures and components
- Accompanying software that is sophisticated and versatile and is available for free on the accompanying CD and from the book's Web site (www.cambridge.org/doyle)
- Ability to complement any standard finite element textbook or course in solid and structural mechanics
- The use of simple models to understand complex problems

James F. Doyle is a professor in the School of Aeronautics and Astronautics at Purdue University. His main area of research is in experimental mechanics with an emphasis on the development of a new methodology for analyzing impact and wave propagation in complicated structures. The goal is to be able to extract the complete description of the wave and the dynamic system from limited experimental data. Special emphasis is placed on solving inverse problems by integrating experimental methods with computational methods (primarily finite element–based methods). He is a dedicated teacher and pedagogical innovator. He is the winner of the Frocht Award for Teaching and the Hetenyi Award for Research, both from the Society for Experimental Mechanics. Professor Doyle is a Fellow of the Society for Experimental Mechanics. This is his fifth book, after *Propagation in Structures*, Second Edition; *Static and Dynamic Analysis of Structures*; *Nonlinear Analysis of Thin-Walled Structures*; and *Modern Experimental Stress Analysis*.

Cambridge Aerospace Series

Editors: Wei Shyy and Michael J. Rycroft

Guided Explorations of the Mechanics of Solids and Structures

STRATEGIES FOR SOLVING UNFAMILIAR PROBLEMS

James F. Doyle

Purdue University

CAMBRIDGE UNIVERSITY PRESS
Cambridge, New York, Melbourne, Madrid, Cape Town, Singapore, São Paulo, Delhi

Cambridge University Press
32 Avenue of the Americas, New York, NY 10013-2473, USA

www.cambridge.org
Information on this title: www.cambridge.org/9780521896788

First published 2009

Printed in the United States of America

A catalog record for this publication is available from the British Library.

Library of Congress Cataloging in Publication data

Doyle, James F., 1951–
Guided explorations of the mechanics of solids and structures : strategies for
solving unfamiliar problems / James F. Doyle.
    p.   cm. – (Cambridge aerospace series ; 26)
Includes bibliographical references and index.
ISBN 978-0-521-89678-8 (hardback)
1. Structural engineering – Problems, exercises, etc.   2. Structural engineering –
Mathematics.   3. Engineering mathematics.   4. Finite element method.
5. Geometry, Solid.   I. Title.   II. Series.
TA640.4.D69   2009
624.1076 – dc22   2009002801

ISBN   978-0-521-89678-8 hardback

This book is dedicated to ALL
who have shared the ARC
at the LBC.

Contents

GUIDED EXPLORATIONS OF THE MECHANICS OF SOLIDS AND STRUCTURES

Introduction

> Thus, for a complete investigation of dynamical systems, we require not only a computer and the direct integration methods. These provide no more than an ideal computer laboratory in which an arbitrary number of experiments can be performed, yielding an immense data flow. We require, in addition, certain principles according to which the data may be evaluated and displayed, thus giving an insight into the astonishing variety of responses of dynamical systems.
>
> J. Argyris and H.-P. Mlejnek [3]

Suppose the existence of a very powerful computer, so powerful that it can execute any given command such as *build* a cantilever beam, *excite* the beam with this force history, *record* the velocity histories, and the like. Suppose, further, that it cannot answer questions such as *what* is elasticity? *why* is resonance relevant to vibrations? *how* are vibration and stiffness related? Here then is the interesting question: With the aid of this powerful computer, how long would it take a novice to discover the law governing the vibration of structures? The answer, it would seem, is never, unless the novice is a modern Galileo.

Now suppose we add a feature to the computer; namely, access to a vast bibliographic database (in the spirit of Google or Wikipedia) that can respond to such library search commands as *find* every reference to vibrations, *sort* the find according to the type of structure, *report* only those citations that combine experiment with analysis, and so on. With this database the process of discovery will be much more rapid.

How can an engineer with a powerful finite element (FE) program and modicum of background solve an unfamiliar problem? More specifically, how can this engineer construct a sequence of commands to the program that will eventually lead to a solution? The current generation of FE programs is like the preceding powerful computer but without the reference library – they can solve (almost) any problem posed to them, but they cannot pose the question itself. That is the role, the very difficult role, of the engineer. The role of the teacher is akin to the reference library, the teacher gives the engineer access to the ideas, insights, and methods of previous generations of engineers. Thus the ideal situation for a novice engineer would be to have a powerful computer, a huge library on-line, and a personal professor always ready to answer questions and suggest solutions. Of course, this is not realistic, which brings us to the purpose of this book. This book introduces a set of strategies and tools whereby a novice can leverage a given knowledge base to solve an unfamiliar problem. These strategies are elucidated within a computer simulation program called QED, which is also introduced as part of this book.

Computer Simulation and Accelerated Experiences

Consider experimental studies, for example, as a means to understanding an unfamiliar problem. An experiment by its nature is a single realization – a single geometry, material, or load case; multiple test cases and examples are just not economically feasible. The same can be said of classroom examples if the time devoted to them is to remain reasonable. But engineers, being introduced to something new, something unfamiliar, need to see other examples as well as variations on the given examples. For instance, in the stress analysis of a symmetrically notched specimen, these are some logical questions to ask:

- What if the notches are bigger or smaller?
- What if there is one instead of two notches?
- What if the notches are closer or farther apart?
- What if the material is changed?
- What if instead of a notch there is a crack?
- What if the clamped boundary has some elasticity?

These questions are too cumbersome and expensive to answer experimentally but are very appropriate for a computer simulation program.

As computer modeling became easier to use and faster to run, this opened up the possibility of understanding problems through parameter variation and visualizing the results. Whereas computer programs once sufficed to provide numbers – discrete solutions of "stress at a point" and the like – now complete simulations, which present global behavior, trends, and patterns, can be presented and sensitivity studies analyzing the relative importance of geometric and material parameters evaluated. Being able to observe phenomena and zoom in on significant parameters, making judgments concerning those that are significant and those that are not, greatly enhances the depth of understanding.

This understanding can come only through experiences, and here the simulation program can help considerably to accelerate the process of accumulating experiences.

Strategies for Solving Unfamiliar Problems

We take as given that the FE program can solve the problem in the sense that once it is properly posed the program can generate the requisite numbers. Our meaning of solution is not the generation of these numbers, but rather, understanding the formulation of the problem and understanding the significance of the generated numbers. But problems are not solved (in our sense) in a knowledge vacuum; the engineer must bring some background knowledge, understanding, and experience to bear on the problem. The objective here is to systematically leverage or migrate this background to the problem at hand. Furthermore, the FE program through the simulation program is to be part of this learning process.

Assume that a preliminary search on the Web has not yielded any useful hits; then six possible strategies for solving unfamiliar problems have been identified and are listed as follows.

I. Establish a basic knowledge base by duplicating a textbook example for the theory.
II. Perform a parameter sensitivity study.
III. Start with a known solution to a different (but somewhat related) problem and slowly morph the problem to the one of interest.
IV. Start with the given complex problem and remove features so as to identify those features that do and do not affect essential behaviors.
V. Identify separate features of the complex problem, make a model for each, and understand their behavior. Then construct a compound model that combines multiple features so as to understand their interaction.
VI. Construct a simple analytical model that contains the essential features of the problem.

The first strategy is the most basic in that there must be some knowledge base before the unfamiliar problem can be tackled. The power of duplicating a textbook example should not be underestimated because the very process of constructing the FE model often highlights aspects not apparent in the theoretical analysis. The second strategy is always available because it treats the problem as a black box; the objective is to identify the significant parameters from which a simpler reduced model can be constructed. Strategies III and IV are the converse of each other and share the idea of *morphing*, that is, slowly changing a given problem into another. Strategy V explores complexity in a problem. In a sense, the ability to construct a simple analytical model that has the essential features of a complex problem is testament to an understanding of the complex problem, and thus Strategy VI is often the capstone of an analysis.

A further word about *simple analytical models*. There was a time when simple models were the only available analytical tool to study complex problems. With the advent of computers, and especially the development of FE codes, almost any complex problem can now be tackled with as much sophistication as needed. However, when trying to understand a complex problem by using the preceding strategies, it becomes necessary to make many computer runs, and voluminous data can easily be generated that are quite difficult to assimilate. Ironically, this power and versatility of current FE programs have renewed the interest in constructing simple models – not as solutions per se but as organizational principles for seeing through the numbers produced by the codes. Another facet often overlooked is that simple models provide the language with which to describe things and therefore help the thinking process itself. That is, their solutions are more generic or archetypal and therefore more generally applicable.

Outline of the Book

QED is the simulation program written specifically to help implement the preceding strategies for solving problems in the mechanics of solids and structures. Chapter 1 gives an overview of QED; this is not a manual for its use, rather it tries to give an insight into the design and organization of the simulation environment and its connection to the mechanics formulation of problems. Chapter 2 considers some basic problems in the analysis of structures under static loads; the primary considerations are those of stiffness, stress, and equilibrium. Chapter 3 considers dynamic effects in the form of structural vibrations; this is complemented in Chapter 4 with the study of transient dynamic responses, in particular, wave-propagation responses. Chapter 5 introduces some aspects of the nonlinear behavior of structures; the examples discussed include large deflections, large strains, and nonlinear material behaviors. Chapter 6 refines the concept of equilibrium (both static and dynamic) while exploring stability of the equilibrium. The book concludes with Chapter 7, which discusses the construction of simple analytical models; in the process, this chapter also delves deeper into the basic concepts of mechanics introduced in Chapter 1.

The explorations are designed under the assumption that the explorer has little specific analysis background. Thus the first exploration in each chapter is introductory in nature and covers the basic concepts of the problem. As much as possible, these are made visual without overburdening analyses. The subsequent explorations in each chapter then delve deeper into the topics. Each exploration is given as much detail as possible to accomplish the data collection tasks without necessarily explaining why. Means of data analysis are given at the end of each exploration, often with a simple model for comparison. Partial results from the data analysis are presented for guidance – providing complete results would overly prejudice the sense of exploration and discovery.

The ostensible goal of the explorations is to introduce various topics in the mechanics of solids and structures; the deeper goal, however, is to develop the strategies for solving unfamiliar problems, and for this reason the range of topics is both broad and deep. That is, a narrow focus on topics (vibrations only, say) would not engender the sense of unfamiliarity that is sought; similarly, the inclusion of the concept of wave dispersion requires the use of important strategies and tools that would otherwise be overlooked.

The explorations collect data, observe trends, foster experiences, and so on, but they do not explain why. Chapter 7, with the development of simple analytical models, in part, tries to explain why, but obviously must be limited in scope. Each exploration is therefore accompanied by some pertinent further readings that can be used to extend the knowledge base and motivate variations on the explorations. Purposely, the QED explorations are like physical experiments, and so many of these readings refer to physical experimental studies.

1 QED the Computer Laboratory

Structures are to be found in various shapes and sizes with various purposes and uses. These range from the human-made structures of bridges carrying traffic, buildings housing offices, airplanes carrying passengers, all the way down to the biologically made structures of cells and proteins carrying genetic information. Figure 1.1 shows some examples of human-made structures. Structural mechanics is concerned with the behavior of structures under the action of applied loads – their deformations and internal loads.

The primary function of any structure is to support and transfer externally applied loads. It is the task of structural analysis to determine two main quantities arising as the structure performs its role: internal loads (called *stresses*) and changes of shape (called *deformations*). It is necessary to determine the first in order to know whether the structure is capable of withstanding the applied loads because all materials can withstand only a finite level of stress. The second must be determined to ensure that excessive displacements do not occur – a building, blowing in the wind like a tree, would be very uncomfortable indeed even if it supported the loads and did not collapse.

Modern structural analysis is highly computer oriented. This book takes advantage of that to present QED, which is a learning environment that is simple to use but rich in depth. The QED program is a visual simulation tool for analysis. Its intent is to provide an interactive simulation environment for understanding a variety of problems in solid and structural mechanics. This chapter gives an overview of QED and the underlying mechanics and programs. The subsequent chapters give detailed instructions on using QED to solve mechanics problems; here we concentrate on installing QED and providing a big picture overview of the program and its environment. To put all this in perspective, the first section begins with an overview of structural mechanics.

The design of QED is such that it isolates the user from having to cope with the full flexibility of the underlying enabling programs. Although it is not necessary to understand these programs in order to run QED, a working knowledge of them is helpful in explaining some of the choices available in QED, and therefore a section is devoted to summarizing the capabilities of these programs.

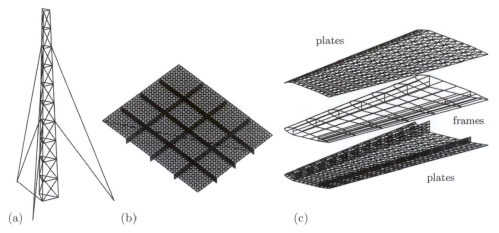

Figure 1.1. Some types of structures: (a) skeletal space truss with cable stays, (b) reinforced shell panel, (c) exploded view of an aircraft wing section.

The explorations in the following chapters utilize a variety of geometries, a variety of analyses, embedded in a variety of learning strategies. This chapter ends with two charts presenting all the explorations arranged according to these aspects.

1.1 Brief Overview of the Mechanics of Structures

This section briefly introduces the primary concepts in the mechanics of solids and structures. The intention is to identify and highlight the main components necessary in the solution of a general structures problem.

Types of Structures

Structures that can be satisfactorily idealized as line members are called *frame* or *skeletal* structures. An example is shown in Figure 1.1(a). Usually the members are assumed to be connected either by frictionless pins or by rigid joints. The members of a frame can support bending (in any direction) as well as axial and torsional loads.

Folded-plate and *shell* structures look very different from their frame counterparts because they are made of two-dimensional (2D) distributed material. For example, a *plate* is an extended continuous body in which one dimension is substantially smaller than the other two and is designed to carry both moments and transverse loads. Figure 1.1(b) shows an example of a plated structure. Shells also support in-plane loads, an action referred to as the *membrane* action.

Real structures are usually constructed as combinations of frame and shell members. Figure 1.1(c) shows an example of an aircraft wing comprising shell and frame members.

When a structural component does not fit into any of the previous special categories, we refer to it as a *solid*. All structural models are approximations of the

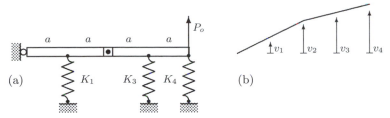

Figure 1.2. Simple structural model with external applied load P_o: (a) two linked rigid bars attached to springs, (b) compatibility of displacements.

three-dimensional (3D) solid; one of the preoccupations in this book is the assessment of the validity of the simpler structural models.

Structural Analyses

There are three fundamental concepts in the mechanics of solids: *deformation and compatibility*, *stress and equilibrium*, and *material behavior*. Each of these is elaborated on here and then made explicit through analyzing the relatively simple structure of Figure 1.2. This structure consists of horizontal rigid bars pinned at the left end and the center and attached to vertical springs.

Even this relatively simple structure has many unknowns; for example, there are the indicated displacements and the forces in the springs and attachment points. Additional steps are therefore required for effecting a solution, and these too are demonstrated.

I. Deformation and Compatibility
Loaded structures undergo deformations (or small changes in shape) and, as a consequence, material points within the structure are displaced to new positions. The principle of *compatibility* is assumed to apply to all of the structures encountered in this book. That is, it is assumed that, if a joint of a structure moves, then the ends of the members connected to that joint by the same amount, consistent with the nature of the connection. For the pin-jointed members of Figure 1.2, compatibility means that the ends of the members meeting at a joint undergo equal translation. For the rigid bar, it means that all points along the bar have the same rotation.

There are four indicated displacements in Figure 1.2(b); these are referred to as *degrees of freedom* or DoFs. We distinguish between *unconstrained* (i.e., unknown) and *constrained* (i.e., known in some way) DoFs. In the present case, the bars are rigid and therefore impose the constraint that the allowable displacements along each bar must form a straight line. Because the rigid bars are articulated, then only two independent displacements are required for describing the deflections. That is, the four indicated displacements are related to each other by

$$v_1 = \frac{a}{2a}v_2 = \tfrac{1}{2}v_2, \qquad v_3 = v_2 + \frac{a}{2a}[v_4 - v_2] = \tfrac{1}{2}v_2 + \tfrac{1}{2}v_4. \qquad (1.1)$$

Figure 1.3. Free bodies of the simple model: (a) left bar, (b) right bar.

We take v_2 and v_4 as the two (unconstrained) DoFs and write the other two as functions of these.

Perhaps needless to say, an analysis of the deformation alone does not provide a solution to a mechanics problem when loads are applied.

II. Stress and Equilibrium

Loads can be regarded as being either *internal* or *external*. External loads consist of applied forces and moments and the corresponding reactions. The applied loads usually have preset values, whereas the reactions assume values that will maintain the equilibrium of the structure. Internal forces are the forces generated within the structure in response to the applied loads. These internal forces give rise to internal stresses.

If the spring K_3 in Figure 1.2 is detached from the bar, then its effect is replaced with that of a force resisting the deflection. If the second bar is removed from the springs and the joint, then the effect of these too is replaced with forces, as shown in Figure 1.3(b). This is called a *free body* or a *free-body diagram*.

The free-body diagrams of the separated bars are shown in Figure 1.3. For each free body, the (vector) sum of the applied loads (both forces $\hat{F}$ and moments $\hat{M}$) must be zero. That is,

$$\sum \hat{F} = 0, \qquad \sum \hat{M} + \sum \hat{r} \times \hat{F} = 0.$$

Note that the assumed initial direction of the reactions is not important; however, at separated joints they must be equal and opposite, as shown in the figure for reaction R_2. The forces F_i are due to the deformation of the springs, and their sense is made consistent with the displacements v_i. Summing the forces vertically and the moments about the left end gives, respectively,

$$R_0 - F_1 + R_2 = 0, \qquad -F_1 a - R_2 2a = 0,$$
$$R_2 - F_3 - F_4 + P_o = 0, \qquad -F_3 a - F_4 2a + P_o 2a = 0. \tag{1.2}$$

This problem is *statically indeterminate* because equilibrium alone is insufficient for a solution (four equations, five unknowns: R_0, F_1, R_2, F_3, F_4). Generally, it should be expected that structural problems are statically indeterminate, and compatibility and material relations must be incorporated to effect a solution.

III. Material Behavior

For illustrative purposes, we restrict ourselves to materials that are *linearly elastic*. By elastic we mean that the stress–strain (or force–deflection) curve is the same for both loading and unloading. The restriction to linear behavior allows us to use the very important concept of superposition.

For our simple structure of Figure 1.2, the only components exhibiting elasticity are the springs. Linear springs have a force–deflection relation described by

$$P = Ku,$$

where P is the applied load, u is the relative deflection of the ends, and K is the spring stiffness. The force in each spring is given by

$$F_1 = K_1 v_1 = \tfrac{1}{2} K_1 v_2, \qquad F_3 = K_3 v_3 = \tfrac{1}{2} K_3 v_2 + \tfrac{1}{2} K_3 v_4, \qquad F_4 = K_4 v_4. \quad (1.3)$$

These are additional equations but in terms of the already identified unknowns. Hence we are now in a position to solve the problem.

IV. Solution Procedure
A total of nine unknowns were previously identified: four displacements, three spring forces, and two reactions. It is possible to set up nine simultaneous equations for the nine unknowns; however, it is more judicious to first identify the minimum set of unknowns, and this is generally done in terms of the unconstrained DoFs of the structure. As previously discussed, these have been identified as the displacements v_2 and v_4. Thus it should be possible to establish a minimum set of $[2 \times 2]$ simultaneous equations.

Get from the second equilibrium equation that $R_2 = -\tfrac{1}{2} F_1$ and combine with the first equilibrium equation to get $R_0 = +\tfrac{1}{2} F_1$. Substitute these into the remaining equilibrium equations and replace F_i in terms of the displacements and spring stiffnesses to get

$$\begin{bmatrix} \tfrac{1}{4} K_1 + \tfrac{1}{2} K_3 & \tfrac{1}{2} K_3 + K_4 \\ \tfrac{1}{2} K_3 a & \tfrac{1}{2} K_3 a + 2 K_4 a \end{bmatrix} \begin{Bmatrix} v_2 \\ v_4 \end{Bmatrix} = \begin{Bmatrix} P_o \\ 2a\, P_o \end{Bmatrix}. \quad (1.4)$$

Observe that this is a stiffness relation analogous to that of a simple spring, but this generalized spring has two DoFs. Furthermore, observe that this spring stiffness is a combination of the individual spring stiffnesses plus the geometry.

This equation can now be solved. For example, suppose $K_1 = K_3 = K_4 = K$; then

$$K \begin{bmatrix} \tfrac{3}{4} & \tfrac{3}{2} \\ \tfrac{1}{2} a & \tfrac{5}{2} a \end{bmatrix} \begin{Bmatrix} v_2 \\ v_4 \end{Bmatrix} = \begin{Bmatrix} P_o \\ 2a\, P_o \end{Bmatrix} \quad \Rightarrow \quad v_2 = -\frac{4 P_o}{9K}, \quad v_4 = +\frac{8 P_o}{9K}.$$

This solution indicates that the second bar rotates clockwise, forcing the first bar to deflect downward.

V. Postprocessing
The solution procedure solved for just the minimum set of unknowns, but of course other quantities such as the spring forces and reactions may also be of interest. In the

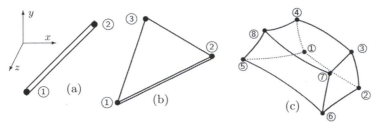

Figure 1.4. Some finite elements: (a) frame, (b) shell, (c) solid. The numbers in circles are the node numbers.

general structures problem, the stresses, strains, and so on may also be of interest. All of these are obtained as part of the postprocessing of the displacements.

It is clear, for example, how the other displacements are obtained from Equations (1.1) and the spring forces from Equations (1.3).

General Structural Analyses

Although the previous example is relatively simple, it nonetheless identified the major components of a structural analysis. The key stage is that of Equations (1.4), which is the minimum set of simultaneous equations to determine the unknowns of the problem.

We write this equation as

$$[K]\{u\} = \{P\},$$

where $[K]$ is the $[N \times N]$ *structural stiffness matrix*, $\{u\}$ is the $[N \times 1]$ *vector of degrees of freedom* used to describe the deformation of the structure, $\{P\}$ is the $[N \times 1]$ *vector of applied loads*, and N is the minimum number of DoFs. All (linear static) structural problems can be represented by this equation; the complexity of the problem in terms of geometry, materials, loads, and so on is reflected in the size of N.

A complex structure is conceived as a collection of components. In the finite element (FE) method, the smallest component is the element – Figure 1.4 shows three such elements. The process of assemblage to form $[K]$ is essentially the summing of the stiffnesses of the individual elements. Almost any general structure can be modeled conveniently and efficiently by a combination of the indicated elements.

Before we discuss these elements in a little more depth, a word about how the deformation of a general structure, modeled with FEs, is described in terms of its DoFs. The number of possible displacement components at each node is known as the *nodal DoF*; the nodal DoFs for different structural types are shown in Table 1.1. From this table, it is clear how the frame and shell structures share common types of DoFs. This is what allows them to be combined to form complex structures such as thin-walled shell structures reinforced with frame members.

Table 1.1. *The nodal DoFs for different structural types: u, v, w are translations and* ϕ_x, ϕ_y, ϕ_z *are rotations*

Structure	Dimension	u	v	w	ϕ_x	ϕ_y	ϕ_z	Global #
Rod	1D	✓						11
Beam	1D		✓				✓	12
Shaft	1D				✓			13
Truss	2D	✓	✓					21
Frame/membrane	2D	✓	✓				✓	22
Grille/plate	2D			✓	✓	✓		23
Truss	3D	✓	✓	✓				31
Frame/shell	3D	✓	✓	✓	✓	✓	✓	32
General structure	3D	✓	✓	✓	✓	✓	✓	33
Solid	3D	✓	✓	✓				111

I. Frame Element

A 3D frame element, as shown in Figure 1.4(a), has six DoFs at each node:

$$\{u\} = \{u,\ v,\ w,\ \phi_x,\ \phi_y\,\phi_z\}^T.$$

In local coordinates, this has three deformation behaviors. There is a rod action with axial displacement $u(x)$ and force $F(x)$ related by

$$\{u\} = \{u\}^T, \qquad F(x) = EA\frac{\partial u}{\partial x}.$$

There are two beam actions: the bending moment $M(x)$ and shear force $V(x)$. The corresponding nodal DoFs are the rotation $\phi_z(x)$ (or the slope of the deflection curve) and the transverse displacement $v(x)$. These are related by

$$\{u\} = \{v,\ \phi_z\}^T, \qquad M(x) = EI\frac{\partial^2 v}{\partial x^2}, \qquad V(x) = -EI\frac{\partial^3 v}{\partial x^3}.$$

There is also a bending about the y axis, giving rise to similar relationships in terms of w and ϕ_y. Finally, there is twisting $\phi_x(x)$ about the longitudinal axis that is related to the torque $T(x)$ by

$$\{u\} = \{\phi_x\}^T, \qquad T(x) = GJ\frac{\partial \phi_x}{\partial x}.$$

In these relations, EA, EI, and GJ are the axial, bending, and torsional stiffnesses, respectively; E and G are Young's and shear moduli, and A, I, and J are the area, moment of inertia, and polar moment of inertia, respectively. Full details on the matrix implementation for frame structures can be found in References [29, 34].

II. Shell Element

Flat-faceted three-noded triangular elements, such as shown in Figure 1.4(b), are suitable for modeling shells because they can be conveniently mapped to form irregular shapes. Three-noded triangular elements have six DoFs at each node:

$$\{u\} = \{u,\ v,\ w,\ \phi_x,\ \phi_y\,\phi_z\}^T.$$

In local coordinates, a loaded segment of a shell supports both in-plane (membrane) and out-of-plane (flexural) actions. The in-plane behavior is analogous to that of a plane 2D frame. Thus at each node there are three DoFs:

$$\{u\} = \{u,\ v,\ \phi_z\}^T.$$

A good element (introduced and documented in the paper by Bergan and Felippa [8]) correctly implements the drilling DoF (ϕ_z) and therefore makes it suit-able for 3D application. The material behavior is represented as a case of plane stress by

$$\begin{Bmatrix} \sigma_{xx} \\ \sigma_{yy} \\ \sigma_{xy} \end{Bmatrix} = \frac{E}{1-v^2} \begin{bmatrix} 1 & v & 0 \\ v & 1 & 0 \\ 0 & 0 & (1-v)/2 \end{bmatrix} \begin{Bmatrix} \epsilon_{xx} \\ \epsilon_{yy} \\ \gamma_{xy} \end{Bmatrix},$$

where v is Poisson's ratio related to the Young's and shear moduli by $v = E/2G - 1$.

The out-of-plane behavior is analogous to that of a plane 2D grille. That is, each node has the DoFs

$$\{u\} = \{w,\ \phi_x,\ \phi_y\}^T.$$

The rotations are related to the deflection by

$$\phi_x = \frac{\partial w}{\partial y}, \qquad \phi_y = -\frac{\partial w}{\partial x}.$$

A good element, now called the discrete Kirchhoff triangular (DKT) element, was first introduced by Stricklin, Haisler, Tisdale, and Gunderson in 1968 [118]. It has been widely researched and documented as being one of the more efficient flexural elements (see Reference [6]). The material behavior is represented as a case of plane stress.

III. Solid Element

A 3D solid is usually modeled with hexahedral elements, as shown in Figure 1.4(c). The Hex20 version has an additional 12 midedge nodes that allow the element to be curved, as indicated in the figure.

Each node has three DoFs:

$$\{u\} = \{u,\ v,\ w\}^T.$$

Inside the element there are six components of stress and strain, and these are re-lated through the material relation

$$\{\sigma\} = [D]\{\epsilon\},$$

where $[D]$ is a $[6 \times 6]$ matrix of material coefficients.

Formulations for its use are given in a number of texts, two of which are References [5, 26]. This element performs very well even when used to construct thin shells.

Structural Analysis Programs

It is clear that, for a general structures problem, it would be unrealistic to establish $[K]$ manually. Structural analysis programs and FE programs in particular automate this process for general systems. The user provides information about the geometry and so on, and the programs provide essentially three services: first, they form $[K]$; second, they solve the simultaneous equations; and third, they provide convenient postprocessing facilities.

The explorations in structural analyses of this book use a program called QED. It is FE based, but one of its special attributes is that it has a good collection of preparameterized models. That is, the input chores have been reduced to a minimum so that the user can quickly generate the model and immediately go to the analysis and postprocessing stages. The rest of the sections in this chapter gives details about QED and its organization; although valuable, these are not essential for its use. However, the next section, which details the instructions for installing and running QED, is absolutely essential.

1.2 Installing and Running QED

These are the instructions for copying QED from the CD, setting it up, and then running it. The target machines have WinNT or WinXP as their operating systems.

Copying and Installing QED

The basic plan is to create a working directory (or folder) and then copy the entire contents of the CD into that directory. It is simplest to do these operations using a COMMAND prompt window.

1. Open a COMMAND prompt window: If it is not on the *Start/Programs/Accessories* list, then click *Start/Run*, type CMD, and click *OK*.
2. Move to the root directory [C:] by typing
        cd\
3. Create a working directory off the root as
        mkdir qed
4. Move to the working directory:
        cd qed
5. Copy everything off the CD (assuming it is in drive D:):
        copy d:*.*
6. Run QED by typing
        qed
   The opening screen should appear and should look like Figure 1.5 but in color.
7. Quit immediately by [pressing q].

Figure 1.5. Opening screen for QED.

QED is now configured to know that all the executables are in the working directory

    `c:\qed` .

IMPORTANT note on file location! If you choose to copy the files to a directory other than `c:\qed`, use a text editor (e.g., Notepad but not Word) and edit the file <<`qed.cfg`>> to replace the line

        `c:\qed`

with

        `?:\???`   (new absolute address)

Save and exit. Make sure to do this while **QED** is not running.

    Every time **QED** is run, it should be launched from the command line in the working directory as done here; do not use Windows Explorer to launch it because **QED** will be confused as to where the working directory and associated files are located.

    If some colors do not appear or the opening screen was blank, then from

Control panel / Appearance and themes / Change screen resolution / Settings

set the color quality to medium (16 bit). On some machines it may be necessary to run **QED** in *compatibility mode*. To do this, launch Windows Explorer and right click on the **QED** program. Click

Properties / Compatibility / Run this program . . .

and choose Windows 95 from the drop-down menu.

The next introductory tour of QED covers additional aspects of configuring QED and the underlying programs. The CD contains a QED manual called *Qedman.pdf*, and this gives further details on configuring QED. In particular, it details how the size of the QED window and the size of fonts can be adjusted by editing the `<<qed.cfg>>` file.

Introductory Tour of QED

Launch QED by typing qed on the command line. The opening screen of Figure 1.5 shows that there are three main functions: Creating models, Analyzing problems, and Viewing the results. We briefly tour each of these.

All interaction with QED is through pressing the highlighted key or the leftmost symbol (number/letter) on the menus. This operation will be typeset as, for example. [press q] to quit. [Press c] to get the Create screen of Figure 1.7 in Section 1.3. This is the collection of preparameterized models. We will go deeper into one of these presently. [Press qa] to quit and get the Analyze screen of Figure 1.17 in Section 1.3. The analyses range from linear statics to nonlinear dynamics. [Press qv] to get the View screen of Figure 1.18 in Section 1.3. These are some of the ways that the results of the analyses can be postprocessed.

[Press qc] to return to the models section. We create a circular cylinder shell by using the preset parameters. [Press c] (for closed cross sections) to get a screen similar to Figure 1.15 in Section 1.3. Creating a model entails specifying the geometry, the boundary conditions, and the type of loads. With these specified, a FE mesh can then be rendered.

The instructions in the explorations are in the form of an enumerated list; we now begin by using that format.

1. [Press g] to get the geometry parameters menu. As is seen, the cylinder (which is [0:geometry=1(=circular)]) already has specified parameters. To change the length, for example, [press 1] to be given the input dialog box at the bottom. To change the length to 18 units, type 18 followed by a carriage return (or enter). The carriage return signals the end of the data input. Observe that the model is immediately updated. Instructions such as these will be typeset as

    [1:length =18]

    in the remainder of this tour as well as in the explorations of the following chapters.

2. [Press q] then [press 1] ("ell") to enter the Loads section. Place a vertical load at the end with

    [0:type =1],
    [3:Z_pos =18],
    [5:Y_load =100],

    all other loads and positions are zero. The cylinder axis is z, and this places the concentrated load at the node nearest to the coordinates $x = 0$, $y = 0$, $z = 18$.

3. [Press qb] to enter the boundary conditions (BCs) section. We accept the default BCs of completely fixed at the left end ([8:L_bc =000000]) and completely free at the right end ([9:L_bc =111111]). The six numbers correspond to the six DoFs (three translations and three rotations) at a general node in a 3D shell.

4. [Press qm] to enter the Mesh section. Mesh with <18> modules through the length and <32> modules around the hoop direction. That is,
[0:Global_ID =32],
[1:length mods =18],
[2:hoop mods =32],
[3:V/H bias =1],
[7:Z_rate =18];
all specials can be set as zero.
Render the mesh by [pressing s]. A small Command Prompt should pop up and show some scrolling. What happened is that QED launched the mesh-generating program GenMesh as a (temporary) child process.

5. The mesh should appear displayed in the right panel. The current view of the cylinder is from the end. For clearer viewing, set the orientation as $\phi_x = +25$, $\phi_y = +35$, $\phi_z = 0$. To do this, [press o] followed by [pressing X] and [pressing Y] multiple times. [Press q] to display the new orientation.
The surfaces of the shell are color coded and given tag numbers. If desired, the display of these surfaces can be toggled on/off (1/0) by [pressing t 4 0], for example. The current tag toggles are displayed at the upper left.

6. If no mesh is displayed, this means one of two possibilities: Either GenMesh did not run or GenMesh could not complete its task. On the command line, type
dir *.log
There should be a <<genmesh.log>> file. If there isn't, then GenMesh did not run. In that case, QED was not configured as instructed earlier. In particular, the location of the executables is not where QED expects to find them; check the comment after the exclamation **IMPORTANT note on file location**. If there is a <<genmesh.log>> file, then the script sent to GenMesh was incomplete. In all probability, the parameters were not set as previously instructed; review the parameter input instructions.

This concludes the introductory tour of QED. It is now possible to proceed to the explorations of the next five chapters. If a deeper understanding of QED is desired, then the next few sections should be read. If an even deeper understanding of the programs underlying QED is desired, then the manual *Qedman.pdf* should be read.

1.3 Overview of QED

A FE analysis can be divided into three separate stages. The *preprocessing* stage allows the model to be created; that is, the geometry is defined, the BCs imposed, the loads applied, and the mesh generated. The *processing* stage allows the model to be analyzed for its static and dynamic responses. The *postprocessing* stage allows the analysis results to be viewed and interrogated in a variety of ways; that is, contour

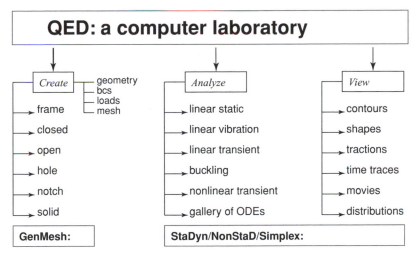

Figure 1.6. Overview of QED and some of its subfunctions. The underlying enabling programs are at the bottom.

plots, deformed shapes, time-history traces, and so on, can be chosen as appropriate. These three stages are referred to as *Create*, *Analyze*, and *View*, respectively, as displayed in the opening screen of Figure 1.5.

Figure 1.6 shows in more detail a schematic of the three functional parts of the program. Its design is such that it isolates the user from having to cope with the full flexibility of the underlying enabling programs and presents each problem in terms of a limited (but richly adaptable) number of choices and combinations. The design philosophy is to make each module very specific but flexible; this helps to fix focus on the significant aspects of behavior. Each problem has a template of properties; therefore consideration need be given only to what needs to be changed.

As shown, QED is an executive for running the underlying programs to achieve the required tasks. These programs can be run directly through their menus, but their operation under QED is by way of driver or *script* files. That is, QED creates the script files to execute GenMesh and StaDyn/NonStaD/Simplex. The name of the script files are

```
GenMesh:                    inmesh,  inmesh#,  insdf
StaDyn/NonStaD/Simplex:  instad,  inpost,  inps
```

These self-explanatory scripts can also be used as templates for running the underlying programs directly.

The next few subsections looks closer at each of the three stages of the FE analysis.

Creating Models

The parameterized models presented by QED and shown in Figure 1.7 are chosen to encompass a range of interesting shapes, loading patterns, and BCs. The models are

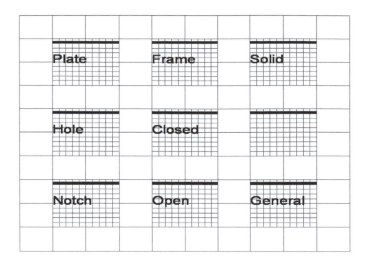

| QED ver 3.54, Oct 2008 | Main KEYs: | q s[nhbfcpw] a e s | C:\pctex\CLASS\546\tutor |

Figure 1.7. Collection of parameterized QED models.

separated into categories based on their geometric configurations. Within each module, parameters can be changed to "morph" the shape from one form to another. For instance, the "plate" module includes the option to define a layered beam or a cantilever curved beam; these share the common geometric characteristic of being simply connected and are thus classified in a single module.

The complete set of parameters for each model is stored in a file called <<---.cfg>>, for example, <<frame.cfg>> for frames. Thus the most recent version of the model is always available for modification. If the model is to be archived, then it is simply a matter of copying <<---.cfg>> to an appropriate new name and copying it back to <<---.cfg>> for reuse.

The distinctions drawn in the model building section are based solely on the shape and geometric characteristics of the model. Within a specific model, changes in the material properties, the applied loads, or the support conditions constitute variations on the standard theme defined by the geometry of the model.

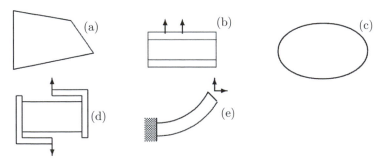

Figure 1.8. Plate geometries: (a) quadrilateral, (b) layered, (c) ellipse, (d) shear test, (e) curved beam.

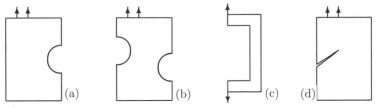

Figure 1.9. Notch geometries: (a) single semicircular, (b) two opposite semicircular, (c) rectangular, (d) edge crack.

I. Plate

This module, in its basic form, is a 2D quadrilateral shape. There are five variations on the standard plate, as shown in Figure 1.8. What each has in common is that they are simply connected regions and are modeled with 2D triangular elements.

Various BCs can be imposed along the plate's edges, including specifying a line of symmetry. Various load types can be applied and include a 3D load vector at a specific point and tractions along the edges.

II. Notch

The notch geometry group has 2D rectangular plates with edge notches or discontinuities. The four variations are shown in Figure 1.9. What they have in common is the high-stress gradients caused by the edge geometric discontinuity; a line crack is a very sharp notch. What they also have in common is that they are simply connected regions and are modeled with 2D triangular elements.

Various BCs can be imposed along the edges. Various load types can be applied and include a 3D load vector at a specific point and tractions along edges.

III. Hole

The hole geometry group has 2D rectangular plates with interior cutouts. The five variations are shown in Figure 1.10. What these have in common is that the regions are multiply connected and are modeled with 2D triangular elements. In addition, the geometry changes give rise to high-stress gradients.

Various boundary conditions can be imposed along the edges. Various load types can be applied and include a 3D load vector at a specific point and tractions along the edges.

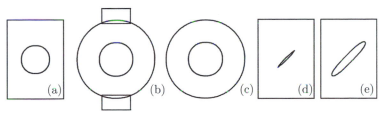

Figure 1.10. Hole geometries: (a) circular, (b) ring with tabs, (c) annulus, (d) crack, (e) elliptical.

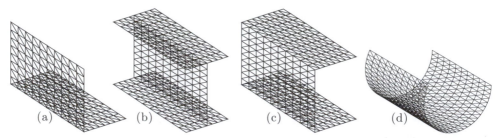

Figure 1.11. Open geometries: (a) angle section, (b) I-beam, (c) rectangular C-channel, (d) circular C-channel.

IV. Open

The open geometry group has 3D thin-walled open structures extruded in the z direction. The four variations are shown in Figure 1.11. What these have in common is that they are folded-plate structures that are modeled with 3D triangular shell elements. They are fully 3D, allowing local buckling of the flange and web crippling to be captured. These structures can have longitudinal and transverse frame reinforcements.

The BCs allow specifying behaviors for complete planes as well as for pivot points. The ends can have reinforcing plates. Loads include an arbitrary point load as well as end resultants.

V. Closed

The closed geometry group has 3D thin-walled enclosed structures extruded in the z direction. The six variations of cylinders are shown in Figure 1.12. What these have in common is that they are closed folded-plate structures and are modeled with 3D triangular shell elements. They are fully 3D, allowing local buckling and

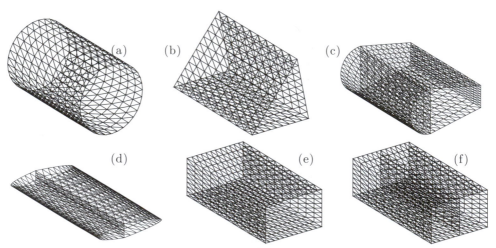

Figure 1.12. Closed geometries: (a) circular, (b) three-stringer box beam, (c) wing section, (d) airfoil section, (e) box beam, (f) two-cell box beam.

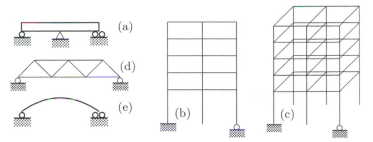

Figure 1.13. Frame geometries: (a) multispan beam, (b) multibay multifloor 2D frame, (c) multibay multifloor space frame, (d) bridge, (e) arch.

crimpling to be captured. These shells can have longitudinal and transverse frame reinforcements.

The BCs allow specifying behaviors for complete planes as well as for pivot points. The ends can have reinforcing plates. Loads include an arbitrary point load as well as end resultants.

VI. Frame

The frame geometry group has 3D multibay multifloor structures. The five variations are shown in Figure 1.13. A variant of (a) has an elastic support and a variant of (c) is a triangulated space truss. What they have in common is that they are skeletal structures modeled with 3D frame elements.

Arbitrary boundary conditions can be imposed at the base. Point and distributed loads can be applied.

VII. Solid

The solid geometry group has 3D objects modeled with Hex20 brick elements. The four variations are shown in Figure 1.14. What these have in common is that they are fully 3D continua and can be modeled with nonlinear elastic, rubberlike elastic, and elastic-plastic constitutive relations.

Arbitrary BCs can be imposed on the edges as well as specify a plane of symmetry. Point and distributed loads can be applied.

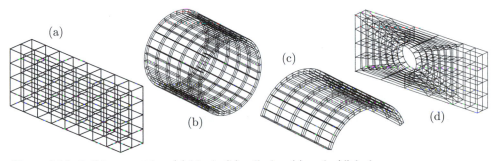

Figure 1.14. Solid geometries: (a) block, (b) cylinder, (c) arch, (d) hole.

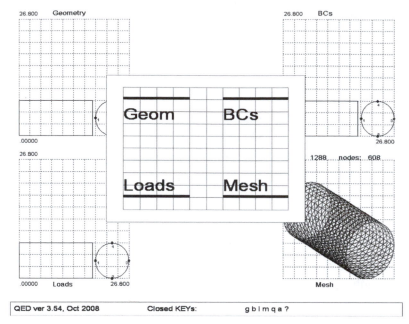

Figure 1.15. The creation of each model entails specifying the geometry, specifying the BCs, specifying the applied loads, and rendering the FE mesh.

VIII. General

The general geometry group allows construction of models that have not already been parameterized. The basic shapes available for change are four-sided plane, arbitrary plane, arbitrary 3D thin-wall z extrusion, and arbitrary 3D frame. The input paradigm is that of forms, which makes specifying the parameters straightforward. The BCs, loads, and material properties are specified via the structure data file (SDF) form.

The general geometry group also allows merging of meshes, thus facilitating the construction of complex models.

Structure Data File

The complete description of the structure and its material properties is kept in a separate file, referred to as the *Structure Data File* (SDF). The file is called <<qed.sdf>>. This is generated from the mesh file <<qed.msh>> by adding information about the material properties, BCs, and loads (see Figure 1.15).

I. Boundary Conditions

A 3D frame member or shell element has six DoFs at each node:

$$\{u\} = \{u, \; v, \; w; \; \phi_x, \; \phi_y \, \phi_z\}^T,$$

as indicated earlier. The default for a DoF is that it is on; thus the BCs are used to fixed or turn off the DoF. This is done by specifying the six-character string for each boundary node. A pinned condition for a 3D frame, for example, is specified as

000111

A 3D solid element has just the three DoFs, $\{u\} = \{u, v, w\}^T$.

II. Loads

A 3D frame member or shell element has six components of applied load at each node:

$$\{P\} = \{P_x, P_y, P_z; T_x, T_y, T_z\}^T.$$

When tractions (force per unit area) are specified, they are converted to nodal loads. A 3D solid element has just the three forces $\{P\} = \{P_x, P_y, P_z\}^T$.

Each geometry allows the imposition of a point load specified at an (x, y, z) location; in this case, the nearest node will be automatically located. Some geometries allow the specification of special loads appropriate for that geometry.

III. Material Properties and Units

The algorithms and source code used do not have a built-in system of units and do not utilize any dimensional conversion constants. Therefore any consistent system of units may be used for input, and the corresponding calculated results will be in the same units. For example, if the Young's modulus is specified as 10000 ksi (kilo-pounds per square inch), the reported values of force will be in kips.

The material properties are kept in a file named <<qed.mat>>. The format for this file is

```
MAXMAT                ::number of materials
Mat #1
E      G    RHO        ::Young's mod, shear mod, density
Mat #2
E      G    RHO        ::Young's mod, shear mod, density
:      :    :
:      :    :
end
```

where RHO is the mass density ρ, input as W/g in common units. Editing this file is the mechanism for specifying arbitrary material properties. However, this file should not be edited while QED is running, otherwise QED will overwrite the changes on exiting.

Of course, to make the explorations physically meaningful, the material and dimensional quantities should also be meaningful; generally, the numbers given correspond to the common set of units [pound (lb), inch (in), second (s)]. The

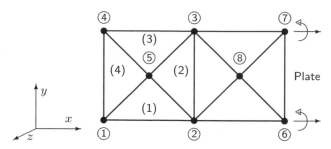

Figure 1.16. Simple plate structure modeled with eight triangular elements. Node numbers are in circles, element numbers are in parentheses.

fundamental units of the SI system are [meter (m), kilogram (kg), second (s)]. The conversion factors between the two systems are

1 m = 39.37 in,   1 in = 25.4 mm,
1 pound-mass = 453.59 g,   1 kg = 2.2 pound-mass.

The initial geometry for each explored model is generally given in both units.

IV. Example Structure DataFile
The following example SDF is for the simple in-plane plate structure shown in Figure 1.16.

```
8-element plate                            ::header
22
1 1 1 1
end
8                                          ::element connectivity
1    4    1    2    5
2    4    2    3    5
3    4    3    4    5
4    4    4    1    5
5    4    2    6    8
6    4    6    7    8
7    4    7    3    8
8    4    3    2    8
end
1                                          ::material #s
1   8   1
end
8                                          ::node coords
1       0.00      0.0    0.0
2       2.00      0.0    0.0
```

```
3      2.00        2.0      0.0
4      0.00        2.0      0.0
5      1.00        1.0      0.0
6      4.00        0.0      0.0
7      4.00        2.0      0.0
8      3.00        1.0      0.0
end
4                                            ::boundary condns
1     0    0    0    0  0  0
2     1    0    0    0  0  0
3     1    0    0    0  0  0
4     0    1    0    0  0  0
end
2                                            ::applied loads
6     1000.0       0.0     0.0 0.0 0.0 -500.0      0.0
7     1000.0       0.0     0.0 0.0 0.0  500.0      0.0
end
1                                            ::material props
1     10.0e6 4.0e6 1.00      2.50e-4  1     1 1.5 0.5
end
0                                            ::specials
end
```

The QED manual details the format for SDFs. Observe that the data are arranged in groups associated with the main information types. The data file is in free format with blanks or commas used as separators; thus any text editor can be used to make changes to the file if desired.

For complicated structures, a need may arise for visual verification of the input data; this is provided by PlotMesh. This is not a graphics environment for data input but a simple means of visually verifying the structure input data file. It can be used to view the same file used by StaDyn/NonStaD and Simplex, and the intermediate mesh files generated by GenMesh. To view a mesh, type `plotmesh` at the command line and [press 0] to change the mesh file name.

Analyzing the Problem

Generally, the same collection of analysis procedures can be applied irrespective of how the model was created. In this subsection, the seven types of analyses available as shown in the screen capture of Figure 1.17 are briefly described.

The complete set of parameters for each analysis and algorithm is stored in a file called <<analysis.cfg>>. If the analysis is to be archived, then it is simply a matter of copying this file to an appropriate new name and copying it back to <<analysis.cfg>> for reuse.

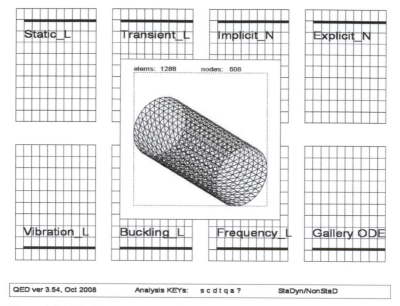

Figure 1.17. Various ways to analyze the problem.

I. Static

This is the most basic type of problem. It solves the linear system

$$[K_E]\{u\} = \chi_1\{P\} + \chi_2\{G\},$$

where $[K_E]$ is the structural *elastic* stiffness matrix, $\{P\}$ is the vector of applied loads, $\{G\}$ is the vector of gravity loads, and $\{u\}$ is the vector of unknown nodal DoFs. The distribution of applied loads $\{P\}$ is specified in the SDF, and gravity effects are specified through the specials group. Both loads can be scaled through the parameters χ_1 and χ_2.

The effect of prestress is estimated by solving the problem

$$[[K_E] + \chi[K_G]]\{u\} = \{\delta P\},$$

where $[K_G]$ is the *geometric* stiffness matrix estimated from the applied loads specified in the SDF. In this case, the applied load $\{\delta P\}$ is a point load specified as part of this type of analysis. The amount of prestress can be scaled through the parameter χ.

II. Buckling

This linear analysis solves the eigenvalue problem

$$[[K_E] + \lambda[K_G(P)]]\{\hat{u}\} = 0.$$

The geometric stiffness matrix $[K_G]$ is estimated from the applied loads specified in the SDF, and λ is the unknown eigenvalue. Either subspace iteration or Jacobi rotations can be used as the solution algorithm. The number of modes included in the output is selectable.

The value of λ reported is to be interpreted as a scaling factor applied to the actual load distribution specified in the SDF. Thus a value of less than unity would mean that the currently applied load system will cause a buckling failure. A negative value would mean that the current loads will not cause failure; sometimes it may be interpreted as meaning that buckling will occur if all the applied loads are reversed.

III. Vibration

This linear analysis solves the eigenvalue problem

$$[[K_E] + \chi[K_G] - \lambda\lceil M \rfloor]\{\hat{u}\} = 0, \qquad \lambda \equiv \omega^2,$$

where $\lceil M \rfloor$ is the structural *mass* matrix, and ω is the vibration frequency in radians per second. The geometric stiffness matrix $[K_G]$ is estimated from the applied loads specified in the SDF, and the amount of prestress can be scaled through the parameter χ. The eigenvalues reported are those of the frequency; an imaginary value will indicate an instability. Either subspace iteration or Jacobi rotations can be used as the solution algorithm.

IV. Transient

This linear analysis uses the unconditionally stable implicit integration scheme of Newmark to solve the system of equations

$$\lceil M \rfloor\{\ddot{u}\} + \lceil C \rfloor\{\dot{u}\} + [K_E]\{u\} = \chi_1\{P\} f_1(t) + \chi_2\{G\} f_2(t),$$

where $\lceil C \rfloor$ is the structural *damping* matrix. The number of time steps and time increments is selectable. A history of the loadings, $f_i(t)$, must be supplied in the form of a time–load file, and this acts as a time scaling on the applied load distribution $\{P\}$. The time history will be interpolated as needed.

The effect of prestress is estimated by solving the problem

$$\lceil M \rfloor\{\ddot{u}\} + \lceil C \rfloor\{\dot{u}\} + [[K_E] + \chi[K_G]]\{u\} = \{\delta P\} f(t).$$

The geometric stiffness matrix $[K_G]$ is estimated from the applied loads (treated as static and preexisting) specified in the SDF. The amount of prestress can be scaled through the parameter χ. In this case, the applied load $\{\delta P\}$ is a point load specified as part of this type of analysis and can vary in time.

V. Forced Frequency

This solves the linear system

$$[[K_E] + i\omega\lceil C \rfloor - \omega^2\lceil M \rfloor]\{\hat{u}\} = \{\hat{P}\}.$$

This is solved over a range of frequencies (ω) to get the frequency-response function (FRF). The frequency increment and number of increments can be specified. Multiple loads can be applied but they will all be in-phase.

The effect of prestress is estimated by solving the problem

$$\left[[K_E] + \chi[K_G] + i\omega\lceil C \rfloor - \omega^2\lceil M \rfloor\right]\{\hat{u}\} = \{\delta\hat{P}\}.$$

The geometric stiffness matrix $[K_G]$ is estimated from the applied loads (treated as static and preexisting) specified in the SDF. The amount of prestress can be scaled through the parameter χ. In this case, the applied load $\{\delta P\}$ is a point load specified as part of this type of analysis.

VI. Nonlinear Implicit

This nonlinear analysis uses the implicit integration scheme of Newmark with Newton–Raphson iterations to solve the system of equations

$$\lceil M \rfloor \{\ddot{u}\} + \lceil C \rfloor \{\dot{u}\} = \chi_1\{P\} f_1(t) + \chi_2\{G\} f_2(t) - \{F\},$$

where $\{F\}$ is the vector of element nodal forces. The number of time steps and time increments is selectable. A history of the loadings, $f_i(t)$, must be supplied in the form of a three-column time–load file, and this acts as a time scaling on the applied loadings. The time history will be interpolated as needed. A vibration eigenanalysis can be performed at selected intervals during the nonlinear deformation.

This analysis (in conjunction with the Solid models) supports three nonlinear material types: elastic, rubber (nearly incompressible) elastic, and elastic-plastic.

VII. Nonlinear Explicit

This nonlinear analysis uses the conditionally stable central-difference explicit integration scheme to solve the system of equations

$$\lceil M \rfloor \{\ddot{u}\} + \lceil C \rfloor \{\dot{u}\} = \chi_1\{P\} f_1(t) + \chi_2\{G\} f_2(t) - \{F\}.$$

The number of time steps and time increments is selectable. A history of the loadings, $f_i(t)$, must be supplied in the form of a three-column time–load file, and this acts as a time scaling on the applied loadings. The time history will be interpolated as needed.

VIII. Gallery of Nonlinear Ordinary Differential Equations (ODEs)

This analysis section is a collection of nonlinear equations in parameterized form. These are 1DoF or 2DoF systems and are detailed in Section 7.7. The Runge–Kutta scheme is used for the time integration. The arrangement is a condensed version of the format for QED itself.

Viewing the Results

Generally, the same collection of postprocessing views can be applied irrespective of how the model was created or how it was analyzed. Of course, not all postprocessing facilities are appropriate for all analyses; for example, time traces would have little meaning for a static analysis. Similarly, contours have no meaning for a 3D solid unless it is sectioned to expose a plane. Figure 1.18 shows the four choices available.

The complete set of parameters for each postanalysis is also stored in the file <<analysis.cfg>>. If the postanalysis is to be archived, then it is simply a matter of

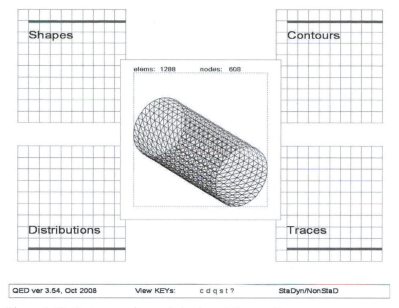

Figure 1.18. Postprocessing and viewing of the results.

copying this file to an appropriate new name and copying it back to `<<analysis.cfg>>` for reuse.

I. Deformed Shapes

The deformed shape is the most basic type of information provided by a FE analysis. The overall behavior of the model is seen at a glance, and it provides a fast and reliable means of judging whether the loads were applied correctly and the effect of the boundary conditions. A vector version of the plot superposes the vector displacement to give a sense of deformed motion.

A sequence of deformed shapes can be made into a *movie*. Vibration and buckled mode shapes can be animated by use of a time sinusoidal function.

II. Contours

Contours of displacement, strain, and stress for plated shells are displayed in a six-window layout, that is,

$$u, v, w;\ \phi_x, \phi_y, \phi_z,$$

$$\epsilon_{xx}, \epsilon_{yy}, 2\epsilon_{xy};\ \kappa_{xx}, \kappa_{yy}, 2\kappa_{xy},$$

$$\sigma_{xx}, \sigma_{yy}, \sigma_{xy};\ M_{xx}, M_{yy}, M_{xy}.$$

For each of these results, all six components are presented simultaneously for comparison. Closer examination of each can then be requested by filling the screen with a single component's plot. For eigenvalue and transient problems, contours can be displayed for each mode or snapshot.

There are also special contours corresponding to photoelastic isochromatic and isoclinic fringes, energy contours, Tresca and von Mises yield criteria, to name a few.

III. Traction Distributions

Two variations of traction distributions are given, one for the in-plane 2D models and one for the general 3D frame structures.

The 2D version is very effective for viewing what is happening within the body. Free-body cuts at any inclination can be sliced through the model to expose the tractions along the exposed boundary. The 3D version is very effective for showing the shear-force and bending-moment distributions in framed structures.

IV. Traces

In some cases, the most appropriate form of presentation of results is as a set of time or pseudo-time traces. For example, the displacement or stress history at a point can be plotted against time for wave-propagation and load incremental problems. This is also relevant to the FRFs that are plotted against frequency.

DiSPtool, the digital signal processing tool, is the stand-alone utility for interrogating traces. At its basic level, it can conveniently plot multiple-response columns against time. But it can also do transforms to and from the frequency domain as well as filter signals in various ways.

1.4 Supporting Programs

QED runs on top of the mesh-generating and analysis programs. It is not necessary to understand these programs in order to run QED, but a working knowledge of them will help explain some of the structure of QED and some of the choices presented in it. The specific operation of these programs is detailed in the manual *Qedman.pdf* on the CD. Their operation is transparent under QED except for the small command prompt window that pops up.

Mesh Generating with GenMesh

When the structural problem becomes large, it is essential to have an automatic scheme for the generation of an input data file; this not only removes the drudgery of making the file, but helps to ensure the model's integrity. The purpose of the program GenMesh is to create SDFs for use by StaDyn/NonStaD/Simplex.

The main capabilities of GenMesh involve generating meshes and performing the following executive functions:

- Creating 2D arbitrary shapes
- Creating 3D frame and shell structures and Solids
- Creating 3D solid objects
- Remapping a given mesh
- Merging multiple meshes
- Adding loads and BCs to a given mesh

The program can be menu driven, but its operation under QED is by way of driver or script files. That is, QED creates the script files and then executes GenMesh as a child process. The names of the script files are <<inmesh>> for the mesh generation and <<insdf>> for adding material properties, BCs, and loads.

Analyses with StaDyn/NonStaD and Simplex

StaDyn/NonStaD is the FE analysis program for frames and thin-walled shells; Simplex is the FE analysis program for 3D solids.

The program **StaDyn** is capable of doing the linear static and dynamic analyses of 3D structures formed from a combination of frame members and folded plates; **NonStaD** is the nonlinear version of **StaDyn**. Their main capabilities involve determining the member displacements and loads and structural reactions for

- Static loading
- Stability eigenanalysis
- Vibration eigenanalysis
- Forced-frequency loading
- Applied transient load history

The programs can be menu driven, but their operation under QED is by way of script files. That is, QED creates the script files to execute them as child processes. The name of the script file for analysis is <<instad>> and <<inpost>> for postanalysis.

The program **Simplex** is capable of doing the linear–nonlinear static–dynamic analyses of 3D solid models. It can handle nonlinear constitutive relations in the form of nonlinear elasticity, rubber elasticity, and plasticity.

As these programs proceed, they create a number of files, among which are

```
stadyn.cfg        stadyn.stf        stadyn.dis
stadyn.log        stadyn.mas        stadyn.snp
stadyn.out        stadyn.mat        stadyn.tm4
stadyn.dyn        stadyn.geo
stadyn.mon        stadyn.lod
```

The LOG file echoes all the input responses as well as having some extra information that might prove useful during postanalysis of the results. The OUT and DYN files are the usual locations for output. The second column of files is associated with various system matrices such as the stiffness matrix and the mass matrix – these binary files are left on the disk in case they may be of value for some other purpose. The last column of files is binary output files to be read as part of the postprocessing.

Each executable program has a file named <<***.CFG>>. This is the configuration file and has in it, for example for <<stadyn.cfg>>,

```
1000000        ::MaxStorage
   5000        ::MaxElem
   5000        ::MaxNode
```

```
          5000    400        ::MaxForce
             1        ::ilump
            16        ::itermax
          1.0000000E-010      ::rtol
@@ DATE:    9- 1-2007      TIME: 11: 2
```

This allows the run-time setting of the dimensions of the arrays. If a memory allocation failure occurs, this is the place to make the adjustments.

1.5 QED Guided Explorations

The following charts lay out the collection of guided explorations in the form of grids. The vertical axis of the first chart is the type of structure, and the horizontal axis is the type of computer analysis available in QED. An effort was made to distribute the explorations over a wide combination of structure types and analysis types without undue duplication or repetition. However, because the explorations are expected to be mostly independent and stand-alone, some duplication was permitted.

The vertical axis of the second chart comprises the sections of this book, and the horizontal axis is an estimate of the dominant strategies used. It is apparent that not all strategies are equally represented – it seems that the three most common deal with expanding the knowledge base (I), performing a sensitivity study (II), and constructing a simple analytical model (VI). Nonetheless, the other strategies are very important because they provide additional techniques for data collection that are sometimes indispensable. A variety of problem types was sought that could use these strategies.

What is not reflected in these charts are the analyses that use the Gallery of ODEs and the use of the signal analysis program DiSPtool. The sections that most use these are

Gallery	5.4, 5.6, 6.5
DiSPtool	3.1, 3.4, 4.5

Not all the geometries and analyses available in QED are utilized in the presented explorations. The ones chosen, however, will give a fair idea how to create the other geometries and run the other analyses.

QED guided explorations: structures and analyses

	Static	Vibration	Buckling	Transient	Nonlinear
Plate					
Quadri	2.2, 2.4	3.2, 3.3	6.2	4.2, 4.5	5.1
Ellipse	○	○	○	5.6	○
C_curve	2.5	○	○	○	○
Notch					
Rect_notch	2.5	○	○	○	○
Hole					
Circle	○	○	○	4.2	○
Ring	2.5	○	○	○	○
Annulus	2.5	○	○	5.6	○
Ellipse	2.6	○	○	○	○
Crack	2.6	○	○	○	○
Closed					
Circular	○	3.2, 3.6	○	○	○
Wing	2.2, 2.4	○	○	○	○
Airfoil	2.4	3.3	○	○	○
Triangle	2.4	○	○	○	○
Box beams	6.2	○	6.2	○	○
Open					
Angle	2.4	○	○	○	○
I-Beam	2.4, 6.2	○	6.2	○	6.2
C-channel	2.4, 6.2	○	6.2, 6.3	○	6.2
Frame					
1D Beam	2.3	3.1, 6.1	3.1, 6.1	4.1, 5.5	5.5, 6.6
2D Frame	2.3	6.1	6.1	4.3	○
2D Bridge	2.3, 6.3	○	6.3	○	6.3
2D Arch	○	6.5	6.5	4.3	6.5
Solid					
Block	2.1, 2.2	3.5	5.3, 6.4	○	5.1, 6.4
Cylinder	○	3.6	○	○	○
Arch	○	3.5	6.5	○	6.5
Hole	○	○	○	○	5.2
General					
2D Arbit	2.6	○	○	4.6	○

QED guided explorations: sections and strategies

	I	II	III	IV	V	VI
Static						
2.1	√	√	○	○	○	√
2.2	○	√	√	○	○	○
2.3	√	○	○	○	○	√
2.4	○	√	√	○	○	○
2.5	√	○	√	○	○	√
2.6	√	√	○	○	○	√
Vibration						
3.1	√	○	○	○	○	√
3.2	○	√	√	○	○	○
3.3	○	√	√	○	○	○
3.4	√	√	○	○	○	√
3.5	○	√	○	○	○	√
3.6	√	√	○	○	○	√
Wave						
4.1	√	○	○	○	○	√
4.2	√	○	○	√	○	○
4.4	○	√	○	√	○	○
4.3	○	√	○	○	○	√
4.5	○	○	○	√	○	○
4.6	√	○	√	○	○	√
Nonlinear						
5.1	√	○	○	○	√	√
5.2	√	○	○	○	√	○
5.3	√	○	○	○	○	○
5.4	√	√	○	○	○	○
5.5	○	√	○	√	○	√
5.6	○	√	○	○	○	○
Stability						
6.1	√	○	√	○	√	○
6.2	○	√	√	○	○	√
6.3	○	√	○	○	√	○
6.4	○	√	√	○	○	○
6.5	√	○	○	√	○	√
6.6	√	○	○	√	○	√

2 Static Analysis

There are two important concepts in the design of structures to withstand loads. One is *stiffness*, which relates to the ability of a structure to maintain its shape under load; the other is *stress*, which relates to the fact that all structural materials can withstand only a certain level of stress without failing. Stiffness is a global structural concept, whereas stress is a local concept.

The stiffness properties of structural members are greatly affected by their cross-sectional properties; this is especially true of thin-walled members. For example, Figure 2.1(a) shows a C-channel fixed at one end with an upward load applied at the other end along the vertical wall. What is interesting is that this load causes a counterclockwise rotation as shown and not a clockwise rotation as might be expected. The reason is because the shear center (the center of twist) is to the left of the wall.

Figure 2.1(b) shows an example of stress distribution in a bar with a hole. Changes in local geometry can cause significant changes in stress, giving rise to what are called stress concentrations. These are clearly visible around the edge of the hole.

The explorations in this chapter consider the stiffness properties of various structures and the stress distributions in some common components. The first and second explorations establish the stiffness properties of basic structural components and some thin-walled 3D structures. The third exploration examines the equilibrium behavior of beam and frame structures consisting of uniform members. The fourth exploration considers the stress distributions in thin-walled structures. The fifth and sixth explorations look at the role played by local geometry in determining the stress distributions. Of particular interest in these is the distinction between a stress singularity and a stress concentration.

A static analysis is typically done to determine the stress distributions in structures. In general, however, structural problems are *statically indeterminate*; that is, equilibrium alone is insufficient for solving the force and moment distributions, and recourse must be made to incorporating the deformations and constitutive relation simultaneously. The equilibrium of a general linear structural system is represented by the FE equations

$$[K_E]\{u\} = \chi_1\{P\} + \chi_2\{G\},$$

35

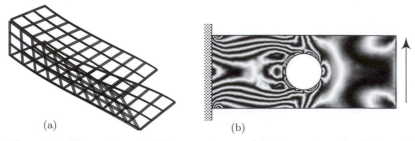

Figure 2.1. Illustrations of stiffness and stress: (a) deformation of a C-channel, (b) stress distribution in a bar with a hole.

where $[K_E]$ is the structural stiffness matrix, $\{P\}$ is the vector of applied loads, $\{G\}$ is the vector of gravity loads, and $\{u\}$ is the vector of unknown nodal DoFs. The stiffness properties of structures are affected by the presence of a prestress – an obvious example is a pretensioned cable. The effect of prestress is estimated by solving the problem

$$\big[[K_E] + \chi[K_G]\big]\{u\} = \{\delta P\},$$

where $[K_G]$ is the geometric stiffness matrix estimated from the preexisting loads. In these analyses, the applied load $\{\delta P\}$ is a point load specified as part of this type of analysis.

Many of the simple models are based on replacing 3D solid components with a reduced dimensional model. The third exploration shows how some statically indeterminate structural problems can be usefully replaced with statically determinate ones.

2.1 Deformation of Structural Members

In this section, we explore the stiffness and stress behaviors of some basic structural members. A cantilevered bar as shown in Figure 2.2 is loaded such that it undergoes axial stretching, transverse bending, and axial torsion; the objective is to determine how the deformation, stress distribution, and stiffness of each of these modes depend on the structural parameters.

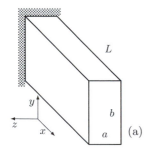

Figure 2.2. Solid bar: (a) geometry (b) single Hex20 solid element showing node positions.

Part I. Axial Loading

We begin with a sensitivity study of the axial behaviors of members. The sensitivity study entails changing the height and length parameters and recording displacement, strain, and stress data. The objective of the data collection and analysis is to use the strain and stress information to construct a simple model to explain the axial stiffness behavior.

Launch QED from the command line in the working directory.

CREATING THE MODEL

1. Create an aluminum bar of dimensions [10 × 2 × 2] using the [Solid:block] geometry. That is,
   [0:geometry =1],
   [1:X_length =10],
   [2:Y_length =2],
   [3:Z_length =2],
   with all offsets being zero. Choose linear elastic material # 1 (aluminum) by
   [a:mat# =1],
   [b:constit s/e =1].
   The yield properties are not relevant to this analysis.

2. Use BCs of [0:type=1(=ENWS)] to fix [000000] the west side with all others being free [111111]. That is,
   [0:type =1],
   [4:sequence=snew],
   [5:east =111111],
   [6:north =111111],
   [7:west =000000],
   [8:south =111111].

3. Use Loads of [0:type=1(=near xyz)] to apply an axial load $P_x = 10000$ at the end of the bar. That is,
   [0:type =1],
   [1:X_pos =10],
   [2:Y_pos =1],
   [3:Z_pos =1],
   [4:X_load =10000].
   Ensure all other loads are zero.

4. Mesh using <10> elements along the length and <2> in the other directions. That is,
   [0:Global_ID =111],
   [1:X_dirn =10],
   [2:Y_dirn =2],
   [3:Z_dirn =2],
   [d:red_int=3];
   the other [specials] can be set to zero.
   [Press s] to render the mesh.

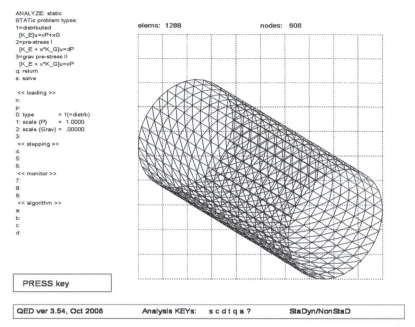

Figure 2.3. Opening screen for QED static analysis.

For clearer viewing, [press o] to set the orientation as $\phi_x = 25$, $\phi_y = -30$, $\phi_z = 0$.

COLLECTING THE DATA

1. Analyze the bar for its linear Static response; the opening screen should be as shown in Figure 2.3. Set
   [0:type =1],
   [1:scale{P} =1.0],
   [2:scale{Grav} =0.0].
   [Press s] to perform the analysis.
   Before exiting the analysis section, look in the file <<stadyn.out>> to see the complete displacement field. (Any convenient text editor can be used for this.)

2. View the deformed Shapes with
   [1:vars =1],
   [2:xMult =500],
   [5:rate =0].
   [Press s] to show the Shapes. Zoom on the second panel by [pressing z2] to see how highly distorted the bar is in the vicinity of the load point.

3. Return to the Loads section and use [0:type=2(=pressure)] to apply a uniform pressure on the end face. That is,
   [0:type =2],
   [1:face =1],
   [4:press =2500]. (This is t_x.)

This will give the same resultant force as before.

Remesh and redo the Static analysis.

4. View the deformed Shapes with

    [1:vars =1],
    [2:xMult =500],
    [5:rate =0].

    Observe that the end face remains almost flat. Observe that the deformation at each cross section is essentially uniform with no distortion (change of angles) of the elements. Observe that there is a small transverse contraction.

5. View time Traces of the responses. For static problems, the time trace comprises the zero load step and the maximum load step. For displacements at the center end, set

    [0:style type =1],
    [1:vars =1],
    [2:xMult =1],
    [5:rate =0],
    [6:node/IP =308],
    [7:elem# =1].

    (The utility **PlotMesh** for viewing <<qed.sdf>> files can be used to find the node of any point by [pressing nx] and inputting the coordinates.)

    [Press s] to show the results. Observe (from the numbers above the plots) that the response is predominantly the axial ([Var:U]) displacement.

    Record the maximum value.

    Change the node number to

    [6:node/IP =158],

    which corresponds to the center midlength. Observe that the displacement is approximately half the tip value.

    View the nodal strains with

    [1:vars =14].

    (A complete list of viewable variables can be seen by [pressing h] for the help menu.) Observe that the axial strain ([Var:Exx]) is an extension whereas the transverse strains ([Var:Eyy] and [Var:Ezz]) are contractions and equal. Also observe that the shear strains are negligible.

    Record the maximum value of [Var:Exx] and [Var:Eyy].

    View the nodal stresses with

    [1:vars =24].

    Observe that the stress is predominantly the axial ([Var:Sxx]) stress.

    Record the maximum value of ([Var:Sxx]).

6. Return to the Geometry section to change the height of the bar to [10 × 2 × 4]. That is, set

    [1:X_length =10],
    [2:Y_length =2],
    [3:Z_length =4].

    Remesh and redo the static analysis.

Record the axial displacement at center end and axial stress at center mid-length.

7. Return to the Geometry section to change the length of the bar to $[20 \times 2 \times 2]$. That is, set

[1:X_length =20],
[2:Y_length =2],
[3:Z_length =2].

Remesh and redo the static analysis.

Record the axial displacement at center end and axial stress at center mid-length.

ANALYZING THE DATA

1. Make a table with all the data converted to stiffnesses according to

$$K_{11} = \frac{t_x A}{u_L},$$

where t_x is the applied traction (pressure).

2. Make separate plots of stiffness against width and length. Comment on the sensitivity of the stiffness to each parameter.

3. A simple model can be constructed by first assuming that the bar is in uniaxial stress. That is,

$$\sigma_{xx} = E\epsilon_{xx} = E\frac{\partial u}{\partial x}, \qquad \epsilon_{yy} = \epsilon_{zz} = -\nu\epsilon_{xx}, \qquad F = \int_A \sigma_{xx} dA = EA\frac{\partial u}{\partial x},$$

with other stresses and strains being zero. In this, E and ν are the Young's modulus and Poisson's ratio, respectively. Also, $\nu = E/(2G) - 1$, where G is the shear modulus.

How well do the observed stresses and strains agree with these assumptions?

How well is the axial strain estimated as $\epsilon_{xx} = u_L/L$, where u_L is the displacement at the tip?

4. In the style of Section 7.4, assume a Ritz function for the displaced shape as $u(x, y, z) = [1 - x/L]u_L$. This gives the strain energy and stiffness as, respectively,

$$\mathcal{U} = \frac{1}{2}\int_V \sigma_{xx}\epsilon_{xx} \, dV = \frac{1}{2}\int_V E\left[\frac{\partial u}{\partial x}\right]^2 dA \, dx = \frac{1}{2}\frac{EA}{L}u_L^2,$$

$$K_{11} = \frac{\partial^2 \mathcal{U}}{\partial u_L^2} = \frac{EA}{L} = \frac{Eab}{L}.$$

Superpose, as a continuous line, this simple model result on the preceding sensitivity plots.

How well does the simple model capture the behavior of the recorded data?

Part II. Bending Loading
Basically the same geometry as that of Part I is used here, but the loading is transverse, putting the member into bending. The sensitivity study and data analysis are similar to those of Part I (except that the width is also changed), and because the flexural behavior is more complex, the simple modeling is also more complex.

Begin by launching **QED** from the command line in the working directory.

CREATING THE MODEL

1. Create an aluminum bar of dimensions $[10 \times 2 \times 2]$ using the [Solid:block] geometry. That is,
   [0:geometry =1],
   [1:X_length =10],
   [2:Y_length =2],
   [3:Z_length =2],
   with all offsets being zero. Choose linear elastic material # 1 (aluminum) by
   [a:mat# =1],
   [b:constit s/e =1].
   The yield properties are not relevant to this analysis.

2. Use BCs of [0:type=1(=ENWS)] to fix [000000] the west side with all others being free [111111]. That is,
   [0:type =1],
   [4:sequence =snew],
   [5:east =111111],
   [6:north =111111],
   [7:west =000000],
   [8:south =111111].

3. Use Loads of [0:type=1(=near xyz)] to apply a transverse load $P_y = 1000$ at the end of the bar. That is,
   [0:type =1],
   [1:X_pos =10],
   [2:Y_pos =1],
   [3:Z_pos =1],
   [4:X_load =0],
   [5:Y_load =1000],
   [6:Z_load =0].
   Ensure all other loads are zero.

4. Mesh using
   [0:Global_ID =111],
   [1:X_dirn =10],
   [2:Y_dirn =2],
   [3:Y_dirn =2],
   [d:red_int =3];
   the other [specials] can be set to zero.
   [Press s] to render the mesh.

COLLECTING THE DATA

1. Analyze the bar for its linear Static response.
2. View the deformed Shapes with
   [1:vars =1],
   [2:xMult =100],
   [5:rate =0].
   Observe that there is no apparent gross deformation in the vicinity of the load point; this is because the transverse deflections are at least two orders of magnitude larger than occurred in Part I. Consequently, there is no need to apply a uniform traction.
   [Press o] to set the orientation as $\phi_x = 0$, $\phi_y = 0$, $\phi_z = 0$ and zoom the deformed shape. Observe that the deflections of the centerline are vertical only (note that this is an exaggerated linear deformation and not a true large deformation) and that initially vertical lines rotate so as to remain perpendicular to the centerline. Thus the rotation is equal to the slope of the centerline.
3. View time Traces of the responses. For displacements at the center end, set
   [0:style type =1],
   [1:vars =1],
   [2:xMult =1],
   [5:rate =0],
   [6:node/IP =308],
   [7:elem# =1].
   [Press s] to show the results. Observe (from the numbers above the plots) that the response is predominantly the transverse ([Var:V]) displacement.
   Record the maximum value of ([Var:V]).
   Change the node number to
   [6:node/IP =168],
   which corresponds to the top midlength.
   View the nodal strains with
   [1:vars =14].
   Observe that the axial strain ([Var:Exx]) is a contraction whereas the other strains ([Var:Eyy] and [Var:Ezz]) are extensions and equal.
   Record the maximum value of [Var:Exx] and [Var:Eyy].
   View the nodal stresses with
   [1:vars =24].
   Observe that the stress is predominantly the axial ([Var:Sxx]) stress.
   Record the maximum value of this stress.
   Change the node number to
   [6:node/IP =158],
   which corresponds to the center midlength. Observe that the strains and stresses are negligible compared with those on the top surface.
4. Change the geometry of the bar to [10 × 2 × 4], that is,
   [1:X_length =10],
   [2:Y_length =2],
   [3:Z_length =4].

Change the load position to

[1:X_pos =10],
[2:Y_pos =1.0],
[3:Z_pos =2].

Remesh and redo the static analysis.

Record the transverse displacement at center end and axial stress at top mid-length.

5. Change the geometry of the bar to $[10 \times 4 \times 2]$, that is,

[1:X_length =10],
[2:Y_length =4],
[3:Z_length =2].

Change the load position to

[1:X_pos =10],
[2:Y_pos =2.0],
[3:Z_pos =1.0].

Remesh and redo the static analysis.

Record the transverse displacement at center end and axial stress at top mid-length.

6. Change the geometry of the bar to $[20 \times 2 \times 2]$, that is,

[1:X_length =20],
[2:Y_length =2],
[3:Z_length =2].

Change the load position to

[1:X_pos =20],
[2:Y_pos =1.0],
[3:Z_pos =1.0].

Remesh and redo the static analysis.

Record the transverse displacement at center end and axial stress at top mid-length.

ANALYZING THE DATA

1. Make a table with all the data converted to stiffnesses according to

$$K_{22} = \frac{P_y}{v_L}.$$

2. Make separate plots of stiffness against width, height, and length. Comment on the sensitivity of the stiffness to each parameter.

3. A simple model can be constructed by first assuming that the bar is in uniaxial stress. How well do the observed stresses and strains agree with these assumptions?

Also assume that the displacements can be represented by

$$u(x, y, z) = -y\phi_z(x) = -y\frac{\partial v}{\partial x}, \qquad v(x, y, z) = v(x),$$

$$\sigma_{xx} = E\epsilon_{xx} = E\frac{\partial u}{\partial x} = -yE\frac{\partial^2 v}{\partial x^2}, \qquad M = -\int_A \sigma_{xx}\,dA = EI_{zz}\frac{\partial^2 v}{\partial x^2},$$

where ϕ_z is the rotation of the cross section and the second moment of area I_{zz} is given by $\int y^2 \, dA = \frac{1}{12}ab^3$. This indicates a linear distribution of stress and strain on the cross section.

How well do the observed stresses and strains agree with these assumptions?

4. The strain energy is

$$\mathcal{U} = \frac{1}{2}\int_V \sigma_{xx}\epsilon_{xx}\, dV = \frac{1}{2}\int_V Ey^2\left[\frac{\partial^2 v}{\partial x^2}\right]^2 dA\, dx = \frac{1}{2}\int_L EI_{zz}\left[\frac{\partial^2 v}{\partial x^2}\right]^2 dx.$$

In the style of Section 7.4, assume a polynomial Ritz function for the transverse displaced shape; then imposing zero displacement and slope at the fixed end leads to

$$v(x) = \frac{x^2}{L^2}\left[3 - 2\frac{x}{L}\right]v_L + \frac{x^2}{L^2}\left[-1 + \frac{x}{L}\right]L\phi_L$$

where v_L and ϕ_L are the tip displacement and slope, respectively. This gives the strain energy as

$$\mathcal{U} = \frac{EI_{zz}}{2L^3}\left[6v_L^2 - 6Lv_L\phi_L + 2L^2\phi_L^2\right].$$

Because there are 2DoFs, this gives rise to two equilibrium equations:

$$P_y = \frac{\partial \mathcal{U}}{\partial v_L} = \frac{EI_{zz}}{L^3}\left[12v_L - 6L\phi_L\right], \qquad T_z = \frac{\partial \mathcal{U}}{\partial \phi_L} = \frac{EI_{zz}}{L^3}\left[-6Lv_L + 4L^2\phi_L\right].$$

Because there is no applied torque, then $\phi_L = \frac{3}{2}v_L/L$, and the force displacement relation becomes

$$P_y = \frac{3EI_{zz}}{L^3}v_L, \qquad K_{22} = \frac{3EI_{zz}}{L^3} = \frac{Eab^3}{4L^3}.$$

Superpose, as a continuous line, this simple model result on the preceding sensitivity plots.

How well does the simple model capture the behavior of the recorded data?

5. As an alternative simple model, assume a one-parameter Ritz function given by $v(x) = [x^2/L^2]v_L$, which leads to

$$P_y = \frac{4EI_{zz}}{L^3}v_L, \qquad K_{22} = \frac{4EI_{zz}}{L^3} = \frac{Eab^3}{3L^3}.$$

Conjecture why this simple model does not function as well as the previous one.

Part III. Torsional Loading

Again, the same geometries as those of Part I are used here but the loading causes an axial twisting of the member. Because Hex20 elements do not have rotation as a DoF, the torque will actually be applied as two parallel forces forming a couple.

The objective of the data collection and analysis is to perform a sensitivity study to determine the dependence of the stiffness behavior on the structural parameters of length and width and thereby construct a simple model. Begin by launching QED from the command line in the working directory.

CREATING THE MODEL

1. Create an aluminum bar of dimensions $[10 \times 2 \times 2]$ using the [Solid:block] geometry. That is,
   [0:geometry =1],
   [1:X_length =10],
   [2:Y_length =2],
   [3:Z_length =2],
   with all offsets being zero. Choose linear elastic material # 1 (aluminum) by
   [a:mat# =1],
   [b:constit s/e =1].
   The yield properties are not relevant to this analysis.

2. Use BCs of [0:type=1(=ENWS)] to fix [000000] the west side with all others being free [111111]. That is,
   [0:type =1],
   [4:sequence =snew],
   [5:east =111111],
   [6:north =111111],
   [7:west =000000],
   [8:south =111111].

3. Use Loads of [0:type=3(=couple)] to apply a torque of $T_x = 2000$ at the end of the bar. That is,
   [0:type =3],
   [1:X_pos =10],
   [2:Y_pos =2],
   [3:Z_pos =1],
   [4:X_pos =10],
   [5:Y_pos =0],
   [6:Z_pos =1],
   [7:X_load =0],
   [8:Y_load =0],
   [9:Z_load =1000].
   The second force forming the couple is automatically set as the negative of the stated force.

4. Mesh using
   [0:Global_ID =111],

[1:X_dirn =10],
[2:Y_dirn =2],
[3:Y_dirn =2],
[d:red_int =3];
the other [specials] can be set to zero.
[Press s] to render the mesh.
For clearer viewing, [press o] to set the orientation as $\phi_x = 25$, $\phi_y = -30$, $\phi_z = 0$.

COLLECTING THE DATA

1. Analyze the bar for its linear Static response.
2. View the deformed Shapes with
   [1:vars =1],
   [2:xMult =250],
   [5:rate =0].
   Observe that there is no apparent gross deformation in the vicinity of the load points.
   [Press o] to set the orientation as $\phi_x = 0$, $\phi_y = -90$, $\phi_z = 0$, and zoom the deformed shape. Observe that the deflection pattern is like a rigid-body rotation of each cross section. During zooming, shift the display so that the lower left edge is visible. Observe that rotation–displacement is almost linear with position.
3. View time Traces of the responses. For displacements at the load point, set
   [0:style type =1],
   [1:vars =1],
   [2:xMult =1],
   [5:rate =0],
   [6:node/IP =318], (=304] for the other point),
   [7:elem# =1].
   [Press s] to show the results. Observe (from the numbers above the plots) that the response is predominantly the transverse ([Var:W]) displacement.
   Record the maximum value of this displacement.
   View the nodal strains at the top midlength with
   [1:vars =14],
   [6:node/IP =168].
   Observe that the shear strain γ_{xz} ([Var:2Exz]) is the dominant strain.
   Record the maximum value of this strain.
   View the nodal stresses with
   [1:vars =24].
   Observe that the shear stress τ_{xz} ([Var:Sxz]) is the dominant stress.
   Record the maximum value of this stress.
   Change the node number to
   [6:node/IP =158],
   which corresponds to the center midlength. Observe that the strains and stresses are negligible compared with those on the top midsurface.

Change the node number to
[6:node/IP =163],
which corresponds to the edge midlength. Observe that the strains and stresses are negligible compared with those on the top midsurface.
Change the node number to
[6:node/IP =155],
which corresponds to the midside midlength. Observe that the shear strain γ_{xy} ([Var:2Exy]) and the shear stress τ_{xy} ([Var:Sxy]) are the dominant strain and stress.

4. Return to the Geometry section and change the dimensions of the bar to $[10 \times 2 \times 4]$. That is, set
[1:X_length =10],
[2:Y_length =2],
[3:Z_length =4].
Change the load positions to
[1:X_pos =10],
[2:Y_pos =2],
[3:Z_pos =2.0].
[4:X_pos =10],
[5:Y_pos =0],
[6:Z_pos =2.0],
[7:X_load =0],
[8:Y_load =0],
[9:Z_load =1000].
This maintains a constant torque of $T_x = 2000$.
Remesh and redo the Static analysis.
Record the z, w displacement at the load points and stress at the top midlength.

5. Return to the Geometry section and change the dimensions of the bar to $[20 \times 2 \times 2]$. That is, set
[1:X_length =20],
[2:Y_length =2],
[3:Z_length =2].
Change the load positions to
[1:X_pos =20],
[2:Y_pos =2],
[3:Z_pos =1],
[4:X_pos =20],
[5:Y_pos =0],
[6:Z_pos =1],
[7:X_load =0],
[8:Y_load =0],
[9:Z_load =1000].
Remesh and redo the static analysis.
Record the z, w displacement at the load points and stress at the top midlength.

ANALYZING THE DATA

1. Make a table with all the data converted to stiffnesses according to

$$K_{44} = \frac{T_x}{\phi_x}, \qquad \phi_x \approx \frac{w_2 - w_1}{b}.$$

2. Make separate plots of stiffness against width and length. Comment on the sensitivity of the stiffness to each parameter.

3. A simple model can be constructed by first assuming that each section rotates as a rigid body so that the displacements can be represented by

$$u(x, y, z) = 0, \qquad v(x, y, z) = -xz\phi_x(x), \qquad w(x, y, z) = +yz\phi_x(x).$$

This gives the stress and strain states:

$$\gamma_{xy} = \frac{\partial u}{\partial y} + \frac{\partial v}{\partial x} = -z\phi_x, \qquad \gamma_{xz} = \frac{\partial u}{\partial z} + \frac{\partial w}{\partial x} = +y\phi_x, \qquad \text{others} = 0,$$

$$\tau_{xy} = G\gamma_{xy} = -zG\phi_x, \qquad \tau_{xz} = G\gamma_{xz} = +yG\phi_x, \qquad \text{others} = 0.$$

How well do the observed stresses and strains agree with these assumptions?

4. The strain energy is

$$\mathcal{U} = \tfrac{1}{2} \int_V [\sigma_{xy}\gamma_{xy} + \sigma_{xz}\gamma_{xz}] \, dV = \tfrac{1}{2} \int_V G[y^2 + z^2]\phi_x^2 \, dA \, dx = \tfrac{1}{2} \int_L GI_{xx}\phi_x^2 \, dx,$$

where the second moment of area I_{xx} is given by $\int [y^2 + z^2] \, dA = \frac{1}{12}[ab^3 + a^3b]$. In the style of Section 7.4, assume as a Ritz function $\phi_x(x) = [1 - x/L]\phi_L$, where ϕ_L is the tip rotation; this gives the strain energy and stiffness as, respectively,

$$\mathcal{U} = \tfrac{1}{2}\frac{GI_{xx}}{L}\phi_L^2, \qquad K_{44} = \frac{\partial \mathcal{U}}{\partial \phi_L} = \frac{GI_{xx}}{L} = \frac{G[ab^3 + a^3b]}{12L}.$$

Superpose, as a continuous line, this simple model result on the preceding sensitivity plots.

How well does the simple model capture the behavior of the recorded data? Conjecture about how the simple model may be improved.

Partial Results

Partial results for the bending stiffness of Part II are shown in Figure 2.4. Note that the bending stiffness of the reference case is the same in each case.

Background Readings

The mechanical behavior of simple structural members is treated in texts such as References [19, 44, 128]. The stiffness properties of these components are covered

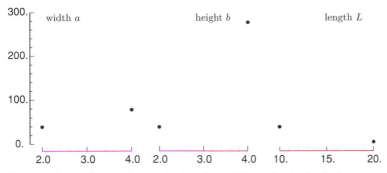

Figure 2.4. Partial results for the bending stiffness (scaled) of a bar.

in texts such as References [18, 29, 102, 129, 134]. The Hex20 element is covered in References [5, 26, 130].

2.2 Stiffness Behavior of Thin-Walled Structures

Engineering structures get their stiffness properties from two main sources. The first is the elastic modulus of the material used; for example, for a given cross section, a steel member is three times as stiff as the aluminum member. The second is the shape of the member and in particular the shape of the cross section. The interplay of these two is interesting when weight is a design consideration. Because steel is three times as heavy as aluminum then, for a given weight, an aluminum member can have three times the cross-sectional area, which for most stiffness properties makes it more stiff than the steel member.

In this section, we explore the stiffness behaviors of some thin-walled structures with different cross sections as shown in Figure 2.5.

Part I. Cantilevered Flat Plate
We begin by analyzing the relatively simple case of a cantilevered plate with a rectangular cross section. The plate is loaded such that it undergoes transverse bending, in-plane bending, and torsion; the objective of the data collection and analysis is to

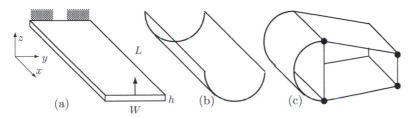

Figure 2.5. Static structures analyzed: (a) cantilevered plate, (b) curved C-channel, (c) wing structure.

use a sensitivity study to determine how the stiffness of each of these modes depends on the structural parameters.

Begin by launching **QED** from the command line in the working directory.

CREATING THE MODEL

1. Create a $[6 \times 2 \times 0.1]$ aluminum plate using the [Plate:quadrilateral] geometry. That is,

   [0:geometry =1],
   [1:X_2 =6],
   [2:Y_2 =0],
   [3:X_3 =6],
   [4:Y_3 =2],
   [5:X_4 =0],
   [6:Y_4 =2],
   [a:mat#=1],
   [b:thick =0.1].

2. Use BCs of [0:type=1(=ENWS)] to fix [000000] the west side with all others being free [111111]. That is,

   [4:sequence =snew],
   [5:east =111111],
   [6:north =111111],
   [7:west =000000],
   [8:south =111111].

3. Use Loads of [0:type=1(=near xyz)] to apply loads of $P_x = 1$, $P_y = 1$, $P_z = 1$, $T_x = 1$, at the end of the plate at the centerline $x = 6$, $y = 1$. That is, set

   [0:type =1],
   [1:X_pos =6],
   [2:Y_pos =1],
   [3:Z_pos =0],
   [4:X_load =1],
   [5:Y_load =1],
   [6:Z_load =1],
   [7:X_mom =1];

   all other loads are zero.

4. Mesh using $<12>$ modules through the length and $<8>$ transverse modules. That is, set

   [0:Global_ID =32],
   [1:X_mods =12],
   [2:Y_mods =8];

   all [specials] can be set to zero. Render the mesh by [pressing s].

   For clearer viewing, [press o] to set the orientation as $\phi_x = -55$, $\phi_y = 0$, $\phi_z = -20$.

COLLECTING THE DATA

1. Analyze the plate for its linear Static response. Set
   [0:type =1],
   [1:scale{P} =100],
   [2:scale{Grav} =0.0].
   Perform the analysis by [pressing s].

2. Before exiting the analysis section, look in the file <<stadyn.out>> to see the complete displacement field. Record the complete displacement line at the load point at node =113. This is the reference data set.

3. View the Contours of displacement with
   [0:type =1],
   [1:vars =1],
   [2:xMult =1],
   [5:rate =0],
   and confirm that the contours are in agreement with expectation. Individual panels can be zoomed by [pressing z #], where the panels are numbered 1 through 6, top row first.

4. Return to the Geometry section and change the dimensions of the plate to [6 × 2 × 0.2]. That is, change
   [b:thick =0.1].
   Remesh and redo the static analysis.
   Record the complete displacements at the load point node =113. This is the sensitivity Case I data set.

5. Change the dimensions of the plate to [6 × 4 × 0.1]. That is, change
   [1:X_2 =6],
   [2:Y_2 =0],
   [3:X_3 =6],
   [4:Y_3 =4],
   [5:X_4 =0],
   [6:Y_4 =4],
   [b:thick =0.1].
   Reposition the load to the end of the plate at the centerline with
   [1:X_pos =6],
   [2:Y_pos =2],
   [3:Z_pos =0].
   Remesh and redo the static analysis.
   Record the complete displacements at the load point node =113. This is the sensitivity Case II data set.

6. Change the dimensions of the plate to [12 × 2 × 0.1]. That is, change
   [1:X_2 =12],
   [2:Y_2 =0],
   [3:X_3 =12],
   [4:Y_3 =2],

[5:X_4 =0],
[6:Y_4 =2],
[b:thick =0.1].
Reposition the load to the end of the plate at the centerline with
[1:X_pos =12],
[2:Y_pos =1],
[3:Z_pos =0].
Remesh and redo the static analysis.
Record the displacements at the load point node =113. This is the sensitivity Case III data set.

ANALYZING THE DATA

1. Make a table of all the data converted to stiffnesses according to

$$K_{11} = \frac{P_x}{u}, \qquad K_{22} = \frac{P_y}{v}, \qquad K_{33} = \frac{P_z}{w}, \qquad K_{44} = \frac{T_x}{\phi_x}.$$

Normalize these stiffnesses by dividing them by their respective reference values.

2. On a single graph and using an appropriate identifying symbol for each stiffness, plot the reference and Case I data against h. Connect like symbols to form the thickness sensitivity plots.

3. On a single graph and using an appropriate identifying symbol for each stiffness, plot the reference and Case II data against W. Connect like symbols to form the width sensitivity plots.

4. On a single graph and using an appropriate identifying symbol for each stiffness, plot the reference and Case III data against L. Connect like symbols to form the length sensitivity plots.

5. Does the dependence on the width, thickness, and length dimensions of the axial stiffness correlate with the dimensional powers $K_{11} \propto EA/L \to W^1 h^1 / L^1$?

6. Does the dependence on the width, thickness, and length dimensions of the in-plane bending stiffness correlate with the dimensional powers $K_{22} \propto E I_{zz}/L^3 \to W^3 h^1 / L^3$?

7. Does the dependence on the width, thickness, and length dimensions of the out-of-plane bending stiffness correlate with the dimensional powers $K_{33} \propto E I_{yy}/L^3 \to W^1 h^3 / L^3$?

8. Does the dependence on the width, thickness, and length dimensions of the twisting stiffness correlate with the dimensional powers $K_{44} \propto G J_{xx}/L \to W^1 h^3 / L^1$?

Part II. Cantilevered Curved C-channel

The plate of Part I is now curved so that it forms a C-channel, as depicted in Figure 2.5(b). In doing so, the positioning of the load for the different modes becomes quite important; otherwise a given load would produce mixed mode effects. In particular, the concepts of centroid and shear center need to be introduced and associated with each loading mode.

The curved plate is loaded such that it undergoes transverse bending, in-plane bending, and torsion; the objective of the data collection and analysis is to first determine the centroid and shear center for the cross section and then determine how the stiffness of each of these modes depends on the structural parameters.

Launch QED from the command line in the working directory.

CREATING THE MODEL

1. Create a curved aluminum C-channel using the [Open:C_chan_circ] geometry. That is,
   [0:geometry =4],
   [1:length =20.],
   [2:radius =2.0],
   [4:angle =180]. (angle of gap)
   Set the the stringer areas as $A = 0.001$ so the reinforcing frame is not constructed. That is,
   [a:mat# =1],
   [b:thick =0.1],
   [d:area =0.001],
   [e:end plate =1].

2. Use BCs of [0:type=1(=L/R pln)] so that the left side is cantilevered and the right side is free. That is,
   [0:type =1],
   [8:L_bc =000000],
   [7:R_bc =111111].

3. Use Loads of [0:type=3(=w/end_frame)] to apply an axial P_z load; the purpose of the frame is to facilitate the placing of the load at an arbitrary position. That is,
   [0:type =3],
   [1:X_pos =0.0],
   [2:Y_pos =0.0],
   with the values
   [4:X_load =0],
   [5:Y_load =0],
   [6:Z_load =1],
   [7:X_mom =0],
   [8:Y_mom =0],
   [9:Z_mom =0].

4. Mesh using < 40 > modules through the length and < 16 > in the hoop direction. That is,
   [0:Global_ID =32],
   [1:length =40],

[2:hoop =16],
[7:Z_rate =40];
all [specials] can be set as zero. Render the mesh by [pressing s].
For clearer viewing, set the orientation ([press o]) as $\phi_x = +35$, $\phi_y = -45$, $\phi_z = 0$.

COLLECTING THE DATA

1. Analyze the channel for its linear Static response. Set
   [0:type =1],
   [1:scale{P} =100],
   [2:scale{Grav} =0.0].
   [Press s] to perform the analysis.
2. View the Contours of displacement with
   [0:type =1],
   [1:vars =1],
   [2:xMult =1],
   [5:rate =0].
   Individual panels can be zoomed by [pressing z #], where the panels are numbered 1 through 6, top row first. For this geometry, the cross section lies in the x, y plane.
   Observe, by zooming, that there is a significant rotation about the x axis ([Var:Rx]) – the axial load is causing a bending action. To avoid this, the axial load must be placed at the *centroid* of the cross section. We will find this position by interpolation by using a second load position. First record the maximum rotation.
3. Return to the Loads section and change
   [1:X_pos =0.0],
   [2:Y_pos =-2.0].
   Because of symmetry, we can assume that the centroid is along the $x = 0$ position.
   Remesh and redo the static analysis. View the displacement contours and record the maximum [Var:Rx] rotation.
   Plot the two values of rotation against vertical position and interpolate to find where the rotation is zero – this is the centroid.
   Return to the Loads section and place the load at the centroid. Remesh and redo the static analysis. View the displacement contours and verify that the [Var:Rx] rotation is negligible.
   Record the maximum axial displacement w [Var:W].
4. Return to the Loads section and change to a P_y loading:
   [4:X_load =0],
   [5:Y_load =1],
   [6:Z_load =0].

Remesh and redo the static analysis. View the displacement contours and observe that the deformation is symmetric.

Record the maximum vertical displacement v [Var:V].

5. Return to the Loads section and change to a P_x loading:

   [1:X_pos =0.0],
   [2:Y_pos =0.0],
   [4:X_load =1],
   [5:Y_load =0],
   [6:Z_load =0].

   Remesh and redo the static analysis. View the displacement contours and observe that there is a significant rotation about the axial direction ([Var:Rz]) – the horizontal load is causing a twisting action. To avoid this, the load must act through the *shear center*. We will find this position by interpolation by using a second load position. First record the maximum rotation.

   Return to the Loads section and change

   [1:X_pos =0.0],
   [2:Y_pos =-4.0].

   Remesh and redo the static analysis. View the displacement contours and record the maximum [Var:Rz] rotation.

   Plot the two values of rotation against vertical position and interpolate to find where the rotation is zero – this is the shear center.

   Return to the Loads section and place the load at the shear center. Remesh and redo the static analysis. View the displacement contours and verify that the [Var:Rz] rotation is negligible.

   Record the maximum horizontal displacement u [Var:U].

6. Return to the Loads section and change to a T_z torque loading:

   [1:X_pos =0],
   [2:Y_pos =0],
   [4:X_load =0],
   [5:Y_load =0],
   [6:Z_load =0],
   [9:Z_mom =1].

   Remesh and redo the static analysis. View the displacement contours and observe that the dominant action is a rotation about the axial direction.

   Record the maximum rotation ϕ_z [Var:Rz].

ANALYZING THE DATA

1. Make a table of all the data converted to stiffnesses according to

$$K_{11} = \frac{P_x}{u}, \qquad K_{22} = \frac{P_y}{v}, \qquad K_{33} = \frac{P_z}{w}, \qquad K_{66} = \frac{T_z}{\phi_z}.$$

2. How well do the data agree with

$$K_{33} = \frac{EA}{L} = \frac{E\pi Rh}{L},$$

$$K_{22} = \frac{3EI_{xx}}{L^3} = \frac{3ER^3h}{L^3}\left[\frac{\pi}{2} - \frac{4}{\pi}\right],$$

$$K_{11} = \frac{3EI_{yy}}{L^3} = \frac{3E\pi R^3h}{2L^3},$$

$$K_{66} = \frac{GJ_{zz}}{L} = \frac{G\pi Rh^3}{3L}.$$

3. How well do the determined centroid and shear center agree with

$$y_c = -\frac{2R}{\pi}, \qquad e = -\frac{4R}{\pi},$$

respectively? Both are measured with respect to the origin.

Part III. Wing Structure

The two previous explorations associated the stiffness with a mode of loading (e.g., torsion or bending); this exploration introduces the concept of a stiffness matrix. Any collection of points on a structure has a stiffness matrix that is related to the other collections of points. For illustrative purposes, consider two identified points 1 and 2; then

$$[K]\{u\} = \{P\} \qquad \text{or} \qquad \begin{bmatrix} K_{11} & K_{12} \\ K_{21} & K_{22} \end{bmatrix} \begin{Bmatrix} u_1 \\ u_2 \end{Bmatrix} = \begin{Bmatrix} P_1 \\ P_2 \end{Bmatrix}.$$

This can also be written as

$$\{u\} = [K^{-1}]\{P\} \qquad \text{or} \qquad \begin{Bmatrix} u_1 \\ u_2 \end{Bmatrix} = \begin{bmatrix} C_{11} & C_{12} \\ C_{21} & C_{22} \end{bmatrix} \begin{Bmatrix} P_1 \\ P_2 \end{Bmatrix}.$$

In these relations, $[K]$ is the stiffness matrix and its inverse $[C]$ is called the compliance matrix. A load applied a single point ($P_1 \neq 0$, $P_2 = 0$, say) causes displacements at all points. It is apparent therefore how the compliance matrix (and hence the stiffness matrix) can be constructed from a series of such one-load tests (e.g., $C_{11} = u_1^1/P_1$ and so on).

The objective of the data collection and analysis is to determine the compliance–stiffness matrix for a general structure using three arbitrary points. An additional objective is to determine if the matrices have any special properties. The model to be used for this purpose is the wing shown in Figure 2.5(c). It will be loaded at different positions at the free end.

Begin by launching **QED** from the command line in the working directory.

CREATING THE MODEL

1. Create a two-spar aluminum wing using the [Closed:wing] geometry; set the dimensions as $L = 10.0$, $x_2 = 6.0$, $y_2 = 0.5$, $y_3 = 3.5$, $y_4 = 4.0$. That is, set the length and coordinates as
   [0:geometry =3],
   [1:length =10.0],
   [6:X_2 =6.0],
   [7:Y_2 =0.5],
   [8:Y_3 =3.5],
   [9:Y_4 =4.0].
   Take the skin and spar thickness as $h = 0.1$ and the stringer areas as $A = 0.3$. That is,
   [a:mat# =1],
   [b:skin =0.1],
   [c:spar =0.1],
   [d:area =0.3],
   [e:end plate =1].

2. Use BCs of [0:type=1(=L/R plane)] to set the whole left plane as fixed and the whole right plane as free. That is, set
   [0:type =1],
   [8:L_bc =000000],
   [9:L_bc =111111].

3. Use Loads of [0:type=1(Near xyz)] to apply a vertical load at the end. That is, set
   [0:type =3],
   [1:X_pos =0.0],
   [2:Y_pos =1.0],
   [3:Z_pos =10],
   [4:X_load =1000],
   [5:Y_load =0];
   [9:Z_mom =0];
   all other loads are set to zero.

4. Mesh using <16> modules through the length and <32> modules around the hoop direction. That is,
   [0:Global_ID =32],
   [1:length mods =16],
   [2:hoop mods =32],
   [3:V/H bias =1],
   [7:Z_rate =16];
   all [specials] can be set as zero. Render the mesh by [pressing s].
   For clearer viewing, set the orientation ([press o]) as $\phi_x = +25$, $\phi_y = +30$, $\phi_z = 0$. If desired, toggle the top skin off by [pressing t 3 0]. Other substructures can similarly be turned on or off.

5. From the `command line`, launch **PlotMesh** and read in (`[press 0]`) the SDF `<<qed.sdf>>`. Find the nearest nodes (`[press nx]`) to the coordinates
   (0, 1, 10),
   (3, 2, 10),
   (6, 3, 10),
   and record them. These will be the monitor nodes and will be referred to as Points 1, 2, and 3, respectively.

COLLECTING THE DATA

1. Analyze the wing for its linear Static response. Set
   `[0:type =1]`,
   `[1:scale{P} =1.0]`,
   `[2:scale{Grav} =0.0]`.
   Perform the analysis by [`pressing s`].
   This is a relatively large mesh, so if **StaDyn** appears not to run, change the memory allocation in `<<stadyn.cfg>>` to a value slightly larger than that suggested in the `<<stadyn.log>>` file.

2. Before exiting the Analysis section, look in the file `<<stadyn.out>>` to see the complete displacement field. Record the horizontal (u) displacement for Point 1, the vertical (v) displacement for Point 2, and the axial (ϕ_z) rotation for Point 3.

3. View the displacement contours with
   `[0:type =1]`,
   `[1:vars =1]`,
   `[2:xMult =1]`,
   `[5:rate =0]`,
   and notice the twisting action. Confirm that the contours are consistent with the recorded data from `<<stadyn.out>>`.

4. Return to the Loads section and change the load values and position to
   `[1:X_pos =3.0]`,
   `[2:Y_pos =2.0]`,
   `[3:Z_pos =10.]`,
   `[4:X_load =0]`,
   `[5:Y_load =1000]`,
   `[9:Z_mom =0]`.
   Remesh and use **PlotMesh** to confirm that the node numbers of the three monitor nodes have not changed.
   Repeat the static analysis and record the horizontal (u) displacement for Point 1, the vertical (v) displacement for Point 2, and the axial (ϕ_z) rotation for Point 3.

5. Return to the Loads section and change the load values and position to
   `[1:X_pos =6.0]`,
   `[2:Y_pos =3.0]`,
   `[3:Z_pos =10.]`,

Figure 2.6. Partial results for the stiffness of a flat plate. Sensitivity plots with respect to width, thickness, and length.

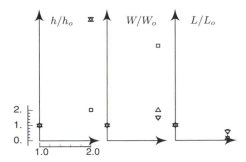

```
[4:X_load =0],
[5:Y_load =0],
[9:Z_mom =1000].
```

Remesh and use **PlotMesh** to confirm that the node numbers of the three monitor nodes have not changed.

Repeat the static analysis and record the horizontal (u) displacement for Point 1, the vertical (v) displacement for Point 2, and the axial (ϕ_z) rotation for Point 3.

ANALYZING THE DATA

1. Form a $[3 \times 3]$ matrix of response divided by load for each load position.
2. Realizing the different orders of magnitude of the compliance terms, is the matrix almost symmetrical? Specifically, is the rotation at position 3 that is due to the vertical load at position 2 equal to the deflection at position 2 that is due to the torque load at position 3?

   If the data show this to be true, then this confirms the *Maxwell reciprocity relation*.
3. Conjecture why the preceding matrix seems to be symmetrical.

Partial Results

The partial results for the sensitivity study of Part I are shown in Figure 2.6. The circle is for K_{11}, the rectangle for K_{22}, the up-triangle for K_{33}, and the down-triangle for K_{33}.

Background Readings

The analysis of thin-walled structures is given in References [30, 87, 93, 99]. The main idea brought out is that, although the structures are geometrically 3D, their behaviors are similar to those of elementary structural members.

2.3 Equilibrium of Beam and Frame Structures

Beam and frame structures are typically statically indeterminate, as discussed in Section 1.1. This makes their solution complicated and difficult to visualize in simple

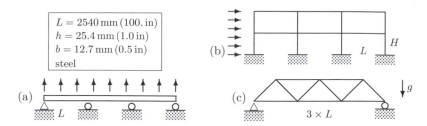

Figure 2.7. Beam and frame structures analyzed: (a) continuous beam with transverse loading, (b) building with lateral load, $H = L/2$, (c) bridge under gravity loading, $H = L/2$. The inset gives the properties of each beam–frame segment.

model terms. The general objective here is to devise a scheme whereby these normally statically indeterminate problems can be usefully approximated as being statically determinate and thus lend themselves to simple modeling.

A typical beam or frame member has member stresses of

$$\sigma_{xx} = \frac{F_x}{A} - \frac{M_z}{I}y, \qquad \sigma_{xy} = \frac{V_y}{A},$$

where F_x is the axial force, V_y is the shear force, M_z is the bending moment, $A = bh$ is the area, and $I = bh^3/12$ is the moment of inertia. Because the moment (bending) stress varies on the cross section, QED actually reports [Var : Sfzz] $= M_z/I$. We explore the distribution of these quantities for the structures shown in Figure 2.7.

Part I. Transverse Loads on Single-Span Beams

We begin with the simplest of structures, that of a single-span beam. The objective of the data collection and analysis is to construct a simple model for solving beam problems and have the potential for generalizing to frames. The key is to understand the characteristics of the force and moment distributions and, in particular, the occurrence of momentless hinges.

Begin by launching QED from the command line in the working directory.

CREATING THE MODEL
1. Create a long slender steel beam of dimensions [100 × 1.0 × 0.5] using the [Frame:1D_beam] geometry. That is,
   [0:geometry =1],
   [1:X_length =100],
   [4:#X_span =1],
   [a:mat#1 =2],
   [b:thick1 =1.0],
   [c:depth1 =0.5].
2. Use BCs of [0:type=1(=AmB)] to cantilever the beam, fixed [000000] at one end and free [111111] at the other. Set
   [0:type =1],

```
[5:end_A =000000],
[6:others =111111],
[7:end_B =111111].
```
3. Use Loads of [0:type=2(=uni_res)] to apply a uniformly distributed load with a given resultant. Set
```
[0:type =2],
[1:side# =1],
[4:X_load =0.0],
[5:Y_load =200.]
```
(applied resultant P_o).

All other loads are set to zero.

This takes the resultant load and divides it equally among all the nodes.
4. Mesh using <50> elements through the length and restrict the motion to being planar. That is,
```
[0:Global_ID =22],
[1:X_elems =50].
```
Set the gravity component as zero:
```
[b?:gravity =0].
```
(The gravity special requires [pressing b] followed by [?] which is to be [x y z] as appropriate.)

Render the mesh by [pressing s]. If necessary, set the orientation of the beam (by [pressing o]) to $\phi_x = 0$, $\phi_y = 0$, $\phi_z = 0$.

COLLECTING THE DATA
1. Analyze the beam for its linear Static response. Set
```
[0:type =1],
[1:scale{P} =1],
[2:scale{Grav} =0.0].
```
Perform the analysis by [pressing s].
2. View the deformed Shapes with
```
[1:vars =1],
[2:xMult =1],
[5:rate =0].
```
[Press s] to show the deformed shape. Individual panels can be zoomed by [pressing z #], where the panels are numbered 1 through 6, top row first. This can also be done while in zoom mode. If necessary, while zooming [press y] a few times to vertically shift the display.

Observe that the shape resulted from a vertical-only deflection. This does not correspond to experience that would expect that there also might be a right-to-left displacement. The anomaly is due to the linear analysis used; a nonlinear analysis, as used in Chapter 5, would indeed exhibit the horizontal displacement. Approximately, a linear analysis of beams is valid for deflections less than the thickness of the beam.

3. View the Distributions of displacements and rotations (slopes) with
   [1:vars =41],
   [2:xMult =1].
   [Press s] to show the distributions.
   Observe that the only displacement is the vertical displacement. [Press z6] to zoom the slopes panel and observe that the free end of the beam has an almost-constant slope.
   View the shear- and moment-stress distributions with
   [1:vars =61],
   [2:xMult =1].
   By zooming, observe that the shear stress [Var:Smxy] is linearly distributed with zero at the free end whereas the moment-stress distribution [Var:Sfzz] is parabolic also zero at the free end.
   Record the maximum shear and moment stresses from the legends.
4. Return to the BCs section and change the boundary conditions to pinned at both ends. That is,
   [0:type =1],
   [5:end_A =000001],
   [6:others =111111],
   [7:end_B =100001].
   Note that end B can move axially. Remesh and rerun the static analysis.
5. View the deformed Shapes with
   [1:vars =1],
   [2:xMult =1],
   [5:rate =0].
   Observe that the deflection is a symmetric upward bow. Although the right end is on rollers, conjecture why there is no horizontal displacement.
6. View the Distributions of displacements and rotations (slopes) with
   [1:vars =41],
   [2:xMult =1].
   Observe that the only displacement is the vertical bow. [Press z6] to zoom the slopes panel and observe that the distribution is antisymmetric.
   View the shear- and moment-stress distributions with
   [1:vars =61],
   [2:xMult =1].
   By zooming, observe that the shear stress is linearly distributed with zero in the center whereas the moment-stress distribution is parabolic with a maximum in the center and zeros at the ends.
   Record the maximum shear and moment stresses from the legends.
7. Return to the BCs section, and change the BCs to fixed. That is,
   [0:type =1],
   [5:end_A =000000],
   [6:others =111111],

[7:end_B =000000].

Remesh and rerun the static analysis.

View the Distributions of shear and moment stresses with

[1:vars =61],

[2:xMult =1].

By zooming, observe that the shear stress is linearly distributed with zero in the center whereas the moment-stress distribution is parabolic with extrema at the ends and in the center.

Record the maximum shear and moment stresses from the legends.

8. A closer look at the moment distribution reveals that it seems to be a combination of two cantilevered beams on the ends and a simply supported beam in between. We will refer to the connection point as a *momentless hinge*.

Estimate the position of the hinges.

Zoom on the shear-stress distribution and observe that the shear is nonzero at the hinge.

9. Return to the BCs section and change the BCs to pinned at one end and fixed at the other. That is,

[0:type =1],

[5:end_A =000001],

[6:others =111111],

[7:end_B =000000].

Remesh and rerun the static analysis.

View the Distributions of shear and moment stresses with

[1:vars =61],

[2:xMult =1].

By zooming, observe that the shear stress is linearly distributed with a maximum at the fixed end whereas the moment-stress distribution is parabolic with a single hinge.

Estimate the position of the hinge.

Record the maximum shear and moment stresses from the legends.

ANALYZING THE DATA

1. Make a table of all the recorded stress and hinge location values.

2. The cantilever and pinned beams are *statically determinate*; that is, equilibrium alone is sufficient for solving the force and moment distributions. A free body cut at an arbitrary section x can then easily establish that the shear-force and bending-moment distributions are

$$\text{cantilever: } V(x) = q_0[L - x], \qquad M(x) = q_0[\tfrac{1}{2}x^2 - Lx + \tfrac{1}{2}L^2],$$

$$\text{pinned: } V(x) = q_0[\tfrac{1}{2}L - x], \qquad M(x) = q_0[\tfrac{1}{2}x^2 - \tfrac{1}{2}Lx],$$

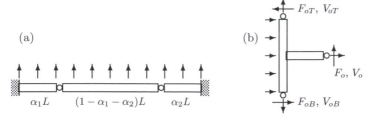

Figure 2.8. Some simple models: (a) model for a beam with elastic boundary conditions, (b) free-body diagram for frame joint. (The circles represent momentless hinges.)

where q_o is the constant load per unit length acting upward; this is calculated from the applied load resultant as P_o/L. The corresponding reported stress distributions are given by

$$\sigma_{mxy}(x) = \frac{V}{A} = \frac{P_o}{LA}[L - x], \qquad \sigma_{mxx}(x) = \frac{M}{I} = \frac{P_o}{LI}[\tfrac{1}{2}x^2 - Lx + \tfrac{1}{2}L^2],$$

$$\sigma_{mxy}(x) = \frac{V}{A} = \frac{P_o}{LA}[\tfrac{1}{2}L - x], \qquad \sigma_{mxx}(x) = \frac{M}{I} = \frac{P_o}{LI}[\tfrac{1}{2}x^2 - \tfrac{1}{2}Lx],$$

with $A = bh$ and $I = bh^3/12$.

How well do the recorded values agree with this analysis?

3. The fixed–fixed and fixed–pinned beams are *statically indeterminate*; that is, equilibrium alone is insufficient for solving the force and moment distributions, and recourse must be made to the deformations and the constitutive relation. This invariably involves solving differential equations. The thrust of our approximate method is to make assumptions that reduce these problems to ones that are statically determinate.

   The simple model in Figure 2.8(a) recognizes that a typical beam segment has one to two momentless hinges and their positions are parameterized by α_1 and α_2. A fixed–fixed beam has $\alpha_1 = \alpha_2 = 0.21$ whereas a fixed–pinned beam has $\alpha_1 = 0.21$, $\alpha_2 = 0$. How well are these supported by the data?

   An application of this simple model gives the bending stresses at the left, center, and right points as,

$$\sigma_{mxx} = \frac{P_o L}{4I}[1 + \alpha_1 - \alpha_2]\alpha_1 \quad = \frac{P_o L}{8I}[1 - \alpha_1 - \alpha_2]^2 \quad = \frac{P_o L}{4I}[1 - \alpha_1 + \alpha_2]\alpha_2.$$

   How well do the recorded values agree with this analysis?

4. When the simple model is used to determine the shear stress, it must be kept in mind that, although the hinge has no moment, it does have a shear force. The estimated shear stress at the beam ends are

$$\sigma_{mxy} = \frac{P_o}{2A}[1 + \alpha_1 - \alpha_2], \qquad = \frac{P_o}{2A}[1 - \alpha_1 + \alpha_2].$$

   How well do the recorded values agree with this analysis?

Part II. Transverse Loads on Continuous Beams

Beams typically have multiple spans as witnessed, say, by a bridge with multiple pylon supports, and are called *continuous* beams. When each span is viewed as a single beam, then its BCs are neither fixed nor free. The objective of the data collection and analysis is to extend the simple model of Part I to continuous beams.

If QED is not already running, begin by launching it from the command line in the working directory.

CREATING THE MODEL

1. Create a long slender three-span steel beam of dimensions [300 × 1.0 × 0.5] using the [Frame:1D_beam] geometry. That is,
   [0:geometry =1],
   [1:X_length =100],
   [4:#X_span =3],
   [a:mat#1 =2],
   [b:thick1 =1.0],
   [c:depth1 =0.5].

2. Use BCs of [0:type=1(=AmB)] to pin the ends and fix the connections. Set
   [0:type =1],
   [5:end_A =000001],
   [6:others =000000],
   [7:end_B =100001].

3. Use Loads of [0:type=2(=uni_res)] to apply a uniformly distributed load with a given resultant. Set
   [0:type =2],
   [1:side# =1],
   [4:X_load =0.0],
   [5:Y_load =600.]. (applied resultant P_o)
   All other loads are set to zero.
   This retains the same load per unit length as in Part I.

4. Mesh using <50> elements through the length of each span and restrict the motion to being planar with no gravity. That is,
   [0:Global_ID =22],
   [1:X_elems =50],
   [b?:gravity =0].
   Render the mesh by [pressing s].

COLLECTING THE DATA

1. Analyze the beam for its linear Static response. Set
   [0:type =1],
   [1:scale{P} =1],
   [2:scale{Grav} =0.0].
   Perform the analysis by [pressing s].

2. View the Distributions of shear and moment stresses with
   [1:vars =61],
   [2:xMult =1].
   By zooming, observe that the moment-stress distributions correspond to the concatenation of pinned–fixed, fixed–fixed, and fixed–pinned beams. Observe the moment jumps at the fixed supports.
   Record the maximum moment stresses in each beam segment.
3. Return to the BCs section and release the middle supports so that they are pinned. That is,
   [0:type =1],
   [5:end_A =000001],
   [6:others =100001],
   [7:end_B =100001].
   Remesh and rerun the static analysis.
   View the Distributions of shear and moment stresses with
   [1:vars =61],
   [2:xMult =1].
   By zooming, observe that the moment-stress distribution is now continuous with no jumps.
   Record the maximum moment stresses in each beam segment. Record the location of the hinges.
4. Return to the BCs section and fix the end supports. That is,
   [0:type =1],
   [5:end_A =000000],
   [6:others =100001],
   [7:end_B =100000].
   Remesh and rerun the static analysis.
   View the Distributions of shear and moment stresses. Observe that the moment-stress distribution is nearly the same in each beam segment.
   Record the maximum moment stresses in each beam segment. Record the location of the hinges.

ANALYZING THE DATA
1. Make a table of all the recorded stress and hinge location values.
2. A continuous beam has neither fixed nor pinned boundary conditions but lies somewhere in between. A usual assumption made is that $\alpha_1 = \alpha_2 \approx 0.2$.
   How well is this assumption supported by all the data?
   Suggest a refined assumption that better reflects the data.
3. Use the simple model with $\alpha = 0.2$ to determine the stresses in the beam.
   How well does the simple model agree with the data?

Part III. Lateral Loads on Buildings
Structures that provide open spaces are typically unbraced (no diagonal supports) and are called *portal frames*. The objective of the data collection and analysis is to further extend the simple models of Parts I and II to these type of frames.

Begin by launching QED from the command line in the working directory.

CREATING THE MODEL

1. Create a single-story steel portal frame using the [Frame:2D_frame] geometry. That is,
   [0:geometry =2],
   [1:X_length =100],
   [2:Y_height =50],
   [4:# X_bays =1],
   [5:#Y_floors =1],
   [7:brace =0],
   [a:mat#1 =2],
   [b:thick1 =1.0], (walls/columns)
   [c:depth1 =0.5],
   [d:mat#2 =2],
   [e:thick2 =1.0], (floors/girders)
   [f:depth2 =0.5].

2. Use BCs of [0:type=1(=AmB)] to pin [000001] the base of the frame. Set
   [0:type =1],
   [5:corner_A =000001],
   [6:others =111111],
   [7:corner_B =000001].

3. Use Loads of [0:type=2(=uni_res)] to apply a uniformly distributed load with a given resultant to the west wall. Set
   [0:type =2],
   [1:side# =4],
   [4:X_load =100].
   [5:Y_load =0.0].
   All other loads are set to zero.

4. Mesh using <50> elements for the floors, <25> elements for the walls and restrict the motion to being planar. That is,
   [0:Global_ID =22],
   [1:X_elems =50],
   [2:Y_elems =25].
   Set all gravity components as zero. Render the mesh by [pressing s].

COLLECTING THE DATA

1. Analyze the frame for its linear Static response. Set
   [0:type =1], [1:scale{P} =1],
   [2:scale{Grav} =0.0].
   Perform the analysis by [pressing s].

2. View the deformed Shapes with
   [1:vars =1],
   [2:xMult =10], (this exaggerates the deformed shape)

[5:rate =0].

[Press s] to show the deformed shape.

Observe the lateral sway of the frame and the distortion of the floor. Also observe the rotations at the joints. Considered as beams, the floor and walls behave as if they have a common connection BC somewhere between fixed and pinned; this is called an *elastic boundary condition*.

3. View the Distributions of stresses with

[1:vars =61],

[2:xMult =1].

By zooming, observe that the axial stress [Var:Smxx] is constant in each member, with opposite signs in the west and east walls and compression in the floor. Observe that the shear stress [Var:Smxy] is linearly distributed in the west wall and constant in the other members, and the moment stress [Var:Sfzz] is parabolic in the west wall and linear in the other members. Observe that each member has only one momentless hinge.

Record the maximum axial, shear, and moment stresses. Estimate and record the locations of the hinges.

4. Return to the BCs section and fix the base of the frame. Set

[0:type =1],

[5:corner_A =000000],

[6:others =111111],

[7:corner_B =000000].

Remesh and rerun the static analysis.

5. View the deformed Shapes with

[1:vars =1],

[2:xMult =10],

[5:rate =0].

[Press s] to show the deformed shape.

Observe that the lateral sway of the frame is considerably less than that of the previous case.

6. View the Distributions of stresses with

[1:vars =61],

[2:xMult =1].

By zooming, observe that the axial stress [Var:Smxx] and shear stress [Var:Smxy] are not too different from those of the previous case. Observe that the moment stress [Var:Sfzz] is parabolic in the west wall and linear in the other members. Observe also that the location of the momentless hinges is no longer at the base of the walls.

Record the maximum axial, shear, and moment stresses at the base of the frame. Estimate and record the locations of the hinges in each member.

7. Return to the Geometry section to create a two-story three-bay building. That is, set

[0:geom =2],

[1:X_length =100],
[2:Y_height =50],
[4:# X_bays =3],
[5:#Y_floors =2].

Go to BCs and fix the base of the frame. Set

[0:type =1],
[5:corner_A =000000],
[6:others =000000],
[7:corner_B =000000].

Go to Loads to set

[4:X_load =200].

This retains the same load per unit length.

Remesh and rerun the static analysis.

8. View the deformed Shapes with

[1:vars =1],
[2:xMult =50],
[5:rate =0].

Observe the lateral sway of the frame and the large distortion of the middle floors.

View the Distributions of stresses with

[1:vars =61],
[2:xMult =1].

By zooming, observe that the axial stress [Var:Smxx] is constant in each member but is largest in the middle floor and decreases west to east. Observe that the shear stress [Var:Smxy] is linear in the west walls and constant in the other members, being larger in the ground walls. Observe that the moment stress [Var:Sfzz] is parabolic in the west walls, linear in the other members, and that there is only one hinge per member.

Record the maximum axial stress in the middle floor and west wall, the shear stress at the midpoint of the ground walls, and the moment stresses at the base of the frame. Estimate and record the locations of the hinges in each member.

ANALYZING THE DATA

1. The key to extending the simple model is to make assumptions that convert the statically indeterminate frame problem into one that is statically determinate. Furthermore, it is the presence of momentless hinges that allows the parameterization of the variable boundary conditions.

For lateral-loading-type problems, three reasonable assumptions are (1) the hinge occurs at the midpoint of each member, (2) the shear in the middle vertical members is twice that of the outside ones, (3) the axial force in the middle vertical members is negligible.

How well are these assumptions supported by the data?

2. With the preceding model in mind and by making horizontal free-body cuts through the hinges to produce free bodies, one that includes just the top floor and one that includes the complete top floor plus half of the bottom floor, it can be shown that the maximum axial and shear stresses in the west wall are

$$\sigma_{mxx_{oT}} = +\tfrac{1}{8}q_o H^2/LA \, , \qquad \sigma_{mxy_{oT}} = -\tfrac{1}{12}q_o H/A,$$

$$\sigma_{mxx_{oB}} = +\tfrac{3}{8}q_o H^2/LA \, , \qquad \sigma_{mxy_{oB}} = -\tfrac{1}{4}q_o H/A.$$

The distributed load is computed from the applied load as $q_o = P_o/2H$. How well do these results agree with the data?

By making a free-body cut through the hinges that surround the middle joint of the west wall (this free body is shown in Figure 2.8), it can be shown that the axial stress in the first middle floor is

$$\sigma_{mxxo} = -\tfrac{2}{3}q_o H/A.$$

How well does this result agree with the data?

3. Conjecture about an appropriate free body to determine the maximum moment in the west wall.
4. Conjecture about ways to improve the simple model.

Part IV. Bridge Under Gravity Loads

The bridge of Figure 2.7(c) supports the loads acting on the deck basically by using the slants to suspend the deck from the top. Thus the dominant member loads are axial tension and compression (which is why the frames are often referred to as trusses).

The objective of the data collection and analysis is to construct a simple model for frames that are dominated by truss actions. Begin by launching **QED** from the command line in the working directory.

CREATING THE MODEL

1. Create a steel bridge using the [Frame:2D_bridge] geometry. That is,
   [0:geometry =4],
   [1:X_length =100],
   [2:Y_height =50],
   [4:# X_span =3],
   [7:brace =0],
   [a:mat#1 =2],
   [b:thick1 =1.0], (vertical/slant)
   [c:depth1 =0.5],
   [d:mat#2 =2],
   [e:thick2 =1.0], (bottom/top cord)
   [f:depth2 =0.5].

2. Use BCs of [0:type=1(=AmB)] to pin the ends. That is, set
   [0:type =1],
   [5:end_A =000001],
   [6:others =111111],
   [7:end_B =100001].
   Note that end B can move horizontally.
3. Mesh using <25> elements for the horizontals, <25> elements for the slants, and restrict the motion to being planar. That is,
   [0:Global_ID =22],
   [1:X_elems =25],
   [2:Y_elems =25].
   Set gravity on in the vertical direction with
   [by:gravity =-386]. (gravity in [in/s^2])
   The gravity special requires [pressing b] followed by [y] and the number. Render the mesh by [pressing s].

COLLECTING THE DATA

1. Analyze the frame for its linear Static response. Set
   [0:type =1],
   [1:scale{P} =0],
   [2:scale{Grav} =1.0].
   This activates only the gravity load. Perform the analysis by [pressing s].
2. View the deformed Shapes with
   [1:vars =1],
   [2:xMult =200], (this exaggerates the deformed shape)
   [5:rate =0].
   Observe that the horizontal members exhibit a sag similar to that of a beam with a uniform load. What is significant is that the bridge as a whole does not exhibit any sag; that is, the deformation of the horizontals (called *cords*) is local. This can be easily confirmed by placing a point load (equal to the total weight of the bridge *W*) at the top middle joint.
3. View the Distributions of stresses with
   [1:vars =61],
   [2:xMult =1].
   By zooming, observe that the axial stress [Var:Smxx] is constant in each member, with tension in the bottom horizontals, compression in the top horizontals, and alternating sign in the slants. Observe that the shear stress [Var:Smxy] and the moment stress [Var:Sfzz] have the appearance of a continuous beam with distributed load and pinned supports at the joints.
   Record the maximum axial, shear, and moment stresses from the legends.
4. Return to the Mesh section and give each member just a single element. That is,
   [0:Global_ID =22],

[1:X_elems =1],
[2:Y_elems =1].
Render the mesh and rerun the static analysis with
[0:type =1],
[1:scale{P} =0],
[2:scale{Grav} =1.0].
View the Distributions of stresses with
[1:vars =61],
[2:xMult =1].
By zooming, observe that the magnitude of the axial stress has changed only slightly but the magnitudes of the shear and moment stress are almost negligible. This has arisen because the loads are now concentrated at the joints. Record the maximum axial stress.

5. Return to the Mesh section and change each member to a truss member that supports only axial loads. That is,
[0:Global_ID =21],
[1:X_elems =1],
[2:Y_elems =1].
Render the mesh and rerun the static analysis. View the Distributions of stresses with
[1:vars =61],
[2:xMult =0.5].
Observe that the only stress is the axial stress, and it has not changed relative to the previous analysis.

ANALYZING THE DATA

1. The key assumption is that the main stiffness of the frame arises from the axial properties of the members. A subsidiary assumption is that bending effects are local.

How well is the first assumption supported by the data?

2. The simple model is essentially that all members are axially loaded only with no moment distributions or moments at joints. The applied loads act only at the joints.

The total weight of the frame is given by

$$W = [5 + 6/\sqrt{2}]\rho A L g = 133.$$

Assume this weight is divided equally among all the joints; then, by using a free body of the leftmost joint, it can be shown that the axial forces in the horizontal (F_{h1}) and slanted (F_{s1}) members lead to the stresses,

$$\sigma_{mxxh1} = +\frac{5}{14} W/A, \qquad \sigma_{mxxs1} = -\frac{5\sqrt{2}}{14} W/A.$$

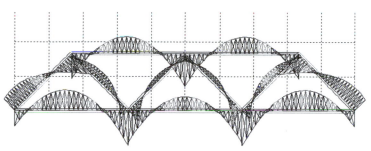

Figure 2.9. Partial results for a bridge under gravity loading. Moment-stress distributions.

How well do these results agree with the data?

By using a free body of the upper leftmost joint, it can be shown that the axial forces in the horizontal (F_{h2}) and slanted (F_{s1}, F_{s2}) members lead to the stresses,

$$\sigma_{mxxh1} = -\frac{8}{14}W/A, \qquad \sigma_{mxxs1} = -\frac{5\sqrt{2}}{14}W/A, \qquad \sigma_{mxxs2} = +\frac{3\sqrt{2}}{14}W/A.$$

How well do these results agree with the data?

3. Conjecture how the distributed load may be more accurately represented.
4. For a bridge of given geometry and material, the weight and stiffness are dominated by the member cross-sectional area. Conjecture how the local (flexural) deflections of the members can be decreased without changing the area.

Partial Results

Partial results for the moment distributions in the bridge of Part IV are shown in Figure 2.9.

Background Readings

References [53, 76] are very good introductions to civil engineering structural analysis. Both have many photographs of various types of structures and both have chapters devoted to the approximate analysis of indeterminate structures.

2.4 Stress Analysis of Thin-Walled Structures

Section 2.2 explored the stiffness behavior of thin-walled structures; this companion section explores various aspects of stresses in thin-walled structures. An essential distinction to be brought out is that between open cross sections (I-beam, C-channel, etc.) and closed cross sections (cylinder, airfoil, etc.); the shear flows (shear stresses) are very different and consequently so too are the torsional responses.

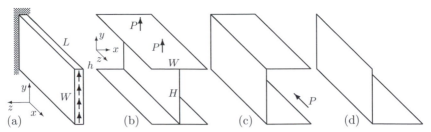

Figure 2.10. Static structures analyzed: (a) cantilevered plate, (b) I-beam, (c) C-channel, (d) angle.

2.4.1 Open Section Thin-Walled Structures

Figure 2.10 shows some examples of open cross sections. The objective of the explorations is to determine how the stresses in these examples depend on the structural and geometrical parameters.

For easy reference, the vertical panel is usually referred to as the *web*, and the horizontal panel is usually referred to as the *flange*.

Part I. Cantilevered Plate

We begin with the simple cross section of a narrow rectangle. The loading is such as to cause in-plane bending, in-plane shearing, and an axial twisting. The objective of the data collection and analysis is to determine the sensitivity of the stresses to the geometric parameters of length, thickness, and width. That is, our primary interest is establishing the dimensional dependencies $L^\alpha h^\beta W^\gamma$ of the stresses.

Launch **QED** from the command line in the working directory.

CREATING THE MODEL

1. Create an aluminum plate of dimensions $[10 \times 4 \times 0.2]$ using the [Plate:quadrilateral] geometry. That is,
   [0:geometry =1],
   [1:X_2 =10],
   [2:Y_2 =0],
   [3:X_3 =10],
   [4:Y_3 =4],
   [5:X_4 =0],
   [6:Y_4 =4],
   [a:mat# =1],
   [b:thick =0.2].
2. Use BCs of [0:type=1(=ENWS)] to fix [000000] the west side with all others being free [111111]. That is,
   [0:type =1],
   [4:sequence =snew],

```
[5:east =111111],
[6:north =111111],
[7:west =000000],
[8:south =111111].
```

3. Use Loads of [0:type=2(=traction)] to apply tractions on the east side. That is,

```
[0:type =2]
[1:side# =2],
[5:Y_trac =100], (traction t_y)
[7:X_mom =100]. (distributed moment m_x)
```

Ensure that all other loads are zero. The distributed moment causes a torsional effect about the x axis.

4. Mesh using <40> modules through the length and <16> transverse modules. Set

```
[0:Global_ID =32],
[1:X_mods =40],
[2:Y_mods =16].
```

Render the mesh by [pressing s].
By [pressing o], ensure the orientation is set as $\phi_x = 0$, $\phi_y = 0$, $\phi_z = 0$.

COLLECTING THE DATA

1. Analyze the plate for its linear Static response. Set

```
[0:type =1],
[1:scale{P} =1.0],
[2:scale{Grav} =0.0].
```

Perform the analysis by [pressing s].

2. View the Distributions of tractions with

```
[1:vars =1],
[2:xMult =5],
[4:X_cur =0],
[5:Y_cur =0],
[6:Z_cur =0],
[7:T_cur =90],
[8:% inc =5].
```

[Press s] to show the distribution of tractions.

The keys [xX yY tT] interactively change the position and orientation of the free-body cut. Place a vertical cut about midway along the plate and observe the almost-linear variation of normal traction from top to bottom, being zero at the center – this is the membrane-bending action. Observe the parabolic variation of shear traction from top to bottom, being maximum at the center – this is the membrane-shearing action.

3. View the Contours of stress with

```
[0:type =1],
[1:vars =21],
```

[2:xMult =1],

[5:rate =0],

and confirm that the contours are in agreement with expectation. Individual panels can be zoomed by pressing z #, where the panels are numbered 1 through 6, top row first.

4. Record the σ_{xx} ([Var:Smxx]) stress at the top midlength, the τ_{xy} ([Var:Smxy]) stress at the center midlength, and the flexural shear σ_{xy} ([Var:Sfxy]) stress at the center midlength. This is the reference data set.

   If there is difficulty in interpolating the stresses, then the [2:xMult=1.0] parameter can be changed to put a contour at the desired location. Alternatively, fixed scales can be used to bracket the estimated value.

5. Return to the Geometry section and change the dimensions of the plate to $[10 \times 8 \times 0.2]$. That is,

   [1:X_2 =10],

   [3:X_3 =10],

   [4:Y_3 =8],

   [6:Y_4 =8],

   [b:thick =0.2].

   Remesh and redo the static analysis. Record the three stresses; this is the sensitivity Case I data set.

6. Return to the Geometry section and change the dimensions of the plate to $[10 \times 4 \times 0.4]$. That is,

   [1:X_2 =10],

   [3:X_3 =10],

   [4:Y_3 =4],

   [6:Y_4 =4],

   [b:thick =0.4].

   Remesh and redo the static analysis. Record the three stresses; this is the sensitivity Case II data set.

7. Return to the Geometry section and change the dimensions of the plate to $[20 \times 4 \times 0.2]$. That is,

   [1:X_2 =20],

   [3:X_3 =20],

   [4:Y_3 =4],

   [6:Y_4 =4],

   [b:thick =0.2].

   and remesh. Remesh and redo the static analysis. Record the three stresses; this is the sensitivity Case III data set.

ANALYZING THE DATA

1. Scale each stress to the corresponding reference data set (the reference data set becomes unity). Make a table of all the scaled values.

2. On a single graph and using an appropriate identifying symbol for each stress, plot the reference and Case I data against W. Connect like symbols to form the width sensitivity plots.

3. On a single graph and using an appropriate identifying symbol for each stress, plot the reference and Case II data against h. Connect like symbols to form the thickness sensitivity plots.

4. On a single graph and using an appropriate identifying symbol for each stress, plot the reference and Case III data against L. Connect like symbols to form the length sensitivity plots.

5. A simple model for beam bending gives the bending stress as

$$\sigma_{xx} = \frac{My}{I} \longrightarrow \frac{(t_y hWL/2)(W/2)}{\frac{1}{12}hW^3} = \frac{3t_y L}{W} \qquad \text{or} \qquad \sigma_{xx} \propto L^1 h^0 / W^1.$$

Does the dependence of the bending stress on the width, thickness, and length dimensions correlate with these dimensional powers?

6. A simple model for the shear stress in a beam is

$$\tau_{xy} = \frac{V}{A} \longrightarrow \frac{(t_y hW)}{hW} = t_y \qquad \text{or} \qquad \tau_{xy} \propto W^0 h^0 L^0.$$

Does the dependence of the shear stress on the width, thickness, and length dimensions correlate with these dimensional powers?

7. A simple model for twisting of a rectangular cross section gives the shear stress as

$$\tau_{xy} = \frac{2Tz}{J} \longrightarrow \frac{2(m_x hW)(h/2)}{\frac{1}{3}Wh^3} = \frac{3m_x}{h} \qquad \text{or} \qquad \tau_{xy} \propto W^0 L^0 / h^1.$$

Does the dependence of the twisting shear stress on the width, thickness, and length dimensions correlate with these dimensional powers?

8. For the reference data set only, how well do the three recorded stresses agree with

$$\sigma_{mxx} = \frac{My}{I} = \frac{3t_y L}{W}, \qquad \tau_{mxy} = \frac{V}{A} = t_y, \qquad \tau_{fxy} = \frac{2Tz}{J} = \frac{3m_x}{h},$$

respectively?

Conjecture why the membrane shear stress τ_{mxy} seems to be significantly off relative to the other comparisons.

Part II. Symmetrically Loaded I-Beam

The I-beam is put on two supports and two transverse loads are applied; all these are placed symmetrically with respect to the center so that each half of the beam is essentially cantilevered. The objective of the data collection and analysis is to determine the effect of material distribution in the cross section on the stress distributions.

Begin by launching QED from the command line in the working directory.

CREATING THE MODEL

1. Create an aluminum I-beam using the [Open:I_beam] geometry; set the dimensions as $L = 10$, $W = 1.5$, $H = 3$. That is, set
   [0:geometry =2],
   [1:length =10.],
   [2:width =1.5],
   [3:depth =3.0].
   Take the flange thickness as $t_f = 0.001$, the web thickness as $t_w = 0.20$, and the stringer areas as $A = 0.001$, and do not attach an end plate. That is,
   [a:mat# =1],
   [b:flange =0.001],
   [c:web =0.2],
   [d:area =0.001],
   [e:end plate =0].

2. Use BCs of [0:type=2(=2_supp)] so that the beam has two support points near the center. That is,
   [0:type =2)],
   [6:Z_sup1 =4.5],
   [7:Z_sup2 =5.5].

3. Use Loads of [0:type=2(=2_vert)] to apply vertical loads near the ends. That is,
   [0:type =2],
   [1:Z_pos1 =0.5],
   [2:Z_pos2 =9.5],
   with the values
   [4:Y_load1 =100],
   [5:Y_load2 =100].

4. Mesh using $<40>$ modules through the length, $<6>$ for the flange, and $<12>$ for the web. That is,
   [0:Global_ID =32],
   [1:length =40],
   [2:flange =6],
   [3:web =12],
   [7:Z_rate =40];
   all [specials] can be set as zero. Render the mesh by [pressing s].
   For clearer viewing, set the orientation ([press o]) as $\phi_x = +20$, $\phi_y = +65$, $\phi_z = 0$. If desired, toggle the top flange off by [pressing t 3 0]. Other substructures can similarly be turned on or off.

COLLECTING THE DATA

1. Analyze the I-beam for its linear Static response.
   Set
   `[0:type =1]`,
   `[1:scale{P} =1.0]`,
   `[2:scale{Grav} =0.0]`.
   [`Press s`] to perform the analysis. If StaDyn appears not to run, change the memory allocation in `<<stadyn.cfg>>` to a value slightly larger than that suggested in the `<<stadyn.log>>` file.

2. View the stress Contours with
   `[0:type =1]`,
   `[1:vars =21]`,
   `[2:xMult =1]`,
   `[5:rate =0]`.
   Observe the bending action of the σ_{xx}, (`[Var:Smxx]`) stress and the almost-parabolic distribution (through the height) of the τ_{xy}, (`[Var:Smxy]`) shear stress in the web.

3. Record the normal membrane σ_{xx} (`[Var:Smxx]`) stress at the top center and the τ_{xy} (`[Var:Smxy]`) shear stress at the center of either the left or right webs.
   If there is difficulty interpolating the stresses, then the `[2:xMult=1.0]` parameter can be changed to put a contour at the desired location.

4. Return to the Geometry section and change the flange and web thicknesses to
   `[b:flange =0.1]`,
   `[c:web =0.1]`;
   this leaves the cross-sectional area unchanged.
   Remesh and redo the static analysis. Record the normal membrane stress at the top center and shear stress at the center of either the left or right webs.

5. Return to the Geometry section and change the geometry to a C-channel with
   `[0:Geometry =3]`.
   Remesh and redo the static analysis. Record the maximum normal membrane stress at the center top and the shear stress at the center of either the left or right webs. Observe that the σ_{xx} stress goes from compression to tension in the top flange.

ANALYZING THE DATA

1. Assign the numbers 1, 2, and 3 to the three cases respectively; this will be the independent variable for plotting purposes.
2. On a single graph, plot the recorded normal stress for each case.
3. A simple model for the bending action of the I-beam is

$$\sigma_{xx} = \frac{My}{I} \quad \longrightarrow \quad \sigma_{\max} = \frac{24P}{H^2 t_w[1 + 6(W/H)(t_f/t_w)]}.$$

Superpose this relation on the plot and connect the points.

How well do the model and data agree?

Conjecture why the C-channel case is significantly off in the comparison.

4. On a single graph, plot the recorded shear stress for each case.

5. A simple model for the shear stress in the I-beam is

$$\tau_{xy} = \frac{V}{A} \quad \longrightarrow \quad \tau_{max} = \frac{P}{Ht_w}.$$

Superpose this relation on the plot and connect the points.

How well do the model and data agree?

Conjecture why the C-channel case is significantly off in the comparison.

Part III. Cantilevered Angle Section

When dealing with members of arbitrary cross section, it is essential to distinguish among the concepts of centroid, shear center, neutral axis, and principal axis. The objective of the data collection and analysis is to determine each of these for an angle cross section.

Begin by launching QED from the command line in the working directory.

CREATING THE MODEL

1. Create an aluminum angle section using the [Open:angle] geometry; set the dimensions as $L = 20$, $W = 4$, $H = 4$. That is, set
   [0:geometry =1],
   [1:length =20],
   [2:width =4.0],
   [3:depth =4.0].
   Take the flange thickness as $t_f = 0.4$, the web thickness as $t_w = 0.4$, the stringer areas as negligible with $A = 0.001$, and attach an end plate. That is,
   [a:mat# =1],
   [b:flange =0.4],
   [c:web =0.4],
   [d:area =0.001],
   [e:end plate =1].

2. Use BCs of [0:type=1(=L/R pln)] so that the left side is cantilevered and the right side is free. That is,
   [0:type =1)],
   [8:L_bc =000000],
   [7:R_bc =111111].

3. Use Loads of [0:type=3(=w/end_frame)] to apply an axial P_z load; the purpose of the frame is to facilitate the placing of the load at an arbitrary position. That is,
   [0:type =3],
   [1:X_pos =2.0],
   [2:Y_pos =2.0],
   with the values

```
[4:X_load =0],
[5:Y_load =0],
[6:Z_load =1],
[7:X_mom =0],
[8:Y_mom =0],
[9:Z_mom =0].
```

4. Mesh using <40> modules through the length and <16> in the hoop direction.
   That is,
   ```
   [0:Global_ID =32],
   [1:length =40],
   [2:flange =8],
   [3:web =8],
   [7:Z_rate =40];
   ```
   all [specials] can be set as zero. [Press s] to render the mesh.
   For clearer viewing, set the orientation ([press o]) as $\phi_x = +35$, $\phi_y = -45$,
   $\phi_z = 0$.

COLLECTING THE DATA

1. Analyze the channel for its linear Static response. Set
   ```
   [0:type =1],
   [1:scale{P} =100],
   [2:scale{Grav} =0.0].
   ```
   [Press s] to perform the analysis.
2. View the Contours of stress with
   ```
   [0:type =1],
   [1:vars =21],
   [2:xMult =1],
   [5:rate =0].
   ```
   Observe, by zooming and noting the legend, that there is a significant membrane axial stress (σ_{xx},[Var:Smxx]) that varies from tension on the outside to compression at the joint – the axial load is causing a bending action. To avoid this, the axial load must be placed at the *centroid* of the cross section. We find this position by interpolation by using a second load case. First record the maximum and minimum σ_{xx} stresses.
3. Return to the Loads section and change
   ```
   [1:X_pos =0.0],
   [2:Y_pos =0.0].
   ```
   Because of symmetry along the 45° line, we can assume that the centroid is along this line. Remesh and redo the Static analysis. View the stress contours and observe that now the membrane axial stress (σ_{xx},[Var:Smxx]) varies from compression on the outside to tension at the joint. Record the maximum and minimum stresses.
   Plot the values of stress against position and interpolate to find where all stresses are the same – this is the centroid. Record these coordinates.

Return to the Loads section and place the load at the centroid. Remesh and redo the Static analysis. View the stress contours and verify that the [Var:Smxx] stress is almost uniform. Record this stress in the middle section. (If the contours are too sparse to interpolate, View the Distributions with [1:vars =93] and move the cursor to the point of interest to display all the stresses at that point.)

4. Return to the Loads section and change to a T_x moment loading:

[1:X_pos =2.0],
[2:Y_pos =2.0],
[4:X_load =0],
[5:Y_load =0],
[6:Z_load =0],
[7:X_mom =1],
[8:Y_mom =0].

Remesh and redo the Static analysis. View the stress Contours and observe by zooming that the (σ_{xx},[Var:Smxx]) stress has a bending action but it is not symmetric about the $y = 2$ axis. It appears that the single moment about the x axis is causing bending about two axes.

View the Distributions with
[1:vars =93]
to interrogate the stresses at individual points. [Press s] to display the structure and then the [x/X y/Y z/Z] keys to move the cursor. Record where, at midlength, the (σ_{xx},[Var:Smxx]) stress is zero in both the flange and web. A line connecting these two points is called the *neutral axis*; that is, the line along which the bending stress is zero for this loading.

Return to the Loads section and change to a T_y moment loading:
[7:X_mom =0],
[8:Y_mom =-1].

Remesh and redo the Static analysis. View the Distributions and record where, at midlength, the (σ_{xx},[Var:Smxx]) stress is zero in both the flange and web. Observe that this neutral axis is different than the previous case.

5. Return to the Loads section and change to a P_y loading:

[1:X_pos =2.0],
[2:Y_pos =2.0],
[4:X_load =0],
[5:Y_load =1],
[6:Z_load =0],
[7:X_mom =0],
[8:Y_mom =0].

Remesh and redo the Static analysis. View the stress contours and observe, by zooming, that the membrane stresses are as expected for a cantilevered beam. Also observe the significant value of the flexural (σ_{xy},[Var:Sfxy]) stress. From Part I, this is to be associated with a torsional effect – it appears that the transverse load is causing a torsional stress. To avoid this, the transverse load must act through the *shear center* of the cross section. We find this position by interpolation by using a second load case. First record the (σ_{xy},[Var:Sfxy]) stress in the middle section.

Return to the Loads section and change
[1:X_pos =-2.0],
[2:Y_pos =2.0].
Remesh and redo the Static analysis. View the stress contours and record the (σ_{xy},[Var:Sfxy]) stress in the middle section.

Plot the values of stress against position and interpolate to find where the stress is zero – this is the x-location of the shear center. Record this value.

Return to the Loads section and place the load at the shear center. Remesh and redo the Static analysis. View the stress contours and confirm that the (σ_{xy},[Var:Sfxy]) stress is now insignificant.

ANALYZING THE DATA

1. The coordinates of the centroid of an angle section are

$$x_c = \frac{W^2}{2[H + W]}, \qquad y_c = \frac{H^2}{2[H + W]}.$$

How well do the measured coordinates compare with these?

2. Draw the centerline of the angle cross section and indicate the position of the centroid. Superpose the two recorded neutral axes on this drawing.
   Do the neutral axes go through the centroid?
   Conjecture why it might be expected that the neutral axes go through the centroid.

3. The *principal axis* of a cross section is the axis for which the moment and neutral axis are parallel. Compute the orientation (angle) of the two neutral axes and plot them against the orientation of the corresponding moment (treated as 0° and −90°, respectively). Connect the two points. Superpose a line drawn at −45°. Determine where the two lines intersect; this is the estimate of the orientation of the principal axes.
   The orientation of the principal axes for the symmetric angle section is ±45°. How well does the measured estimate agree with this?

4. The shear center for cross sections having a single intersecting point is the intersecting point itself. How well does the measured shear center agree with this?
   Conjecture about the location of the shear center for an I-beam.
   Conjecture again why the C-channel case in Part II is significantly off in the comparisons.

Partial Results

Partial results for the sensitivity study of Part I are shown in Figure 2.11(a). The stresses for the I-beam and C-channel of Part II are shown in Figure 2.11(b).

Partial results for determining the cross-sectional properties of Part III are shown in Figure 2.12.

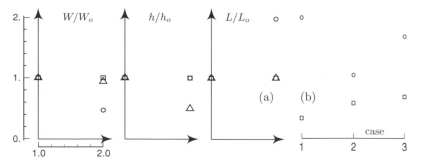

Figure 2.11. Partial results for the stress sensitivity study: (a) flat plate, (b) open cross-section beams.

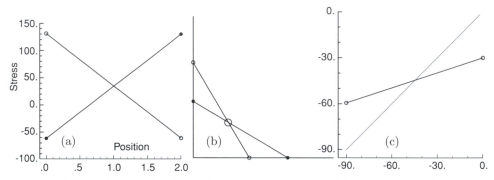

Figure 2.12. Partial results for the angle section: (a) centroid, (b) neutral axes, (c) principal orientation.

2.4.2 Closed Thin-Walled Structures

Some examples of closed cross sections are shown in Figure 2.13. The wing and airfoil are referred to as being two-celled (or multicelled) whereas the box beam is single-celled. Both the wing and the box beam have longitudinal stiffeners called *stringers*.

Each structure will be loaded differently so that different aspects of the stress analysis can be explored.

Part I. Bending of a Wing Structure with Stringers

The wing is put in bending by a moment applied about the horizontal axis. The objective of the data collection and analysis is to compare the stress distributions with those of a simple model that emphasizes the dominant role of the stringers.

Begin by launching QED from the command line in the working directory.

CREATING THE MODEL

1. Create an aluminum two-spar wing using the [Closed:wing] geometry; set the dimensions as $L = 100$, $x_2 = 40$, $y_2 = 4$, $y_3 = 16$, $y_4 = 20$. That is, set
    [0:geometry =3],
    [1:length =100],

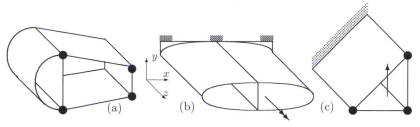

Figure 2.13. Cantilevered thin-walled structures: (a) bending of a wing, (b) airfoil with torsional load, (c) three-stringer box beam.

```
[6:X_2 =40],
[7:Y_2 =4],
[8:Y_3 =16],
[9:Y_4 =20].
```
Take the skin and spar thickness as $h = 0.05$ and the stringer areas as $A = 2$ and attach an end plate. That is,
```
[a:mat# =1],
[b:skin =0.05],
[c:spar =0.05],
[d:area =2],
[e:end plate =1].
```

2. Use BCs of [0:type=1(=L/R pln)] to fix the wing at one end and set free at the other. That is,
```
[0:type =1],
[8:L_bc =000000],
[9:R_bc =111111].
```

3. Use Loads of [0:type=3(=end w/fr)] to apply a vertical load by means of an end frame. That is,
```
[0:type =3],
[1:X_pos =0.0],
[2:Y_pos =10],
[5:Y_load =-1000];
```
set all other loads as zero.

4. Mesh using <20> modules through the length and <40> in the hoop direction. That is,
```
[0:Global_ID =32],
[1:length =20],
[2:hoop =40],
[3:V/H bias =1],
[4:Z_rate =20];
```
all [specials] can be set as zero. Render the mesh by [pressing s]. For clearer viewing, set the orientation ([press o]) as $\phi_x = +25$, $\phi_y = +50$, $\phi_z = 0$.

If desired, toggle the top skin and end plates off by [pressing t 3 0] and [t 6 0], respectively. Other substructures can similarly be turned on or off.

COLLECTING THE DATA

1. Analyze the structure for its linear Static response. Set
    [0:type =1],
    [1:scale{P} =1.0],
    [1:scale{Grav} =0.0].
    Perform the analysis by [pressing s]. If StaDyn appears not to run, change the memory allocation in <<stadyn.cfg>> to a value slightly larger than that suggested in the <<stadyn.log>> file.

2. View the stress Contours with
    [0:type =21],
    [1:vars =1],
    [2:xMult =1],
    [5:rate =0].
    We wish to concentrate on the spars (the vertical sections), so remove all skins by ([pressing t 1 0]), ([pressing t 3 0]), and ([pressing t 4 0]). Individual panels can be zoomed by [pressing z #], where the panels are numbered 1 through 6, top row first.
    Observe that the σ_{xx} ([Var:Smxx]) stress is the dominant stress in the spars and except at edges and attachment points, all other stresses are zero.
    Observe that the σ_{xx} ([Var:Smxx]) stress varies from tension at the bottom to compression at the top and is uniform along the length. This is characteristic of a beam in bending. Record the maximum σ_{xx} ([Var:Smxx]) stress in the leading spar.

3. Return to the Geometry section and change the stringer area to
    [d:area =4].
    Remesh and redo the Static analysis. Record the maximum σ_{xx} ([Var:Smxx]) stress in the leading spar.

4. Repeat the analysis and data collection for the additional stringer areas
    [d:area =8, 16].
    Make sure to remesh after each area change.

ANALYZING THE DATA

1. Plot the stress data against area.
2. A simple model for the bending stresses in a thin-walled wing with stringers is to assume that all the bending stiffness arises solely from the stringers. That is,

$$\sigma_{xx} = \frac{My}{I}, \qquad I = \sum_i y_i^2 A_i,$$

where A_i is the area of stringer i and y_i is its distance from the neutral axis (the horizontal line of symmetry).

Superpose, as a continuous line, this model on the preceding plot.
How well does this simple model match the data?

3. Conjecture about the comparison as the area is made smaller.

Part II. Torsion of an Airfoil Structure

The airfoil is twisted by a torque applied along the axis. The objective of the data collection and analysis is to compare the stress distributions with those of a simple model based on constant shear flow in each cell.

Begin by launching **QED** from the command line in the working directory.

CREATING THE MODEL

1. Create an aluminum thin-walled airfoil using the [Closed:airfoil] geometry; set the dimensions as $L = 5$, width $W = 5$, depth $H = 1.0$. That is,

   [0:geometry =4],
   [1:length =5.0],
   [6:X_2 =1.5],
   [7:Y_2 =0.5],
   [8:X_3 =4.0].

   Set the skin and spar thicknesses as $h = 0.062$ and $h = 0.085$, respectively, the stringer areas as $A = 0.001$ (this essentially removes the stringers), and attach an end plate. That is,

   [a:mat# =1],
   [b:skin =0.062],
   [c:spar =0.084],
   [d:area =0.001],
   [e:end plate =1].

2. Use BCs of [0:type=4(=pivot)] so that the end points are in pivots placed at the center of the airfoil. That is,

   [0:type =4],
   [6:X_pos =2.0],
   [7:Y_pos =0.0],
   [8:L_bc =000000],
   [9:R_bc =000001].

3. Use Loads of [0:type=3(=end w/fr)] to apply an axial torque load by means of a rigid frame attached to the end plate. That is,

   [0:type =3],
   [1:X_pos =2.0],
   [2:Y_pos =0.0],
   [9:Z_mom =100];

   all other loads are zero.

4. Mesh using <15> modules through the length and <40> modules around the hoop direction. That is,

   [0:Global_ID =32],
   [1:length =15],

[2:hoop =40],
[3:V/H bias =1],
[4:Z_rate =15];
all [specials] can be set as zero.
Render the mesh by [pressing s]. For clearer viewing, set the orientation ([press o]) as $\phi_x = +25$, $\phi_y = +30$, $\phi_z = 0$. If desired, toggle the end plate off by [pressing t 6 0]. Other substructures can similarly be turned on or off.

COLLECTING THE DATA

1. Analyze the airfoil for its linear Static response. Set
   [0:type =1],
   [1:scale{P} =1.0],
   [2:scale{Grav} =0.0].
   Perform the analysis by [pressing s].
2. View the displacement Contours with
   [0:type =1],
   [1:vars =1],
   [2:xMult =1],
   [5:rate =0].
   For clearer viewing, surfaces may be toggled on or off by [pressing t # 1/0]. Observe that the dominant action is a rotation about the z axis. Observe the nonuniform axial (w, [Var: W]) displacement.
   View the stress contours with
   [1:vars =21]
   and observe that the axial stress (σ_{xx}, [Var:Smxx]) varies along the length, being a maximum at the corners and zero at midlength. Observe that the shear stress (σ_{xy}, [Var:Smxy]) is almost uniform in each cell.
   Record the maximum shear stress.
3. Return to the Geometry section and decrease the height of the airfoil by two. That is, set
   [7:Y_2 =0.25].
   Remesh and repeat the Static stress analysis.

ANALYZING THE DATA

1. Plot the shear-stress data against height.
2. A simple model [35] for the shear flows in a two-celled section is

$$q_1 = \frac{T_z[A_1 a_2 + A_2 a_{12}]}{2[A_1^2 a_2 + 2A_1 A_2 a_{12} + A_2^2 a_1]},$$

$$q_2 = \frac{T[A_2 a_1 + A_1 a_{12}]}{2[A_1^2 a_2 + 2A_1 A_2 a_{12} + A_2^2 a_1]},$$

where A_i is the area of the cell and

$$a_1 = \oint_1 \frac{ds}{t}, \qquad a_{12} = \int_{12} \frac{ds}{t}, \qquad a_2 = \oint_2 \frac{ds}{t}$$

are circumference lengths. The shear stress is obtained from the shear flow by

$$\tau_i(s) = q_i / h(s),$$

where s is the position around the cell circumference and $h(s)$ is the thickness at the point of interest. As a further approximation, assume both cells are the same and the overall shape is elliptical; then

$$q_1 = q_2 = \frac{T_z}{2A} = \frac{2T_z}{\pi WH}, \qquad \tau = \frac{2T_z}{\pi WHh}.$$

Superpose, as a continuous line, this model on the preceding plot.
Is the behavior of the shear stresses in agreement with this simple model?

3. Conjecture about the role of the spar in the torsional behavior of the airfoil.

Part III. Bending of a Thin-Walled Beam

The box beam is put in bending by a vertical load applied at the end. The objective of the data collection and analysis is to determine the effect of the stringers and load position on the stress behavior of the structure.

Launch **QED** from the command line in the working directory.

CREATING THE MODEL

1. Create an aluminum thin-walled shear beam using the [Closed:triangle] geometry; set the dimensions to $L = 40$, $x_2 = 20$, $y_2 = 0$, $x_3 = 20$, $y_3 = 20$. That is,
   [0:geometry =2],
   [1:length =40.],
   [6:X_2 =20.],
   [7:Y_2 =0.0],
   [8:X_3 =20.],
   [9:Y_3 =20.].
   Take the skin thickness as $h = 0.1$, the stringer areas as $A = 2.0$, and attach an end plate. That is,
   [a:mat# =1],
   [b:skin =0.1],
   [d:area =2.0],
   [e:end plate =1].

2. Use BCs of [0:type=1(=L/R pln)] to set the left plane as fixed and the right plane as free. That is,
   [0:type =1],
   [8:L_bc =000000],
   [9:R_bc =111111].

3. Use Loads of [0:type=3(=end w/fr)] to apply a vertical load at the coordinates $x = 0$, $y = 2$, by means of a rigid frame attached to the end plate. That is,
   [0:type =3],

     `[1:X_pos =0.0]`,
     `[2:Y_pos =10.]`,
     `[5:Y_load =1000]`;
     set all other loads as zero.

4. Mesh using $<28>$ modules through the length and $<48>$ modules around the circumference. That is,
     `[0:Global_ID =32]`,
     `[1:length =28]`,
     `[2:hoop =48]`,
     `[3:V/H bias =1]`,
     `[4:Z_rate =28]`;
     set all `[specials]` as zero. Render the mesh by `[pressing s]`.
     For clearer viewing, toggle the slanted face off by pressing `[t 3 0]`, and set the orientation as $\phi_x = +25$, $\phi_y = +35$, $\phi_z = 0$.

COLLECTING THE DATA

1. Analyze the structure for its linear Static response. Set
     `[0:type =1]`,
     `[1:scale{P} =1.0]`,
     `[2:scale{Grav} =0.0]`.
     Perform the analysis by `[pressing s]`.
     If StaDyn appears not to run, change the memory allocation in `<<stadyn.cfg>>` to a value slightly larger than that suggested in the `<<stadyn.log>>` file.

2. View the displacement Contours with
     `[0:type =1]`,
     `[1:vars =1]`,
     `[2:xMult =1]`,
     `[5:rate =0]`.
     Observe the twisting action. Record the maximum vertical (v, `[Var:V]`) deflections at the stringers and the maximum axial (ϕ_z, `[Var:Rz]`) twist.
     View the stress Contours with
     `[1:vars =21]`.
     Observe that, except at corners, the shear stress (σ_{xy}, `[Var:Smxy]`) is nearly uniformly distributed in each segment, but the stress that is due to bending (σ_{xx}, `[Var:Smxx]`) varies on the cross section.
     At the midlength location $z = 20$, record the bending stress at the apexes and the shear stress at the panel center. (If necessary, the surfaces can be toggled on or off by pressing `[t # 0]` where the bottom, vertical, and slanted sides are numbered 1, 2, 3, respectively.)

3. Repeat the analysis and data collection with the load at positions
     `[1:X_pos =0, 5, 10, 15, 20]`.
     Make sure to remesh after each load change.

4. Return to the Geometry section and change the stringer area to $A = 0.2$. That is,

[d:area =0.2].
Remesh and redo the complete Static analysis and data collection.

ANALYZING THE DATA

1. Plot all deflections against load position using a different symbol for each stringer area case.
   What is the effect of stringer area on the deflections?
2. Plot the rotations against load position using a different symbol for each stringer area case.
   Obtain the shear center – the load position that does not cause any rotation.
   What is the effect of stringer area on the shear center?
   How well does the location compare with $e = 0.707\,W$?
   Conjecture why it is expected that the shear center does not depend on the stringer area.
3. Plot the bending stresses against load position using a different symbol for each stringer area case.
   What is the effect of load position on this stress?
   What is the effect of stringer area on this stress?
   Estimate the bending stresses if the load was applied at the shear center; how well do these stresses agree with

$$\sigma_{xx} = \left[-(x - x_c) + 2(y - y_c) \right] \frac{V[L - z]}{W^2},$$

   where (x_c, y_c) is the area centroid?
4. Plot the shear stress against load position using a different symbol for each stringer area case.
   What is the effect of load position on this stress?
   What is the effect of stringer area on this stress?
   Estimate the shear stresses if the load was applied at the shear center; how well do these stresses agree with

$$\sigma_{xy1} = -\frac{\tau_o}{2 + \sqrt{2}}, \qquad \sigma_{xy2} = \frac{\tau_o}{\sqrt{2}}, \qquad \sigma_{xy3} = -\frac{\tau_o}{2 + \sqrt{2}}, \qquad \tau_o \equiv \frac{V}{Wh},$$

   where the subscript numbers refer to the panel?

Partial Results

Partial results for the stresses in the wing of Part I are shown in Figure 2.14.

Background Readings

Analytical treatments for the stresses in thin-walled structures are given in References [87, 93, 99, 123]. Because the walls are thin, the state of stress is essentially that of plane stress, and this simplifies the analyses. An essential distinction brought out

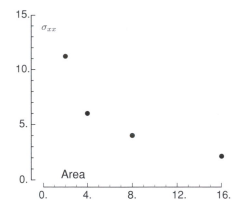

Figure 2.14. Partial results for the stresses in a wing.

is that between closed cross sections (cylinder, airfoil, etc.) and open cross sections (I-beam, C-channel, etc.); the shear flows are very different and consequently so too are the torsional responses.

2.5 Stress Analysis of a Ring

A FE analysis can solve almost any problem with any geometry. What we often need, however, is an understanding of the underlying mechanisms of the problem. That is, it is necessary to see through the numbers produced by the analysis and discern the underlying patterns. This can often be achieved by looking at variations of a given problem.

This exploration addresses the static stress distributions and stress concentrations in a nonuniform specimen. The complexity of the geometry and its influence on the resulting stress patterns are investigated through a series of simple models with increasing geometric complexity. The final (or general) model to be understood is that of an annulus with tabs in which the load is applied as shown in Figure 2.15.

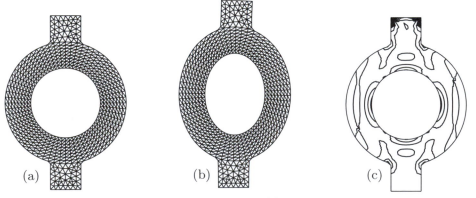

Figure 2.15. Stress analysis of a ring: (a) mesh, (b) exaggerated ($\times 150$) deformed shape, (c) contours of von Mises stress.

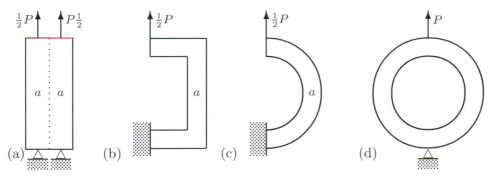

Figure 2.16. Sequence of geometries: (a) plate, (b) rectangle C, (c) circular C, (d) annulus.

The variations of geometries are shown in Figure 2.16. The complexity is in terms of the effect of the radius and the influence of the bending at sections not just on the horizontal cross section.

Part I. General Problem

We begin with the full ring model. The objective of the data collection is to identify significant stress features of the ring and to record a data set for assessing the simpler models.

Launch QED from the command line in the working directory.

CREATING THE MODEL

1. Create a complete aluminum ring model with tabs using the [Hole:ring tab] geometry. That is,
   [0:geometry =2],
   [1:inner_R =2.0],
   [2:outer_R =3.5],
   [5:X_tab =1.75],
   [6:Y_tab =1.5],
   [a:mat# =1],
   [b:thick =0.25].
2. Use BCs of [0:type=1(bottom)] to set the bottom on rollers. That is,
   [0:type =1]
   [9:bottom =100000].
   For this BC type, the bottom center node is automatically fixed.
3. Use Loads of type [0:type=1(near xyz)] to apply a vertical force on the top tab. That is,
   [0:type =1],
   [1:X_pos =0.0],
   [2:Y_pos =4.0],
   [5:Y_load =-1000],
   with all other loads being zero.

4. Mesh using <64> modules around the hoop, <8> modules radially, and <6> tab modules. That is,

```
[0:Global_ID =22],
[1:hoop divs =64],
[2:radial divs =8],
[3:tab divs =6],
[7:# smooth =0],
[8:max elems =9000];
```

all [specials] can be set as zero.

Render the mesh by [pressing s]. If necessary, reorient the mesh by [pressing o] to set $\phi_x = 0$, $\phi_y = 0$, $\phi_z = 0$.

COLLECTING THE DATA

1. Analyze the ring for its linear Static response. Set

```
[0:type =1],
[1:scale{P} =1.0],
[2:scale{Grav} =0.0].
```

Perform the analysis by [pressing s].

2. View the deformed Shapes with a magnification of 200. That is,

```
[1:vars =1],
[2:xMult =200],
[5:rate =0].
```

Observe how the width shrinks and the height elongates. Individual panels can be zoomed by [pressing z #], where the panels are numbered 1 through 6, top row first.

3. View the Distributions of tractions with

```
[1:vars =1],
[2:xMult =2],
[4:X_cur =0],
[5:Y_cur =0],
[6:Z_cur =0],
[7:T_cur =90],
[8:% inc =5].
```

[Press s] to show the distribution of tractions.

The keys [xX yY tT] interactively change the position and orientation of the free-body cut. Place the cut along the horizontal line of symmetry and observe the variation of normal traction from the inside to outside radius.

4. View the Contours of specials to see where the maximum stresses occur. Set

```
[0:type =1],
[1:vars =51],
[2:xMult =1],
[5:rate =0].
```

Zoom the first panel and observe that, as expected, the stresses are quite large at the tab root, but they are also quite large on the inner radius at the horizontal and vertical lines of symmetry.

View the Contours of stress with
[0:type =1],
[1:vars =21].
Record the extreme values of σ_{yy} along the horizontal line of symmetry; call these values DataSet I.

Part II. Uniform Plate

The simplest model of any loaded structure is just a uniform block for which most geometric irregularities such as notches or holes are ignored. The simple model of Figure 2.16(a) gives an estimate of the stress as

$$\sigma_{yy} = \frac{P}{2A}.$$

This simple model is related to the tabbed ring by $a = R_o - R_i$, $A = ah$, where h is the thickness of the ring.

The objective of the data collection and analysis is to assess this simple model and relate it to the ring problem. Begin by launching QED from the command line in the working directory.

CREATING THE MODEL

1. Create a simple [$3 \times 6 \times 0.25$] aluminum plate using the [Plate:quadrilateral] geometry. That is, set
   [1:X_2 =3],
   [2:Y_2 =0],
   [3:X_3 =3],
   [4:Y_3 =6],
   [5:X_4 =0],
   [6:Y_4 =6],
   [a:mat# =1],
   [b:thick =0.25].
2. Use BCs of [0:type=3(=2_supp)] to give the plate two support points. That is, set
   [0:type =3],
   [1:posn_1 =1.0],
   [2:posn_2 =2.0].
3. Use Loads of [0:type=3(=2_vert)] to apply two symmetric loads on the top surface. That is, set
   [0:type =3],
   [2:posn_1 =1.0],
   [3:posn_2 =2.0],
   [5:valu_1 =500],
   [6:valu_2 =500].
4. Mesh using $< 12 \times 24 >$ modules by setting
   [0:Global_ID =22],
   [1:X_mods =12],

[2:Y_mods =24];
with all [specials] as zero. Render the mesh by [pressing s]. If necessary, reorient the mesh by [pressing o]) to set $\phi_x = 0$, $\phi_y = 0$, $\phi_z = 0$.

COLLECTING THE DATA

1. Analyze the plate for its linear Static response with
   [0:type =1],
   [1:scale{P} =1.0],
   [2:scale{Grav} =0.0].
   Perform the analysis by [pressing s].
2. View the deformed Shapes with
   [1:vars =1],
   [2:xMult =200],
   [5:rate =0].
   Observe the severe deformation in the vicinity of the loads and supports.
3. View the stress Contours with
   [0:type =1],
   [1:vars =21],
   [2:xMult =1],
   [5:rate =0].
   Zoom on the σ_{yy} ([Var:Smyy]) stress and observe that the stress concentrations are near the points of applied load and support. Observe that it is necessary to move to a location approximately one width from the applied load for the stress distribution to become nearly uniform. This can also be seen through the traction distributions.
   Record the value of stress in the middle of the model.

ANALYZING THE DATA

1. Do the observed stress contours agree with the simple model?
   Comment on what ways they are different.
2. Compare the recorded value of stress with that of the simple model. How well do they agree?
   Compare the recorded value of stress with that of DataSet I. How well do they agree?
3. Although this simple model is a very crude model of a ring, it nonetheless can always be used as an order-of-magnitude estimate of the average stresses. Conjecture on situations for which this could be useful.

Part III. Rectangular C Model

This simple model, shown in Figure 2.16(b), introduces bending effects in addition to the axial loading. The normal traction distribution on the narrow section of the C is estimated by

$$\sigma_{yy} = \frac{F}{A} + \frac{Mx}{I} = \frac{P}{2A}\left[1 - \frac{x}{R}\frac{AR^2}{I}\right],$$

where $R = (R_o + R_i)/2$ is the horizontal distance to the middle of the cross section, x is the distance from this point, and $I = ha^3/12$ for the narrow section.

The objective of the data collection and analysis is to assess this simple model and relate it to the ring problem. Begin by launching QED from the command line in the working directory.

CREATING THE MODEL

1. Create an aluminum C specimen with rectangular corners using the [Notch:rect_notch] geometry. That is, set
   [0:geometry =3],
   [1:X_dim1 =3.5],
   [2:Y_dim1 =7.0], (outer dimensions)
   [6:X_dim2 =2.0],
   [7:Y_dim2 =4.0], (inner dimensions)
   [a:mat# =1],
   [b:thick =0.25].
2. Use BCs of [0:type=3(=fixed)] to set the bottom left side as fixed. That is, set
   [0:type =3].
3. Use Loads of [0:type=1(near xyz)] to apply 500 vertically at the top. That is, set
   [0:type =1],
   [1:X_pos =0.0],
   [2:Y_pos =7.0],
   [5:Y_load =500],
   with all other loads being zero.
4. Mesh using a total of <100> modules around the circumference. That is,
   [0:Global_ID =22],
   [5:tot_mods =100],
   [7:# smooth =5],
   [8:max elems =2000];
   all [specials] can be set as zero.

COLLECTING THE DATA

1. Analyze the plate for its linear Static response with
   [0:type =1],
   [1:scale{P} =1.0],
   [2:scale{Grav} =0.0].
   [Press s] to perform the analysis.
2. View the deformed Shapes with
   [1:vars =1],
   [2:xMult =10].
   [5:rate =0].
   Note that the multiplication factor is smaller than the previous case; why? Observe how the C shape is bending.

3. View the stress Contours with
   [0:type =1],
   [1:vars =21],
   [2:xMult =1],
   [5:rate =0].
   Observe that, aside from the severe stress concentrations at the inner corners
   and load points, the primary action is the nonuniform σ_{yy} ([Var:Smyy]) stress in
   the vertical leg; it goes from tension on the inside to compression on the outside.
   Record the maximum stress on the cross section.

ANALYZING THE DATA

1. The scale factor for observing the deformed shape in this part is significantly
   smaller than in Part II; why?
2. The maximum stress in this part is significantly higher than in Part II; why?
3. Do the stress contours agree with what is expected from the simple model?
   Does the recorded stress agree with that of the simple model?
4. How close are the recorded values of stress to the values of DataSet I?

Part IV. Circular C Model

This simple model, shown in Figure 2.16(c), adds the effect of curved bending to the
previous model. The distribution of stress at the midsection is no longer linear, but
parabolic, and approximated by

$$\sigma_{yy} = \frac{F}{A} + \frac{M}{AR} - \frac{M}{I}\frac{x}{(1-x/R)} = \frac{P}{2A}\left[2 - \frac{x/R}{(1-x/R)}\frac{AR^2}{I}\right],$$

where R is the middle radius of the cross section, x is the radial distance from this
line, and $I = ha^3/12$ for the section.

   The objective of the data collection and analysis is to assess this simple model
and relate it to the ring problem. Begin by launching **QED** from the command line
in the working directory.

CREATING THE MODEL

1. Create an aluminum semicircular C model using the [Plate:C_curve] geometry
   and dimensions similar to those of the original ring. That is,
   [0:geometry =5],
   [1:rad_in =2.0],
   [2:rad_out =3.5],
   [3:angle =180],
   [a:mat# =1],
   [b:thick =0.25].
2. Use BCs of [0:type=4(=set T/B)] to set the left bottom side as fixed. That is,
   [0:type =4],
   [8:bottom =000000].

3. Use Loads of [0:type=1(=near xyz)] to apply 500 vertically at the top. That
   is, set
   [0:type =1],
   [1:X_pos =0.0],
   [2:Y_pos =3.5],
   [5:Y_load =500],
   with all other loads being zero.
4. Mesh using <32> modules around the hoop direction and <8> modules radially, with no smoothing. That is,
   [0:Global_ID =22],
   [1:R_mods =8],
   [2:H_mods =32],
   [4:min R% =1],
   [5:exp =1.0],
   [7:# smooth =0];
   set all [specials] as zero.

COLLECTING THE DATA
1. Analyze the plate for its linear Static response with
   [0:type =1],
   [1:scale{P} =1.0],
   [2:scale{Grav} =0.0].
   [Press s] to perform the analysis.
2. View the deformed Shapes with
   [1:vars −1],
   [2:xMult =10],
   [5:rate =0].
   Observe how the C shape is bending.
3. View the Distributions of tractions with
   [1:vars =1],
   [2:xMult =2],
   [4:X_cur =0],
   [5:Y_cur =0],
   [6:Z_cur =0],
   [7:T_cur =90],
   [8:% inc =5].
   [Press s] to show the distribution of tractions.
   The keys [xX yY tT] change the position and orientation of the free-body cut.
   Place the cut along the horizontal line of symmetry and observe the variation of
   normal traction from the inside to outside radius.
4. View the stress Contours with
   [0:type =1],
   [1:vars =21],

[2:xMult =1],
[5:rate =0].
Observe that this model has removed the high-stress concentrations incurred at the inner corners of the previous model.
Record the edge stresses on the cross section.

ANALYZING THE DATA
1. Do the stress contours agree with what is expected from the simple model? Do the recorded stresses agree with those of the simple model?
2. How close are the recorded values of stress to the values of DataSet I?

Part V. Complete Ring
This model completes the full geometry of the ring and is shown in Figure 2.16(d). A simple model for the stress distribution in a ring [128] is given by

$$\sigma_{\theta\theta} = \frac{P}{2A} \left[\frac{2}{\pi} + \left(\cos\theta - \frac{2}{\pi} \right) \frac{x/R}{(1 - y/R)} \frac{AR^2}{I} \right],$$

where y is the distance from the middle surface and θ is the angle off the horizontal line of symmetry.

The objective of the data collection and analysis is to assess this simple model and relate it to the ring problem. Begin by launching **QED** from the command line in the working directory.

CREATING THE MODEL
1. Create a complete aluminum ring model using the [Hole:annulus] geometry and dimensions of the original ring. That is,
   [0:geometry =3],
   [1:inner_R =2.0],
   [2:outer_R =3.5],
   [a:mat# =1],
   [b:thick =0.25].
2. Use BCs of [1:type=1(=bottom)] to fix the bottom. That is,
   [0:type =1],
   [8:bottom =000000].
3. Use Loads of type [0:type=1(near xyz)] to apply a vertical force of 1000 at the top. That is,
   [0:type =1],
   [1:X_pos =0.0],
   [2:Y_pos =3.5],
   [5:Y_load =1000],
   with all other loads being zero.
4. Mesh using <64> modules around the hoop direction and <8> modules radially. That is,
   [0:Global_ID =22],

[1:hoop divs =64],
[2:radial divs =8],
and all [specials] as zero.
Render the mesh by [pressing s]. If necessary, reorient the mesh by [pressing o] to set $\phi_x = 0$, $\phi_y = 0$, $\phi_z = 0$.

COLLECTING THE DATA

1. Analyze the ring for its linear Static response. Set
   [0:type =1],
   [1:scale{P} =1.0],
   [2:scale{Grav} =0.0].
   [Press s] to perform the analysis.
2. View the deformed Shapes with
   [1:vars =1],
   [2:xMult =100].
   [5:rate =0].
   What is the biggest difference between the present deformed shape and that of the previous model?
3. View the stress Contours with
   [0:types =1],
   [1:vars =21],
   [2:xMult =1],
   [5:rate =0].
   Observe that the distribution in the region along the midsection has not changed significantly from those of Parts III and IV. This leads to the conclusion that, for geometries that are not identical but somewhat similar, the stress distributions may be approximated using results from the simpler geometry. That is, for a relatively thin-walled section, the predominant action is that of bending. On the other hand, when the hole is small, the predominant action is that of a uniform distribution of axial stresses, except in the immediate vicinity of the hole.
   Record the edge stresses on the cross section.

ANALYZING THE DATA

1. Do the stress contours agree with those expected from the simple model?
   Do the recorded stresses agree with the simple model?
2. How close are the edge values of stress to the values of DataSet I?
3. In terms of stress features, in what ways does the complete ring resemble the general problem?
   In terms of stress features, in what ways does the complete ring differ significantly from the general problem?
4. What is the effect of the load tabs in the general problem?

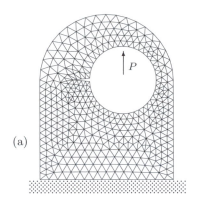
(a)

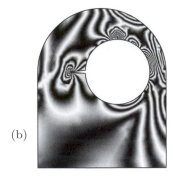
(b)

Figure 2.17. A model with two stress singularities and multiple stress concentrations around the hole: (a) mesh, (b) von Mises stress contours.

Partial Results

Partial results for the complete ring with tabs are shown in Figure 2.15.

Background Readings

There are many good introductory texts to the area of solids and structures; a selection is given in References [19, 47, 128]. More advanced ideas in the theory of elasticity are developed in References [11, 124].

2.6 Stress Concentrations and Singularities

This problem exploration considers the difference between a stress concentration and a stress singularity in a linear elastic body. The stress in the former case is finite, and any refinement of a FE mesh or experimental technique will approach this finite value in the limit. The stress in the latter case, however, is infinite, and any refinement of a FE mesh or experimental technique will approach this infinite value in the limit. Clearly, from a stress analysis point of view, attempting to measure the stress at a singularity would be fruitless; what is needed instead are alternative measures of the loading effect near these stress singularities.

The main singularity we analyze is that associated with cracks, although there are many others such as point loads, any reentrant corner, and contact areas.

Part I. General Problem

We begin by observing the difference between a stress singularity and a stress concentration. The model and loading of Figure 2.17 is used, and the effect of mesh refinement on the computed values of stress is explored. The model used here is not one of QED's preparameterized ones but uses the [General:] forms motif; this section therefore also acts a tutorial for this type of model creation.

Launch QED from the command line in the working directory.

CREATING THE MODEL

1. Create a model similar to Figure 2.17 using the [General:arbitrary] geometry forms. The paradigm is that of forms and the tabs are shown on the top ribbon. Access the form for constructing arbitrary 2D shapes by [pressing fa]. This allows construction of the model geometry; BCs, Loads, and material properties will be specified via the SDF form.

2. All entries are changed by pressing two characters: the first letter of the box followed by the first letter of the line. This will invoke a dialog box for data input; the data entry is completed by [pressing ENTER]. For example, specify the element type by [pressing et], then typing [4] in the dialog box, followed by ENTER. Choose material [m_tag=1] (aluminum) and [name=qed.msh]. A summary of these instructions is

   [et 4],
   [em 1],
   [en qed.msh].

   A colon will be used to separate the command from the input.

3. The geometry is constructed by specifying a collection of control points ([Points of control:]) and connecting them ([Connect pts:]) in various ways. The model of Figure 2.17 has eight control points. The locations of these points are specified by

   [p1: =0.0 3.0],
   [p2: =0.0 0.0],
   [p3: =4.0 0.0],
   [p4: =4.0 3.0],
   [p5: =1.5 3.05],
   [p6: =3.5 3.0],
   [p7: =1.5 2.95],
   [p8: =1.0 3.0].

   Ensure remaining entries are zero.

   Segments are specified through connecting two control points and specifying a line style. The six available line styles are given in the information box in the lower right hand corner. Connect the eight segments as

   [c1: =1 2 1],
   [c2: =2 3 1],
   [c3: =3 4 1],
   [c4: =4 1 -2],
   [c5: =5 6 2],
   [c6: =6 7 2],
   [c7: =7 8 1],
   [c8: =8 5 1].

   Ensure remaining entries are zero.

   Note that the solid material is always to the left of the segment direction. Finally, we need to divide each segment into a number of elements. Divide as

   [ds:smooth =5],

[d1: =1 3 15],
[d2: =4 6 20],
[d3: =7 8 4].

Ensure remaining entries are zero.

Note that ranges of segments are specified and these can overlap. Smoothness of the created mesh is achieved by trying to equalize the areas of neighboring elements.

4. The mesh is rendered by [pressing s] and can be viewed in PlotMesh as file <<qed.msh>>. It should be the same as Figure 2.17.

5. Access the SDF form by [pressing fs]. This allows the specification of BCs, Loads, material Properties and so on. In the [Make SDFile:] block, specify
[m1:in mesh =qed.msh],
[m2:out SDF =qed.sdf],
[m3:Global =22],
[m4:BW_red =1].

The last of these reduces the bandwidth of the stiffness matrix. In the [Proper-ties:] block, specify
[p1:tag_1 =1],
[p2:h_1 =1.0].

Ensure other tags are zero.

The BCs are imposed by isolating the appropriate nodes inside a cylinder. We will impose that the bottom is fixed:
[b1:R =.01],
[b2:X1 =-1.0 0.0 0.0],
[b3:X2 =4.0 0.0 0.0],
[b4:D =000000].

Ensure that the radii for the other cylinders are zero.

Loads can be specified either through a cylinder or as a point load at the nearest node. We impose the latter:
[lb:X =2.5 4.0 0.0],
[lc:P =0 1000 0],
[ld:T =0 0 0].

Ensure that all other loads are zero.

6. The SDF is rendered by [pressing s] and can be viewed in PlotMesh as file <<qed.sdf>>. The model is now ready for analysis.

COLLECTING THE DATA

1. Analyze the model for its linear Static response. Set
[0:type =1],
[1:scale{P} =1.0],
[2:scale{Grav} =0.0].
[Press s] to perform the analysis.

If necessary reorient the mesh by [pressing o] and changing $\phi_x = 0$, $\phi_y = 0$, $\phi_z = 0$.

2. View the displacement Contours, having set
   `[0:type =1]`,
   `[1:vars =1]`,
   `[2:xMult =1]`,
   `[5:rate =0]`.
   Observe the discontinuity of v displacement across the crack. Individual panels can be zoomed by `[pressing z #]` where the panels are numbered 1 through 6, top row first. (This can also be done while in zoom mode.)
   View the stress Contours with
   `[1:vars =21]`.
   There are multiple sites where the stresses are large. To get a combined stress comparison between these sites, view the von Mises stress contours with
   `[1:vars =51]`.
   Zoom the von Mises panel and observe that there are three main areas of stress concentration: the load point, the crack tip, and the right side of the hole.
   Make a note of the maximum stress (from the contour legend) at the three concentration points.

3. Return to the `[General:]` model section and access the arbitrary 2D shapes form by `[pressing fa]`. Change the density (Division) of elements around the hole by specifying
   `[d2: =4 6 40]`.
   Render the mesh by `[pressing s]`.
   Access the SDF form by `[pressing fs]`; then render it by `[pressing s]`. Confirm the change of element density by `[pressing R]` in **PlotMesh**.

4. Analyze the problem as Static. View the von Mises stress contours with
   `[1:vars=51]`.
   Zoom the von Mises panel and observe (from the contour legend) that the maximum stress at the load point has nearly doubled, but that of the stress concentration to the right is unchanged. This can be confirmed by setting the axes scales as fixed with
   `[0:type =2]`,
   `[5:var_max =5e3]`,
   `[6:var_min =0]`.

5. Return to the `[General:]` model section and change the density (Division) of elements by specifying
   `[d2: =4 6 20]`,
   `[d3: =7 8 16]`.
   This restores the original density around the hole but increases it at the crack. Render the mesh by `[pressing s]` then render the SDF by `[pressing fs s]`.

6. Analyze the problem as Static. View and zoom the von Mises stress contours and observe (from the contour legend) that the maximum stress at the crack-tip point has nearly quadrupled.

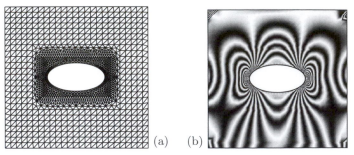

Figure 2.18. Elliptical hole in a plate: (a) mesh, (b) contours of σ_{yy} stress.

ANALYZING THE DATA

1. A stress singularity means that the stresses vary as $1/r^\alpha$, where r is the distance from the singularity point. Conjecture on the similarities and differences between the singularities at the crack tip and load point.

2. Conjecture on how these similarities and differences could be assessed quantitatively.

Part II. Elliptical Holes

From the analysis of an elliptical hole [27, 124] of size $[2a \times 2b]$ in an infinite plate (see Figure 2.18), under uniform remote uniaxial tension σ_{nom}, it is found that the maximum stress at the edge of the hole is

$$\sigma_{\max} = \sigma_{\text{nom}} \left[1 + 2\frac{a}{b}\right] \quad \text{or} \quad \sigma_{\max} = \sigma_{\text{nom}} \left[1 + 2\sqrt{\frac{a}{\rho}}\right], \qquad \rho = \frac{b^2}{a},$$

where ρ is the radius of curvature at the edge. This gives the stress concentration factor (SCF) for a circular hole as 3; but as the radius of curvature becomes very small, the SCF goes to infinity. In other words, for a sharp crack, the stress is infinite even under moderate applied loads. We need a different measure of loading in the vicinity of the crack tip.

Multiply the relation for the maximum stress for the ellipse by $\sqrt{\rho}$ to get

$$\sqrt{\rho}\,\sigma_{\max} = \sigma_{\text{nom}} \left[\sqrt{\rho} + 2\sqrt{a}\right].$$

The quantity $\sqrt{\rho}\,\sigma_{\max}$ has a definite limit $\to 2\sigma_{\text{nom}}\sqrt{a}$ as the root radius approaches zero. This motivates the introduction of a measure of crack-tip loading, called the *stress intensity factor*, defined as

$$K \equiv \tfrac{1}{2}\sqrt{\pi\rho}\,\sigma_{\max} = \sigma_{\text{nom}}\sqrt{\pi a}\left[1 + \frac{1}{2}\sqrt{\frac{\rho}{a}}\right] = K_{\text{crack}}\left[1 + \frac{1}{2}\sqrt{\frac{\rho}{a}}\right].$$

The stress intensity factor K should be thought of as a load measure analogous to stress, but one that also accounts for the very sharp geometric discontinuity at a crack tip.

The objective of the data collection and analysis is to assess this relationship between stress intensity and ellipse radius. Begin by launching **QED** from the command line in the working directory.

CREATING THE MODEL

1. Create an aluminum $[5 \times 5 \times 0.25]$ plate with an elliptical hole using the [Hole:ellipse] geometry. That is,
   [0:geometry =5],
   [1:length 2a =2.0],
   [2:length 2b =1.0],
   [3:X_dim =5.0],
   [4:Y_dim =5.0],
   [5:X_pos =2.5],
   [6:Y_pos =2.5],
   [7:angle =0],
   [a:mat# =1],
   [b:thick =0.25].
2. Use BCs of [0:type=1=(ENWS)] to fix the south boundary. That is, set
   [0:type =1],
   [4:sequence =enws].
   [5:east =111111],
   [6:north =111111],
   [7:west =111111],
   [8:south =000000].
3. Use Loads of [0:type=2(=traction)] to apply a traction on the north edge. That is, set
   [0:type =2],
   [1:side# =3],
   [5:Y_trac =100];
   ensure that all other loads are zero.
4. Mesh using
   [0:Global_ID =22],
   [1:X_mods =25],
   [2:Y_mods =25],
   [3:notch divs =16],
   [7:# smooth =0],
   [8:max elems =9000].
   Set all [specials] as zero and [press s] to render the mesh.
   If necessary, reorient the mesh by [pressing o]) and setting $\phi_x = 0$, $\phi_y = 0$, $\phi_z = 0$. The mesh should look like that in Figure 2.18.

COLLECTING THE DATA

1. Analyze the plate for its linear Static response. Set
   [0:type =1],

```
[1:scale{P} =1.0],
[2:scale{Grav} =0.0].
```
[Press s] to perform the analysis.

2. View the displacement Contours having set
```
[0:type =1],
[1:vars =1],
[2:xMult =1],
[5:rate =0].
```
Confirm that the displacements are continuous and the appropriate symmetries are exhibited. Individual panels can be zoomed by pressing [z #], where the panels are numbered 1 through 6, top row first. (This can also be done while in zoom mode.)

View the stress contours with
```
[1:vars =21].
```
Zoom the σ_{yy} ([Var:Smyy]) panel and record the maximum value from the contour legend; call this σ_{max}.

3. Return to the Geometry section and change the ellipse geometry to
```
[2:length 2b =2].
```
Remesh and redo the Static analysis.

Record σ_{max}.

4. Repeat the analysis for the additional geometry values of
```
[2:length 2b =0.5, 0.25, 0.125]
```
and record the maximum stress. Remember to remesh after each change. It may also be necessary to refine the mesh for the smaller radii.

Observe how the localized region of high stress shrinks as the ellipse becomes sharper.

ANALYZING THE DATA

1. Make a plot of $\sqrt{\rho}\,\sigma_{max}$ against $\sqrt{\rho}$, where $\rho = b^2/a$.
2. Does the stress follow the expected behavior?
3. Conjecture on the behavior for small radius.

Part III. Stress Analysis of Cracks

It is apparent that directly determining the stress state in the vicinity of a crack by using mesh refinement is computationally expensive. Here, instead, the singularity is assumed, and this is related to quantities that are globally easier to compute.

The virtual crack-closure method (VCCM) [62, 63, 108] relates the work done in closing a FE crack to the work done in the continuum crack. This results in

$$K_{\mathrm{I}} = \sqrt{\frac{E P_y(v_t - v_b)}{\Delta a}}, \qquad K_{\mathrm{II}} = \sqrt{\frac{E P_x(u_t - u_b)}{\Delta a}},$$

where Δa is the crack-length extension, P_x and P_y are the crack-tip forces, E is Young's modulus, and u and v are the crack-opening displacements of the nodes

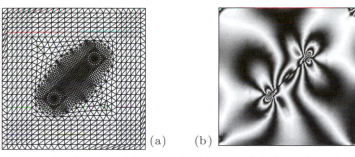

Figure 2.19. Crack at arbitrary angle: (a) mesh, (b) photoelastic fringes.

nearest the crack tip. The subscripts b and t refer to the bottom and top of the crack, respectively. The subscripts I and II refer to the Mode I (opening) and Mode II (shearing) stress intensities, respectively.

The objective of the data collection and analysis is to investigate how K_I and K_{II} change as the orientation of the crack is changed. Begin by launching **QED** from the command line in the working directory.

CREATING THE MODEL

1. Create a $[5 \times 5 \times 0.25]$ aluminum plate with a slanted crack (see Figure 2.19) by using the `[Hole:crack]` geometry. That is,
   `[0:geometry =4]`,
   `[1:length 2a =2.0]`,
   `[3:X_dim =5.0]`,
   `[4:Y_dim =5.0]`,
   `[5:X_pos =2.5]`,
   `[6:Y_pos =2.5]`,
   `[7:angle =0]`.
   `[a:mat# =1]`,
   `[b:thick =0.25]`.
2. Use BCs of `[0:type=1(=ENWS)]` to fix the south boundary. That is, set
   `[0:type =1]`,
   `[4:sequence =ewns]`,
   `[5:east =111111]`,
   `[6:north =111111]`,
   `[7:west =111111]`,
   `[8:south =000000]`.
3. Use Loads of `[0:type=2(=traction)]` to apply a traction on the north edge. That is, set
   `[0:type =2]`,
   `[1:side# =3]`,
   `[5:Y_trac =100]`;
   ensure all other loads are zero.

4. Mesh using
   [0:Global_ID =22],
   [1:X_mods =25],
   [2:Y_mods =25],
   [3:crack divs =16],
   [7:# smooth =0],
   [8:max elems =9000].
   Set all [specials] as zero. Render the mesh by [pressing s].

COLLECTING THE DATA

1. Analyze the plate for its linear Static response. Set
   [0:type =1],
   [1:scale{P} =1.0],
   [2:scale{Grav} =0.0].
   [Press s] to perform the analysis.
2. View the displacement Contours with
   [0:type =1],
   [1:vars =1],
   [2:xMult =1],
   [5:rate =0].
   Confirm that the displacements exhibit the appropriate symmetries and are continuous except at the crack where they show a discontinuity in v.
   View the stress contours with
   [1:vars =21].
   Zoom the σ_{yy} ([Var:Smyy]) panel. Confirm that the remote stress is nearly uniform and approximately equal to the imposed value of 100.
   View the special stress contours with
   [1:vars =51],
   [7:fsig/h =50].
   Make a gray-scale PostScript image of the (pseudo) photoelastic fringes by [pressing g3]. The stored file is named <<fringe.ps>>; use GhostView to view the image. (Note that smaller values of f_σ means more sensitivity.)
   Still within the View/contours section, change the variables for numbers output with
   [1:vars =92].
   [Press s] to see the numbers and record the stress intensity values.
3. Return to the Geometry section and change the orientation of the crack to
   [7:angle =20].
   Remesh and repeat the analysis.
   Record the stress intensity values.
4. Repeat the complete analysis for other values of crack orientation such as
   [7:angle = 0, 20, 40, 60, 80].

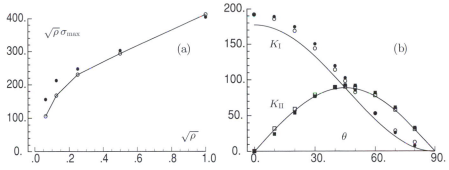

Figure 2.20. Partial results for stress singularities and concentrations: (a) stress behavior for an elliptical hole, (b) stress intensity factor variation with angle.

ANALYZING THE DATA

1. Plot the stress intensity factor data against crack orientation.
2. The stress intensity factors for a crack in an infinite sheet under remote tension σ_o and shear τ_o are, respectively,

$$K_{\mathrm{I}} = \sigma_o\sqrt{\pi a}, \qquad K_{\mathrm{II}} = \tau_o\sqrt{\pi a}.$$

A simple model for the stress intensity variation with angle θ is obtained by resolving the applied stress perpendicular and parallel to the crack to get

$$K_{\mathrm{I}} = \sigma_o\sqrt{\pi a}\,\cos^2\theta, \qquad K_{\mathrm{II}} = \sigma_o\sqrt{\pi a}\,\sin\theta\cos\theta.$$

Superpose this model as a continuous curve. Does the simple model capture the behavior of the crack?

Partial Results

The partial results are shown in Figure 2.20. The two data sets in Figure 2.20(a) are for different mesh densities, whereas in Figure 2.20(b) they correspond to the two crack tips.

Background Readings

Determining the stress concentrations for various structural components has been a very important part of the history of stress analysis. A compendium of results can be found in Reference [100]. Fracture mechanics and stress analysis of cracks are more recent developments; introductory material can be found in References [24, 66], and more advanced ideas are found in Reference [42]. A compendium of stress intensity factors can be found in Reference [106].

3 Vibration of Structures

When a structure is disturbed from its static equilibrium position, a motion ensues. When the motion involves a cyclic exchange of kinetic and strain energy, the motion is called a *vibration*. When this occurs under zero loads, it is called a *free* or *natural* vibration; whereas if the only loads are dissipative, then it is called a *damped* vibration. An example of a damped vibration is shown plotted in Figure 3.1; observe that the amplitudes eventually decreases to zero, but oscillates as it does so.

The explorations in this chapter consider the linear vibrations of structures. The first exploration uses a simple pretensioned cable to introduce the two basic concepts in vibration analyses; namely, that of *natural frequency* and *mode shape*. The second exploration looks at the meaning of mode shape in complex thin-walled structures. The stiffness properties are affected not only by the elastic material properties but also by the level of stress; the third exploration looks at the effect of prestress on the vibration characteristics. A significant insight into linear dynamics can be gained by analyzing it in the frequency domain. The fourth exploration introduces DiSPtool as the tool to switch between the time and frequency domains. Generally, increasing the mass of a structure decreases the vibration frequencies; however, in the presence of gravity, the mass can increase or decrease the stiffness and thereby affect the vibrations differently. The fifth exploration considers this. The final exploration is devoted to the choice of level of modeling and, in particular, the choice of FE.

The dynamics of a linear elastic structure, taking into account the effect of preloadings, is described by

$$[M]\{\ddot{u}\} + [C]\{\dot{u}\} + \big[[K_E] + \chi[K_G]\big]\{u\} = \{P\},$$

where $[M]$ is the structural mass matrix, $[C]$ is the damping matrix, $[K_E]$ is the elastic stiffness, $[K_G]$ is the load-dependent geometric stiffness, χ is the scale on the preloading, and $\{P\}$ are the additional dynamic loads. Free undamped vibrations are described by motions of the form

$$\{u\} = \{\hat{u}\}\sin \omega t, \qquad \{P\} = 0,$$

where ω is the angular frequency. The corresponding vibration eigenvalue problem is described by

$$\big[[K_E] + \chi[K_G] - \lambda[M]\big]\{\hat{u}\} = \{0\}, \qquad \lambda = \omega^2,$$

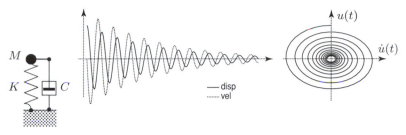

Figure 3.1. Damped response that is due to initial displacement.

where the frequency ω^2 is considered the unknown eigenvalue and the vector $\{\hat{u}\}$ is the eigenvector or mode shape. There are as many eigenvalues and eigenvectors as the system size. A vibration eigenanalysis directly finds the natural frequencies and associated shapes for a general system.

The vibration characteristics of structures can be represented as

$$\text{frequency} \propto \sqrt{\frac{\text{stiffness}}{\text{mass}}} \qquad \text{or} \qquad \omega = \alpha\sqrt{\frac{K}{M}}.$$

In constructing simple models for particular structures, it is often just a matter of identifying an appropriate stiffness and mass to be used.

3.1 Introduction to Vibrations

This section explores some basic aspects of the free vibration of a structure. A particular aspect of interest is the effect of damping on the free vibrations. The pretensioned cable shown in Figure 3.2 is used to illustrate the vibration shapes.

In each case, the cable will be set in motion by plucking, that is, a force is applied slowly to cause a displacement, and then released rapidly. In preparation, create a force file, call it <<force.31>>, with the [time force] contents

```
0.00   0
0.10   1
0.20   1
0.21   0
1.00   0
10.0   0
```

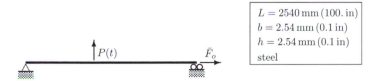

$L = 2540\,\text{mm}\,(100.\,\text{in})$	
$b = 2.54\,\text{mm}\,(0.1\,\text{in})$	
$h = 2.54\,\text{mm}\,(0.1\,\text{in})$	
steel	

Figure 3.2. Effect of damping on the vibration of components. Very thin pretensioned cable.

Part I. Pluck Tests at the Center

A cable, because it has a narrow cross section, has large axial stiffness but very little transverse stiffness. The transverse stiffness comes from pretensioning the cable. QED does not have a preparameterized model of a cable, so in the following discussion, the cable model is achieved by creating a beam of very narrow cross section.

This exploration plucks the cable at the center, and the objective of the data collection and analysis is to determine the frequency and amplitude decay of the ensuing motion. Begin by launching QED from the command line in the working directory.

CREATING THE MODEL

1. Create an axially loaded very slender steel beam of dimensions $[100 \times 0.1 \times 0.1]$ using the [Frame:1D_beam] geometry. That is,
   [0:geometry =1],
   [1:X_length =100],
   [4:X_span =1],
   [a:mat # =2],
   [b:thick_1 =0.1],
   [c:depth_1 =0.1].

2. Use BCs of [0:type=1(=AmB)] to pin both ends. That is,
   [0:type =1],
   [5:end_A =000001],
   [6:others =111111],
   [7:end_B =100001].
   Note that end B can move axially.

3. Use Loads of [0:type=1(=near xyz)] to put the member in axial tension. Set
   [0:type =1],
   [1:X_pos =100.],
   [4:X_load =100.].
   All other loads are set to zero.
   This axial load will be treated as a pretension, and the stiffness of the member will be dominated by the associated geometric stiffness. In this way, the very slender beam will behave essentially as a pretensioned cable.

4. Mesh using <50> elements through the length and restrict the motion to being planar. That is,
   [0:Global_ID =22],
   [1:# X_elems =50].
   Set the damping as $\eta = 0.002$ and all gravity components as zero:
   [a:damping =.002],
   [b?:gravity =0].
   (The gravity special requires [pressing b] followed by [?] which is to be [x y z] as appropriate.)
   Render the mesh by [pressing s]. Set the orientation of the cable to $\phi_x = 0$, $\phi_y = 0, \phi_z = 0$, by [pressing o].

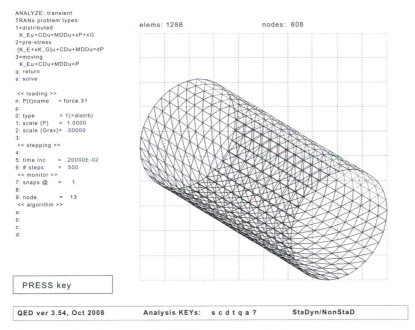

```
ANALYZE: transient
TRANs problem types:                 elems: 1288              nodes: 608
1=distributed
  K_Eu+CDu+MDDu=xP+xG
2=pre-stress
  [K_E+xK_G]u+CDu+MDDu=dP
3=moving
  K_Eu+CDu+MDDu=P
q: return
s: solve

 << loading >>
n: P(t)name    = force.31
p:
0: type        = 1(=distrb)
1: scale {P}   = 1.0000
2: scale {Grav}= .00000
3:
 << stepping >>
4:
5: time inc    = .20000E-02
6: # steps     = 500
 << monitor >>
7: snaps @     =   1
8:
9: node        =   13
 << algorithm >>
a:
b:
c:
d:
```

PRESS key

QED ver 3.54, Oct 2008   Analysis KEYs: s c d t q a ?   StaDyn/NonStaD

Figure 3.3. Opening screen for QED transient analysis.

COLLECTING THE DATA

1. Analyze the problem as linear Transient motion with prestress (see Figure 3.3). That is, set
   [n:P(t)name =force.31],
   [0:type =2],
   [2:scale[K_G] =1],
   [5:time inc =.002],
   [6:#steps =500],
   [7:snap@ =1],
   [9:node =26].
   The prestresses arise from the loads specified in <<qed.sdf>> and gravity. The transient disturbance is specified by [pressing p] to access its submenu. Set
   [1:X_pos =50],
   [5:Y_load =200],
   with all other loads being zero.
   This pluck load is sufficient to cause a geometrically nonlinear problem; however, we will nonetheless do a linear analysis and use this value of load as a scale so the deflections will be visually large.
   [Press q s] to quit the submenu and perform the time integration.
2. View a movie of the deformed Shapes with
   [1:vars =111],
   [2:xMult =1],
   [3:snapshot =1],

[5:rate =0],
[6:size =1.5],
[7:max # =555],
[8:divs =2],
[9:slow =20].

[Press s] to process the snapshots and [press s] again to show the movie. (Note that [8:divs =?], and [9:slow =?] may need to be adjusted for particular computers.) The movie can be replayed by double [pressing s]. Observe that the shape during the motion is essentially a symmetric arc and that all points move in unison.

3. View time Traces of the responses. For displacements at the plucked location set

[0:style type =1],
[1:vars =1],
[2:xMult =1],
[5:rate =0],
[6:node# =26],
[7:elem# =1],
[8:elem type =3].

(The utility PlotMesh for viewing <<qed.sdf>> files can be used to find the node of any point by [pressing nx] and inputting the coordinates.)

[Press s] to show the results; individual panels can be zoomed by [pressing z #], where the panels are numbered 1 through 6, top row first. (This can also be done while in zoom mode.) Observe that the response is predominantly a transverse decaying oscillation.

Store the transverse ([Var:V]) response for further interrogation.

The simplest way to store the response is to zoom on the appropriate panel (this will automatically store the data in a file called <<qed.dyn>>).

4. Launch DiSPtool and go to [View]. [Press p] for the parameters menu and specify

[n:file =qed.dyn],
[1:plot_1 =1 2],
[6:line #1 #2 =1 501].

[Press q] to see the same trace as shown by QED.

[Press z 1] to zoom the panel and then move the cursor by [pressing >]. (The cursor speed can be adjusted up/down by [pressing F/f].)

Record the positive peak amplitudes and their time for at least six peaks after the pluck.

5. Return to the mesh section in QED and change the damping to

[a:damping =0.004].

Remesh, rerun the analysis, and view the traces. Observe the effect of the added damping.

Change the damping a few more times until a value is found that just causes no oscillations. Call this the critical value η_c.

ANALYZING THE DATA

1. Estimate the frequency of vibration by

$$f \approx \frac{1}{t_{i+1} - t_i},$$

where t_i are the recorded times of the peaks. Plot this against average time $(t_{i+1} + t_i)/2$. Estimate an average frequency.

2. Plot the natural log of the amplitude against time. Estimate the slope assuming straight-line behavior.

Does the plot reasonably follow a straight line?

3. A simple model based on Subsection 7.4.I and using the Ritz assumed displacement shape $v(x) = 4v_m[x/L - x^2/L^2]$ (where v_m is the deflection midway along the cable) leads to

$$M\ddot{u} + C\dot{u} + Ku = P,$$

where

$$K = \tfrac{16}{3}\bar{F}_o/L, \qquad C = \tfrac{16}{30}\eta AL, \qquad M = \tfrac{16}{30}\rho AL.$$

This estimates the frequency and damping slope as, respectively,

$$f = \frac{\omega_o}{2\pi} = \frac{1}{2\pi}\sqrt{\frac{K}{M}} = \frac{\sqrt{10}}{2\pi L}\sqrt{\frac{\bar{F}_o}{\rho A}}, \qquad \text{slope} = -\frac{C}{2M} = -\frac{\eta}{2\rho}.$$

Superpose these results as continuous lines on the preceding two plots.
How well do they compare?
How well does the critical value compare with $\eta_c = 2\rho\omega_o$?

4. The static deflection (during the plucking stage) is estimated as $v_m = P_{\max}/K = L[3/16][P_{\max}/\bar{F}_o]$. Speculate on why the frequency and damping results are reasonably close to those of the simple model but this value of static deflection is proportionally quite different.

Part II. Off-Center Pluck Test

The superficial difference of this exploration is that the cable is plucked off center. It turns out, however, that the ensuing motion is significantly more complicated than in Part I. Essentially what happens is that multiple resonances are excited. The objective of the data collection and analysis is to determine these frequencies and the associated vibration shapes.

The same model as in Part I will be used. If QED is not already running, begin by launching it from the command line in the working directory.

COLLECTING THE DATA

1. Return to the mesh section and change the damping back to
   [a:damping =0.004].
   Remesh.

2. In the Transient analysis section, [press p] to access the load submenu and change the location of the pluck to
   [1:X_pos =12],
   [5:Y_load =200],
   with all other loads being zero.
   [Press q s] to perform the analysis.

3. View a movie of the deformed Shapes with
   [1:ivars =111].
   Observe that there is no apparent consistent single shape to the motion, that all points appear to have their own independent motion.

4. View time Traces of the responses at the plucked location by setting
   [1:vars =1],
   [6:node# =7].
   Observe that the transverse response no longer has a smooth decaying harmonic appearance to it.
   Store the transverse response for further interrogation by zooming on the second panel; this will automatically put the data into the file <<qed.dyn>>).

5. Launch DiSPtool and go to [Time:One]. [Press p] for the parameters menu and specify
   [n:file =qed.dyn],
   [1:dt =5e-3],
   [3:# fft pts =1024],
   [4:MA filt # =1],
   [5:MA filt begin =1],
   [6:plot#1#2 =1 1024],
   [w:window =0].
   [Press q r] to compare the raw file with the transformed version. The amplitude spectrum is the panel titled called Amplitude [freq]; zoom on this panel by [pressing z 2] and observe that there are multiple sharp spectral peaks (the large value at zero frequency is not a spectral peak). This shows that the observed time-domain response is a vibration with many simultaneous natural frequencies.
   Record the frequency and amplitude of the first three peaks by moving the cursor [>] to the peak. (The cursor speed can be adjusted up or down by [pressing F/f].)

6. Return to the QED time Traces and change the monitoring node to
   [6:node# =13].
   Show the response and zoom on the second panel.
   Switch to DiSPtool and [press R] to read the new <<qed.dyn>> file and process it. Observe that the values of the spectral peaks have changed slightly.
   Record the frequency and amplitude of the first three peaks.

7. In turn, change the monitoring node to
   [6:node# =19, 26, 33, 39, 45]
   and each time record the frequency and amplitude of the first three peaks.

ANALYZING THE DATA

1. Do the spectral peaks occur at the same frequencies irrespective of monitored location?
2. How well are the three frequencies related through

$$f_n = \frac{\omega_n}{2\pi} = \frac{n\omega_o}{2\pi} = \frac{n}{2\pi}\frac{\sqrt{10}}{L}\sqrt{\frac{\bar{F}_o}{\rho A}},$$

   where n is the peak number?
3. Plot the amplitude data against position associating a different symbol with each frequency set.
   Append the zero-amplitude information at both ends of the cable.
   If the data with a common symbol are connected, is there a pattern to the shapes formed?
   How would the patterns be described?
   If the amplitudes have an ambiguity of sign, can the plots be modified to give a "more natural" pattern to the shape?

Part III. Vibration Eigenanalysis

A vibration eigenanalysis directly finds the resonant frequencies and associated mode shapes. The objective of the data collection and analysis is to correlate the eigenanalysis with the previous analyses.

The same model as in the other parts will be used. If QED is not already running, begin by launching it from the command line in the working directory.

COLLECTING THE DATA

1. Analyze the cable for its Vibration response; the opening screen should look like Figure 3.4.
   Set the scale on the geometric matrix as
   [0:type =1],
   [2:scale[K_G] =1.0].
   The only loads will be those specified in <<qed.sdf>>. Choose
   [8:# modes =16],
   [a:algorithm =1].
   Perform the analysis by [pressing s].
2. View the deformed Shapes.
   For comparative viewing, select the gallery of modes
   [1:vars =51],
   [2:xMult =0.5],
   [3:mode =1],
   [5:rate =0].
   [Press s] to see the gallery. The shapes can be shifted by [pressing x/X y/Y] while zooming.

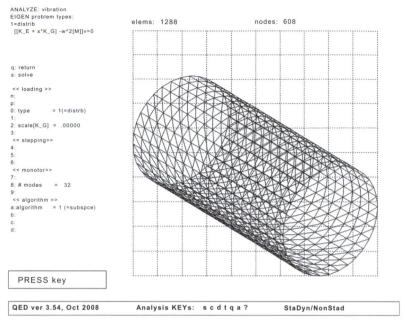

```
ANALYZE: vibration
EIGEN problem types:
1=distrib                        elems: 1288           nodes: 608
  [[K_E + x*K_G] -w^2[M]]v=0


q: return
s: solve

<< loading >>
n:
p:
0: type        = 1(=distrb)
1:
2: scale[K_G]  = .00000
3:
<< stepping>>
4:
5:
6:
<< monotor>>
7:
8: # modes     = 32
9:
<< algorithm >>
a:algorithm    = 1 (=subspce)
b:
c:
d:


PRESS key


QED ver 3.54, Oct 2008      Analysis KEYs:  s c d t q a ?       StaDyn/NonStad
```

Figure 3.4. Opening screen for QED vibration analysis.

Record the first three frequencies ω_i from the second number in the upper left of each panel. Also record (by sketch, comment, or otherwise) the first
three mode shapes.

3. For additional clarity, animate the mode shapes with
   [1:vars =11],
   [2:xMult =1],
   [3:mode =#],
   [5:rate =0],
   [6:size =1.5],
   [7:max# =4],
   [8:divs =40],
   [9:slow =50].
   (Note, again, that [8:divs =?], and [9:slow =?] may need to be adjusted for
   particular computers.)

ANALYZING THE DATA

1. How well do the recorded eigenfrequencies match the frequencies of the spectral peaks and of the simple model?
2. How well do the eigenmode shapes follow the same patterns established in
   Part II?
3. Conjecture on what is missing in Part III that was in the other two parts. (As a
   hint, an eigenanalysis is for the free-undamped-vibration response.)

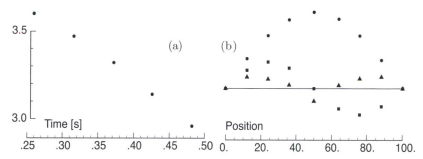

Figure 3.5. Partial results for vibration behavior of a cable: (a) damping, (b) spatial distribution of amplitudes.

Partial Results

The partial results for the damped amplitudes of Part I are shown in Figure 3.5(a). The estimated mode shapes of Part II are shown in Figure 3.5(b); the horizontal line represents the zero position.

Background Readings

References [55, 88, 119, 122] are good introductions to vibrations; References [3, 10, 16] develop the theory more deeply. Structural systems have many modes of vibrations, and the formal approach to deal with these is called *modal analysis*; References [40, 43, 90, 138].

3.2 Modes of Vibration

The two characteristics of a vibration mode are the frequency and the mode shape. The investigations in this section explore the meaning of mode shape. A cantilevered plate is vibrated such that it exhibits transverse bending, in-plane bending, and torsion vibration modes that are easily recognized. A cantilevered shell exhibits a more complex mode-shape pattern, and so an intermediate exploration of a square plate with mixed boundary conditions is introduced to bridge the transition from the cantilevered plate to the shell (see Figure 3.6).

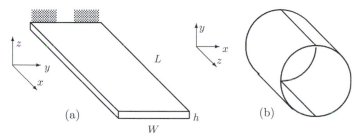

Figure 3.6. Vibration structures analyzed: (a) cantilevered plate, (b) cylindrical shell.

Part I. Cantilevered Plate

We begin with a cantilevered plate. The objective of the data collection and analysis is to use a sensitivity study to show how the frequency of a given mode depends on the structural parameters of length, width, and thickness. Although the frequency changes, the mode shapes remain essentially the same and are used as the identifier to connect the frequencies as the structural parameters are changed.

Launch QED from the command line in the working directory.

CREATING THE MODEL

1. Create an aluminum cantilever plate using the [Plate:quadrilateral] geometry and specify the dimensions as [6 × 2 × 0.1]. That is,

   [0:geometry =1],
   [1:X_2 =6],
   [2:Y_2 =0],
   [3:X_3 =6],
   [4:Y_3 =2],
   [5:X_4 =0],
   [6:Y_4 =2],
   [a:mat#=1],
   [b:thick =0.1].

2. Use BCs of [0:type=1(=ENWS)] to fix [000000] the west side with all others being free [111111]. That is,

   [4:sequence =snew],
   [5:east =111111],
   [6:north =111111],
   [7:west =000000],
   [8:south =111111].

3. No loads need be applied for a vibration eigenanalysis, so set any nonzero loads to zero.

4. Mesh using <12> modules through the length and <8> transverse modules. That is,

   [0:Global_ID =32],
   [1:X_mods =12],
   [2:Y_mods =8];

   set all [specials] as zero.

   [Press s] to render the mesh. For clearer viewing, orient the mesh ([press o]) to set $\phi_x = -55$, $\phi_y = 0$, $\phi_z = -20$. Ensure that the plate surface is toggled on ([press t 1 1]).

COLLECTING THE DATA

1. Analyze the plate for its Vibration response. Set the scale on the geometric matrix as zero and choose the subspace iteration algorithm. That is,

   [0:type =1],
   [2:scale[K_G] =0.0],

```
[8:# modes =24],
[a:algorithm =1].
```
[Press s] to perform the analysis.

2. View the gallery of displacement mode Shapes with
```
[1:vars =51],
[2:xMult =.01],
[3:mode =1],
[5:rate =0].
```
[Press s] to display the gallery. Observe that the modes alternate out-of-plane bending, in-plane bending, and twisting actions. The shapes are best seen with an orientation of about $[-55, 0, -20]$. Individual panels can be zoomed by pressing [z #], where the panels are numbered 1 through 6, top row first.

When two frequencies occur close to each other, as for example the second and third modes, then the reported mode shape may be a combination of the expected separated modes. In these situations the ambiguity is resolved by considering other information such as the modes for slightly changed parameters.

3. Animate the modes with
```
[1:vars =11],
[2:xMult =.01],
[3:mode =1],
[5:rate =0],
[6:size =1.5],
[7:max# =4],
[8:divs =20],
[9:slow =50].
```

4. Record enough information (a brief word description of bending, twisting, etc. plus frequency is adequate) to track the first two out-of-plane bendings, the first in-plane bending, and the first two twisting modes. The frequency (ω rad/s) is the last number in the caption of the gallery. This is the Reference data set.

5. Return to the Geometry section to change the dimensions of the plate to $[6 \times 4 \times 0.1]$ with
```
[1:X_2 =6],
[2:Y_2 =0],
[3:X_3 =6],
[4:Y_3 =2],
[5:X_4 =0],
[6:Y_4 =2],
[b:thick =0.1].
```
Remesh and redo the Vibration analysis.

Record the mode type and frequency for the sequence. This is the sensitivity Case I data set.

6. Return to the Geometry section to change the dimensions of the plate to $[6 \times 2 \times 0.2]$ with
```
[1:X_2 =6],
[2:Y_2 =0],
```

```
[3:X_3 =6],
[4:Y_3 =2],
[5:X_4 =0],
[6:Y_4 =2],
[b:thick =0.2].
```

Remesh and redo the Vibration analysis.

Record the mode type and frequency for the sequence. This is the sensitivity Case II data set.

7. Return to the Geometry section to change the dimensions of the plate to [12 × 2 × 0.1] with

```
[1:X_2 =12],
[2:Y_2 =0],
[3:X_3 =12],
[4:Y_3 =2],
[5:X_4 =0],
[6:Y_4 =2],
[b:thick =0.1].
```

Remesh and redo the Vibration analysis.

Record the mode type and frequency for the sequence. This is the sensitivity Case III data set.

ANALYZING THE DATA

1. Scale all frequency data with respect to their reference value taken from the Reference data set.

2. On a single graph and using an appropriate identifying symbol for each mode, plot the Reference and Case I data against W/W_o. Connect like symbols to form the width sensitivity plots.

3. On a single graph and using an appropriate identifying symbol for each mode, plot the Reference and Case II data against h/h_o. Connect like symbols to form the thickness sensitivity plots.

4. On a single graph and using an appropriate identifying symbol for each mode, plot the Reference and Case III data against L/L_o. Connect like symbols to form the length sensitivity plots.

5. As a simple model, the vibration characteristics of structures can be represented as

$$\text{frequency} \propto \sqrt{\frac{\text{stiffness}}{\text{mass}}} \qquad \text{or} \qquad \omega = \alpha\sqrt{\frac{K}{M}}.$$

For particular structures, it is a matter of identifying the appropriate stiffness and mass for a given mode of vibration.

6. For the out-of-plane bending, assume $K_E = EI_{yy}/L^3$, $M = \rho AL$ so that

$$\omega \quad \longrightarrow \quad \sqrt{\frac{E\frac{1}{12}Wh^3/L^3}{\rho WhL}} \quad \longrightarrow \quad \frac{h}{L^2}.$$

Do the data correlate with the dimensional powers W^0h^1/L^2?

7. For the in-plane bending, assume $K_E = EI_{zz}/L^3$, $M = \rho AL$ so that

$$\omega \longrightarrow \sqrt{\frac{E\frac{1}{12}hW^3/L^3}{\rho WhL}} \longrightarrow \frac{W}{L^2}.$$

Do the data correlate with the dimensional powers $W^1 h^0/L^2$?

8. For the twisting, assume $K_E = GJ_{xx}/L$, $M = \rho I_{xx}L$ so that

$$\omega \longrightarrow \sqrt{\frac{G\frac{1}{3}Wh^3/L}{\rho\frac{1}{12}hW^3L}} \longrightarrow \frac{h}{WL}.$$

Do the data correlate with the dimensional powers $h^1/W^1 L^1$?
Which representation of J_{xx}

$$J_{xx} = \tfrac{1}{3}Wh^3 \quad \text{or} \quad \tfrac{1}{12}hW^3 \quad \text{or} \quad \tfrac{1}{12}Wh^3 + \tfrac{1}{12}hW^3$$

correlates best with the data? Conjecture why.

Part II. Constrained Plate

The cantilevered plate of the previous exploration is now made initially square with the two lateral sides being simply supported. The objective of the data collection and analysis is to establish a connection between the mode shapes as the length of the plate is changed.

Begin by launching **QED** from the command line in the working directory.

CREATING THE MODEL

1. Create an aluminum plate using the [Plate:quadrilateral] geometry and specify the dimensions as [10 × 10 × 0.1]. That is,
   [0:geometry =1],
   [1:X_2 =10],
   [2:Y_2 =0],
   [3:X_3 =10],
   [4:Y_3 =10],
   [5:X_4 =0],
   [6:Y_4 =10],
   [a:mat#=1],
   [b:thick =0.1].

2. Use BCs of [0:type=1(=ENWS)] to fix [000000] the west side, simply support the north and south sides, with the east side being free. That is,
   [4:sequence =ewsn],
   [5:east =111111],
   [6:north =000100],
   [7:west =000000],
   [8:south =000100].

3. No loads need be applied for a vibration eigenanalysis, so set any nonzero loads to zero.

4. Mesh using <30> modules through the length and <30> transverse modules. That is,

```
[0:Global_ID =32],
[1:X_mods =30],
[2:Y_mods =30];
```

set all [specials] as zero.

[Press s] to render the mesh. For clearer viewing, orient the mesh ([press o]) to set $\phi_x = -55$, $\phi_y = 0$, $\phi_z = -20$.

COLLECTING THE DATA

1. Analyze the plate for its Vibration response. Set the scale on the geometric matrix as zero and choose the subspace iteration algorithm. That is,

```
[2:scale[K_G] =0.0],
[8:# modes =32],
[a:algorithm =1].
```

[Press s] to perform the analysis.

2. View the gallery of displacement mode Shapes with

```
[1:vars =51],
[2:xMult =.03],
[3:mode =1],
[5:rate =0].
```

If desired, animate the mode shapes with

```
[1:vars =11],
[2:xMult =.01],
[3:mode =1],
[5:rate =0],
[6:size =1.0],
[7:max# =4],
[8:divs =20],
[9:slow =50].
```

3. The primary information of interest is the number of crests (positive or negative) in the length and transverse directions designated as $[n, m + \frac{1}{2}]$. That is, associate with each mode the number $[n, m + \frac{1}{2}]$, where n is the number of crests in the transverse direction, m is the number of crests in the length direction, and the "$\frac{1}{2}$" signifies a half-wave that is open at the free edge.

Record the frequencies for the following modes:

$[1, 0 + \frac{1}{2}]$, $[1, 1 + \frac{1}{2}]$, $[1, 2 + \frac{1}{2}]$,
$[2, 0 + \frac{1}{2}]$, $[2, 1 + \frac{1}{2}]$, $[2, 2 + \frac{1}{2}]$,
$[3, 0 + \frac{1}{2}]$, $[3, 1 + \frac{1}{2}]$, $[3, 2 + \frac{1}{2}]$.

4. Return to the Geometry section and, in turn, change the length to

```
[1:length =5.0, 15.0, 20.0].
```

Remesh redo the Vibration analysis, and record the mode-shape information for each length.

ANALYZING THE DATA

1. Plot, on a single graph, all the frequency data against length using a particular symbol for a given mode shape.
   Connect data points corresponding to the same mode shape.
2. Do the data indicate that as L becomes large there is no $[m]$ dependence?
   Conjecture on the implications of the answer.
3. Do the large L data vary as $[n]$ or $[n^2]$?
   Conjecture on the implications of the answer.
4. As a simple model, suppose the plate is conceived as a collection of cantilevered beams along the length and simply supported beams across the width. Conjecture on the meaning of the relationship

$$\omega \propto \alpha\left(\frac{m}{L}\right)^2 + \beta\left(\frac{n}{W}\right)^2,$$

   where α and β are proportionality constants.
   Conjecture on why the two terms are additive.

Part III. Cylindrical Shell

The plate of the previous exploration is now curved and seamed to form a cantilevered cylindrical shell. As in Part II, the objective of the data collection and analysis is to establish a connection between the mode shapes as the length of the cylinder is changed.

Begin by launching **QED** from the command line in the working directory.

CREATING THE MODEL

1. Create a thin-walled circular aluminum cylinder using the [Closed:circular] geometry. Set the dimensions as length $L = 12$, radius $R = 2$, thickness $h = 0.1$, and the reinforcing stringer areas as negligible with $A = 0.001$. That is,
   [0:geometry =1],
   [1:length =12.0],
   [2:radius =2.0],
   [a:mat# =1],
   [b:thick_s =0.1],
   [d:area =0.001],
   [e:end plate =0].
   Having no end plate means the cylinder will be open.
2. Use BCs of [0:type=1(=L/R)] to set the left side fixed and right side free. That is, set
   [0:type =1],
   [8:L_bc =000000],
   [9:R_bc =111111].
3. No loads need be applied for a vibration eigenanalysis; confirm that no loads are applied by setting the load as [0:type=1(=near_xyz)] and ensure that no

load frame members are added to the structure. That is, set
[0:type =1],
with all load values set to zero.

4. Mesh using <18> modules through the length and <40> modules around the hoop direction. That is, set
[0:Global_ID =32],
[1:length =18],
[2:hoop =40],
[3:bias V/H =1],
[7:Z_rate =18];
set all [specials] as zero.

[Press s] to render the mesh. For clearer viewing, set the orientation ([press o]) as $\phi_x = +20$, $\phi_y = +30$, $\phi_z = 0$. Ensure that all surfaces are toggled on; if necessary, [press t # 1].

COLLECTING THE DATA

1. Analyze the structure for its Vibration response using the subspace iteration algorithm. Set
[0:type =1],
[2:scale[K_G] =0.0],
[8:# modes =32],
[a:algorithm =1].
[Press s] to perform the eigenanalysis.

2. View the gallery of displacement mode Shapes with
[1:vars =51],
[2:xMult =.01],
[3:mode =1],
[5:rate =0].
For clearer viewing, toggle surfaces on or off by [pressing t # 1/0] where the surfaces are numbered counterclockwise beginning at the bottom.
If necessary for easier identification, animate the mode shapes with
[1:vars =11],
[2:xMult =.01],
[3:mode =1],
[5:rate =0],
[6:size =1.0],
[7:max# =4],
[8:divs =20],
[9:slow =50].

3. The primary information of interest is the number of waves in the hoop direction and in the length direction designated as $[n, m + \frac{1}{2}]$. That is, associate with each mode the number $[n, m + \frac{1}{2}]$, where n is the number of complete sinusoids (include both positive and negative crests) in the hoop direction, m is the

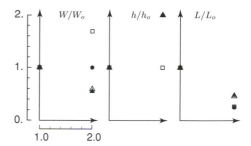

Figure 3.7. Partial results for the vibration of a canti-levered plate. Sensitivity plots with respect to width, thickness, and length.

number of crests (positive or negative) in the length direction, and the "$\frac{1}{2}$" signifies a half-wave that is open at the free edge.

Setting the orientation as $\phi_x = 0$, $\phi_y = 0$, $\phi_z = 0$ will aid in counting n, and setting $\phi_x = 0$, $\phi_y = 90$, $\phi_z = ?$ will aid in counting m.

Record the frequencies for the following modes:

$[1, 0 + \frac{1}{2}]$, $[1, 1 + \frac{1}{2}]$, $[1, 2 + \frac{1}{2}]$,
$[2, 0 + \frac{1}{2}]$, $[2, 1 + \frac{1}{2}]$, $[2, 2 + \frac{1}{2}]$,
$[3, 0 + \frac{1}{2}]$, $[3, 1 + \frac{1}{2}]$, $[3, 2 + \frac{1}{2}]$.

Note that many of the modes have repeated frequencies and only one of them need be recorded.

4. Observe, by animation, that Mode 11 is a torsional mode. Note that when a linear rotational displacement is exaggerated there is an apparent expansion and contraction – ignore this. Record the frequency.

5. Return to the Geometry section and, in turn, change the length to
   [1:length =6.0, 18.0, 24.0].
   Remesh, redo the Vibration analysis, and record the mode shape and frequency information for each length.

ANALYZING THE DATA

1. Plot, on a single graph, all the frequency data against length using a particular symbol for a given mode shape.
   Connect data points corresponding to the same mode shape.
2. Do the data indicate that as L becomes large there is no $[m]$ dependence?
   Conjecture on the implications of the answer.
3. Do the large L data vary as $[n]$ or $[n^2]$?
   Conjecture on the implications of the answer.
4. The $[1, \frac{1}{2}]$ mode is like a global cantilever beam mode. Do the frequencies correlate as $1/L^2$?
5. Do the torsional frequencies correlate as $1/L$?

Partial Results

The partial results for the sensitivity study of Part I are shown in Figure 3.7. Note that the change of parameter (width, height, length) is deemed the independent variable and the frequency is the dependent variable.

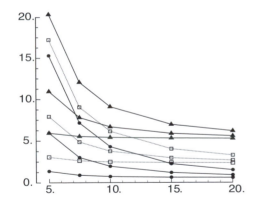

Figure 3.8. Partial results for the vibration of a constrained plate. The effect of length on the modes.

The partial results for the length sensitivity of Part II are shown in Figure 3.8. The interesting question is this: Do the frequencies tend to cluster as the length is increased? And if so, what is the meaning of this?

Background Readings

The vibration frequencies and mode shape are computed with an eigenanalysis. The computational methods for accomplishing this are discussed in depth in References [5, 29, 101]. A compendium of results for plates and shells is given in References [78, 79].

3.3 Prestressed Structures

As observed in Section 3.1, the vibration behavior of a cable is affected by the prestressing. This is actually true of all loaded structures; that is, as the loads change so do the vibration characteristics. The stress effect enters the equations of motion through a stiffness contribution called the *geometric* stiffness. For bulky or stiff structures, such as engine blocks, this effect is usually not significant, but can be quite significant for thin-walled lightweight structures such as a fuselage. The structures analyzed are shown in Figure 3.9.

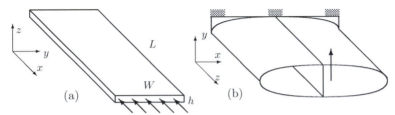

Figure 3.9. Effect of load on the vibration of structures: (a) flat plate, simply supported on all edges, with in-plane membrane loading, (b) cantilevered airfoil with transverse load.

In this section, we explore the connection between the free-vibration behavior of a structure and the applied loads. We also investigate the use of the mode shape as a distinguishing characteristic to describe the vibration.

Part I. Simply Supported Plate with Uniform Load

We begin with a simply supported rectangular plate loaded uniformly in-plane. Such a load is called a membrane load. The objective of the data collection and analysis is to determine the effect of the membrane loading on the transverse vibration characteristics of the plate.

Begin by launching QED from the command line in the working directory.

CREATING THE MODEL

1. Create a rectangular aluminum plate of size $[10 \times 5 \times 0.1]$ using the [Plate:quadrilateral] geometry. That is,
   [0:geometry =1],
   [1:X_2 =10],
   [2:Y_2 =0],
   [3:X_3 =10],
   [4:Y_3 =5],
   [5:X_4 =0],
   [6:Y_4 =5],
   [a:mat# =1],
   [b:thick =0.1].

2. Use BCs of [0:type=1(=ENWS)] to set simply supported boundary conditions on all sides as
   [0:type =1],
   [4:sequence =snew],
   [5:east =110010],
   [6:north =110100],
   [7:west =010010],
   [8:south =100100].
   Note that in-plane motion is allowed.

3. Use Loads of [0:type=2(=traction)] to apply tractions on the east side. That is,
   [0:type =2],
   [1:side# =2],
   [4:X_trac =-1];
   ensure that all other loads are set to zero.

4. Mesh using <20> modules through the length and <10> transverse modules. That is,
   [0:Global_ID =32],
   [1:X_mods =20],
   [2:Y_mods =10],

with all [specials] being zero. Note that [0:Global_ID=32] means the plate can deform in any direction; although the loads are only in-plane, the vibration and buckling will be out-of-plane and the global setting of the DoF should not inhibit this.

[Press s] to render the mesh.

For clearer viewing, [press o] to set the orientation as $\phi_x = -55$, $\phi_y = 0$, $\phi_z = -20$.

COLLECTING THE DATA

1. Analyze the plate for its linear Static response. Set
   [0:type =1],
   [1:scale{P} =1.0],
   [2:scale{Grav} =0.0].
   [Press s] to perform the analysis.
2. View the Contours of displacement with
   [0:type =1],
   [1:vars =1],
   [2:xMult =1],
   [5:rate =0],
   and verify that the dominant action is axial compression.
   Look at the stress contours with
   [1:vars=21],
   and observe (by zooming the panels to see the contour legends) that the membrane stresses are $\sigma_{xx} = -1.0$, $\sigma_{yy} = 0.0$, $\sigma_{xy} = 0.0$, and are uniformly distributed.
3. Analyze the plate for its Vibration response using the subspace iteration algorithm. Set
   [0:type =1],
   [2:scale[K_G] =0.0],
   [8:# modes =16],
   [a:algorithm =1].
   [Press q s] to perform the eigenanalysis.
4. Before exiting the analysis section, record the eigenvalues (frequencies) from the <<stadyn.out>> file.
5. View the deformed Shapes. For comparative viewing, select the gallery of modes with
   [1:vars =51],
   [2:xMult =0.05],
   [3:mode =1],
   [5:rate =0].
   The gallery begins at [3:mode=#]. The deformed shapes can be exaggerated or minimized by changing [2:xMult=] to something other than 1.0 as appropriate.
6. Record (by sketch, comment, or otherwise) the shape of the first 10 modes; these are the load = 0.0 data.

For additional clarity, animate the mode shapes with
[1:vars =11],
[2:xMult =0.05],
[3:mode =#],
[5:rate =0],
[6:size =1.5],
[7:max# =4],
[8:divs =16],
[9:slow =40].
Also View the displacement Contours.

7. Repeat the analysis (including recording the description of the deformed shapes) changing the scale on the geometric stiffness matrix as
[2:scale[K_G] =2000, 4000, 6000, ...]
until at least one of the frequencies becomes imaginary.

ANALYZING THE DATA

1. Plot all the plate data points on a frequency (vertical) against load (horizontal) graph. Represent an imaginary frequency as a negative value. Use a distinguishing symbol to identify the data points according to their mode shape.
2. Connect the data points belonging to a particular mode shape. Observe that some modes cross.
3. Conjecture about the meaning of a zero frequency. Conjecture about the meaning of a negative (imaginary) frequency.
4. A plate is not a beam, but as a simple conceptual model, conceive of the flexural behavior of plates as being similar to that of beams. Then a Ritz single-DoF (SDoF) analysis with $v(x) = v_o[-x/L + x^2/L^2]$ leads to the energies

$$\mathcal{U}_E = \frac{1}{2}\int DW\Big[\frac{\partial^2 v}{\partial x^2}\Big]^2 dx = \frac{1}{2}v_o^2 4DW/L^3,$$

$$\mathcal{U}_G = \frac{1}{2}\int \bar{F}_o\Big[\frac{\partial v}{\partial x}\Big]^2 dx = -\frac{1}{2}v_o^2 P/3L,$$

$$\mathcal{T} = \frac{1}{2}\int \rho A[\dot{v}]^2 dx = \frac{1}{2}\dot{v}_o^2 \rho AL/30,$$

where $DW = Eh^3 W/12(1 - v^2)$ plays a role similar to EI for beams and $-P$ is the axial compressive load. These energies in turn lead to the governing equation for free vibrations as

$$[\rho AL/30]\ddot{v}_o + [4DW/L^3 - P/3L]v_o = 0.$$

This gives an estimate of the frequency behavior as

$$f = \frac{\omega}{2\pi} = \frac{1}{2\pi}\sqrt{\frac{120}{L^4}\frac{DW}{\rho A} - \frac{1}{10}\frac{P}{\rho AL^2}}.$$

The effect of the compressive load is clear. This relation suggests that the frequency squared varies linearly with load; conjecture how might that information be used to predict the buckling collapse load of a structure.

5. Following the final paragraph of Section 7.4, Part II, conjecture about constructing a simple model that takes transverse effects into account.

Part II. Simply Supported Plate with Nonuniform Load

Operating structures typically have nonuniform distributions of loads causing nonuniform stress distributions. Here we explore the effect of nonuniform stresses by applying an end shearing load to the plate of Part I. The objective of the data collection and analysis is similar to that of Part I.

If QED is not already running, begin by launching it from the command line in the working directory.

CREATING THE MODEL

1. Return to the Loads section to apply a shear traction on the east side. That is,
   [0:type =2],
   [1:side# =2],
   [4:X_trac =0],
   [5:Y_trac =1];
   ensure that all other loads are set to zero.

2. Use BCs of [0:type=1(=ENWS)] to set simply supported boundary conditions on all sides but also allow the shear deformation. That is,
   [0:type =1],
   [4:sequence =snew],
   [5:east =110011],
   [6:north =110101],
   [7:west =000011],
   [8:south =110101].

3. Mesh using <20> modules through the length and <10> transverse modules. That is,
   [0:Global_ID =32],
   [1:X_mods =20],
   [2:Y_mods =10],
   with all [specials] being zero.
   [Press s] to render the mesh.

COLLECTING THE DATA

1. Analyze the plate for its linear Static response. Set
   [0:type =1],
   [1:scale{P} =1.0],
   [2:scale{Grav} =0.0].
   [Press s] to perform the analysis.

2. View the Contours of displacement with
   [0:type =1],
   [1:vars =1],
   [2:xMult =1],
   [5:rate =0],
   and verify that the dominant action is an in-plane bendinglike displacement.
   Look at the stress contours with
   [1:vars=21],
   and observe (by zooming the panels to see the contour legends) that the membrane stresses are nonuniformly distributed; in particular, observe that the σ_{xx} ([Var:Smxx]) stress is maximum at the west side and varies from tension to compression south to north. Observe that the σ_{xy} ([Var:Smxy]) shear stress is nearly constant along the length and is parabolic across the width.

3. Analyze the plate for its Vibration response using the subspace iteration algorithm. Set
   [0:type =1],
   [2:scale[K_G] =0.0],
   [8:# modes =16],
   [a:algorithm =1].
   [Press s] to perform the eigenanalysis.

4. View the deformed Shapes. For comparative viewing, select the gallery of modes with
   [1:vars =51],
   [2:xMult =0.04],
   [3:mode =1],
   [5:rate =0].
   Record (by sketch, comment, or otherwise) the shape of the first six modes; these are the load = 0.0 data.
   For additional clarity, animate the mode shapes with
   [1:vars =11],
   [2:xMult =0.05],
   [3:mode =#],
   [5:rate =0],
   [6:size =1.5],
   [7:max# =4],
   [8:divs =16],
   [9:slow =40].
   Also View the displacement Contours.

5. Repeat the analysis (including recording the description of the deformed shapes), changing the scale on the geometric stiffness matrix as
   [2:scale[K_G] =4000, 6000, 8000, 8400, 8500, ...,]
   until at least one of the frequencies becomes imaginary.
   The mode shapes will distort as the load is increased, nonetheless, it should be possible to track the modes from one load to the next. If necessary, decrease the load increment.

ANALYZING THE DATA

1. Plot all the plate data points on a frequency (vertical) against load (horizontal) graph. Represent an imaginary frequency as a negative value. Use a distinguishing symbol to identify the data points according to their mode shape.

2. Connect the data points belonging to a particular mode shape. Observe that some modes cross whereas others come close but do not seem to cross.

3. Are the data trends similar to those of Part I?

4. Distortion of mode shapes can be observed only if there are multiple modes. Conjecture about how the simple model of Part I could be modified to exhibit mode-shape distortion.

Part III. Airfoil Structure

The magnitude of a given load obviously affects the magnitude of the stresses. The position of a given load can also affect the magnitude of the stresses, and this is what we explore here.

A cantilevered thin-walled airfoil structure is loaded transversely and the effect of the load position investigated. The objective of the data collection and analysis is to identify the underlying reason for the stiffness (frequency) going to zero.

Launch QED from the command line in the working directory.

CREATING THE MODEL

1. Create a thin-walled aluminum airfoil using the [Closed:airfoil] geometry; give it the dimensions $L = 10$, width $W = 6$, depth $H = 1$. That is,
   [0:geometry =4],
   [1:length =10.],
   [6:X_2 =3.0],
   [7:Y_2 =0.5],
   [8:X_3 =8.0].
   Set the skin and spar thicknesses as $h = 0.1$ and the stringer areas as negligible with $A = 0.0001$. That is,
   [a:mat# =1],
   [b:skin =0.1],
   [c:spar =0.1],
   [d:area =0.001],
   [e:end plate =1].
   The end plate will have stiffness but negligible mass.

2. Use BCs of [0:type=1(=L/R pln)] to set the left plane fixed and the right plane free. That is,
   [0:type =1],
   [8:L_bc =000000],
   [9:R_bc =111111].

3. Use Loads of [0:type=1(=near_xyz)] to apply a vertical load at the free end. That is, set
   [0:type =1)],

[1:X_pos =3.0],
[2:Y_pos =0.0],
[3:Z_pos =10.0],
[5:Y_load =2300];
set all other loads as zero.

4. Mesh using <20> modules through the length and <40> modules around the hoop direction. That is,
[0:Global_ID =32],
[1:length =20],
[2:hoop =40],
[3:V/H bias =1],
[7:Z_rate =20].
Set all [specials] as zero.
[Press s] to render the mesh. For clearer viewing, set the orientation ([press o]) as $\phi_x = +25$, $\phi_y = +30$, $\phi_z = 0$.
If desired, toggle the end plate off by [pressing t 6 0]. Other substructures can similarly be turned on or off.

COLLECTING THE DATA

1. Analyze the airfoil for its linear Static response. Set
[0:type =1],
[1:scale{P} =1.0],
[2:scale{Grav} =0.0].
[Press s] to perform the analysis.

2. View the displacement Contours with
[0:type =1],
[1:ivars =1],
[2:xMult =1],
[5:rate =0],
and verify that the dominant action is a vertical bending. For clearer viewing, surfaces may be toggled on or off by [pressing t # 1/0].
View the stress contours with
[1:ivars =21]
and observe (by zooming) that the axial membrane stress [Var:Smxx] varies along the length, being a maximum at the root, and has a neutral axis midway up the middle spar such that the top is in compression and the bottom in tension. Record the maximum bending stress at the root and the shear stress [Var:Smxy] at the center of the leading and trailing edges.

3. Analyze the airfoil for its Vibration response using the subspace iteration algorithm. Set
[0:type =1],
[2:scale[K_G] =0.0],
[8:# modes =16],

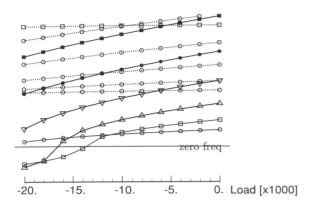

Figure 3.10. Partial results for the vibration behavior of flat plates under axial stress.

[a:algorithm =1].
[Press q s] to perform the analysis.
Before leaving the Vibration analysis section, record the eigenvalues (frequencies) in the <<stadyn.out>> file.

4. Change the prestress scale to
   [2:scale[K_G] =1.0],
   and repeat the analysis. The difference in recorded eigenvalues is due to the effect of the [4:Y_load =2300] load.
   Record the eigenvalues (frequencies) of the first six modes from the <<stadyn.out>> file.
   View the displacement shapes or contours and animate and record a note describing the mode shapes.

5. Return to the Loads section and change the location of the load to
   [1:X_pos =4.0],
   [2:Y_pos =0.0].
   Remesh and repeat the analysis (including the static stress analysis) with
   [2:scale[K_G] =1.0].

6. Change, in turn, the location of the load to
   [1:X_pos =5.0, 6.0].
   Remesh and repeat the analysis (including the static stress analysis) with
   [2:scale[K_G] =1.0].

ANALYZING THE DATA

1. Plot the stresses against load position. What is the effect of the load position on the stresses?
2. Plot the frequencies against load position, connecting those with the same mode shape. How are the modes affected by the load position?
3. Which mode eventually loses all stiffness?

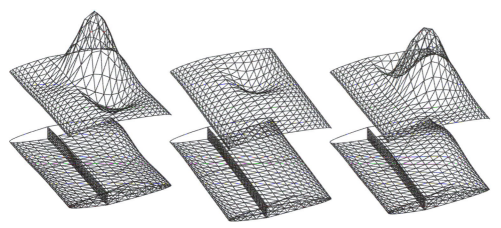

Figure 3.11. Exploded view of the first three mode shapes for the vibrating airfoil.

4. Conjecture why the change of load position caused one of the modes to lose stiffness.

Partial Results

The partial results for Part I are shown in Figure 3.10. It is interesting to observe how some modes are unaffected by the membrane stress. The first three mode shapes for the vibrating airfoil are shown in Figure 3.11. Observe that the large displacements are mostly near the root, similarly to what occurred in Part II.

Background Readings

The presence of axial and membrane stresses can affect the stiffness properties of structures. Their particular contribution is through the *geometric stiffness*; References [18, 30, 102, 129, 134] establish the geometric stiffness matrix for structural members and plates whereas References [5, 26] establish it for the Hex20 element.

3.4 Frequency Analysis of Signals

Significant insights into the dynamic response of structures can be gained by viewing the responses in the frequency domain. This section uses the computer program DiSPtool to explore the relationship between a time-domain waveform and its frequency content.

Figure 3.12 shows the comparison of some signals in both domains. The connection is not obvious, and it takes some practice before an intuitive sense of the connection is developed.

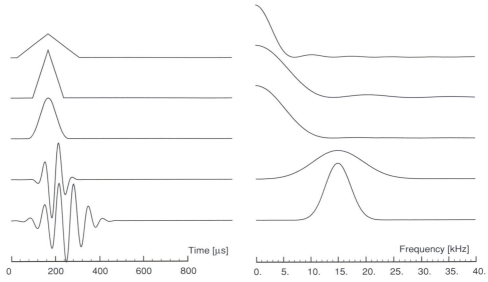

Figure 3.12. Comparison of some pulse-type signals in both the time and frequency domains.

Part I. Connection Between Domain Samplings

The FFT (fast Fourier transform) algorithm takes discretized time-domain data (N samples at a rate of Δt) and transforms them into discretized frequency-domain data (N samples at a rate of Δf). This first exploration establishes the connection between Δt and Δf.

In preparation, use any convenient text editor to create a file called `<<time.100>>` with the two-column `[time amplitude]` contents of a triangular pulse of 100-μs duration:

```
0.0e-6    0
200e-6    0
250e-6    1
300e-6    0
```

This file will be interpolated as needed.

COLLECTING THE DATA

1. Launch DiSPtool from the command line and go to [Time domain, One channel], that is, [press t o].
2. [Press p] to select the parameters input list and change the entries to
    [n:file =time.100],
    [1:dt =1.0e-6], (this is Δt),
    [3:# fft pts =1024], (this is N),
    [4:MA filt# =1],
    [5:MA filt begin =1],

[6:plot#1#2 =1 1024],

[w:window =0].

The abbreviation MA means Moving Average filter – its use will be demonstrated presently. Quit the parameters input by [pressing q].

3. [Press r] to compare the [Raw] input data <<time.100>> to its processed form in the [Window & Smooth] panel. At the moment, the only difference is that the latter is extended to 1023 μs.

4. [Press f] to see the frequency-domain data.

   Observe from the [Log Amplitude] panel that there are amplitudes over the full frequency range. For real-only input signals, the second half of the spectrum is the mirror image of the first half and therefore the only unique information is up to $N/2 + 1$. This frequency value is called the *Nyquist frequency*.

   [Press z1] to zoom a full window view of the amplitude spectrum. [Pressing E/e] gives a horizontal expansion–contraction of the plot while [pressing V/v] gives a vertical expansion–contraction. Points are read by moving the cursor by [pressing >] for left to right and [pressing <] for right to left. The trace is refreshed by pressing [r]. The speed (number of points stepped) of the cursor is increased or decreased by [pressing F/f].

   Record the maximum frequency f and the frequency resolution Δf. The frequency values are in hertz (cycles per second).

5. [Press q p] to access the parameters list, and change in succession

   [1:dt =2.0e-6],

   [1:dt =4.0e-6],

   [1:dt =8.0e-6],

   [1:dt =16.0e-6];

   each time looking at the amplitude spectrum. Observe that the frequency range decreases whereas the frequency resolution gets finer.

   For each case, record the maximum frequency and the frequency resolution.

6. Through the parameters list, change

   [1:dt =1.0e-6],

   and in succession, change

   [4:MA filt# =3],

   [4:MA filt# =5],

   [4:MA filt# =9],

   [4:MA filt# =17],

   [4:MA filt# =33];

   each time look at the amplitude spectrum. Observe that the amplitudes of the high frequencies (disregard components above the Nyquist frequency) are diminished but the dc component is hardly changed.

   [Press d] to look at the difference panels. Observe that the triangle pulse has been smoothed at the corners. Thus, "sharp edges" in a signal are associated with high frequencies and "smoothing a signal" is associated with removing high-frequency content.

ANALYZING THE DATA

1. Plot frequency range f against Δt and frequency resolution Δf against $N\Delta t$. Do the data appear to have a reciprocal relationship?
2. Replot the data as

$$f \text{ vs. } \frac{1}{\Delta t}, \qquad \Delta f \text{ vs. } \frac{1}{N\Delta t}.$$

Are the data reasonably represented by a straight line?
What is the slope of the straight line?
Conjecture on the meaning of the numerical value of the slope.

Part II. Frequency Content of Smoothed Pulses

A triangular pulse of duration ΔT in the time domain has a dominant frequency content ΔF. The objective of the data collection and analysis is to establish an approximate rule relating ΔF to ΔT.

In preparation, use a text editor to create five files (called <<time.#>>, say) with the two-column [time amplitude] contents of a triangular pulse, each of different duration:

0.0e-6 0	0.0e-6 0	0.0e-6 0	0.0e-6 0	0.0e-6 0
200e-6 0	200e-6 0	200e-6 0	200e-6 0	200e-6 0
210e-6 1	220e-6 1	240e-6 1	280e-6 1	360e-6 1
220e-6 0	240e-6 0	280e-6 0	360e-6 0	520e-6 0

Each pulse begins at $t_o = 200\,\mu\text{s}$ and will be extended as needed.

COLLECTING THE DATA

1. Launch DiSPtool from the command line and go to [Time domain, One channel], that is, [press t o].
2. [Press p] to select the parameters input list and change the entries to
   [n:file =time.#],
   [1:dt =1.0e-6],
   [3:# fft pts =1024],
   [4:MA filt# =1],
   [5:MA filt begin =1],
   [6:plot#1#2 =1 1024],
   [w:window =0].
3. For each pulse, without using smoothing, record the frequency at which the amplitude spectrum first becomes zero. We call this the frequency content of the pulse ΔF.
4. If, at any stage, data need to be stored to disk, [press s] and the [STORE Columns] menu for the complete collection of columns is presented.

ANALYZING THE DATA

1. Plot the frequency content ΔF against reciprocal of pulse duration $1/\Delta T$.
2. Superpose, as continuous plots, the relationships $\Delta F = 1/\Delta T$ and $\Delta F = 2/\Delta T$.
3. Which plot best represents the data?

Part III. Frequency Content of Modulated Pulses

The fourth and fifth time traces of Figure 3.12 are examples of modulated pulses. These can be generated by multiplying a finite-width pulse (a triangle, say) by a sinusoid. The interesting feature of these signals is that they are frequency bandlimited.

The objective of the data collection and analysis is to establish an approximate rule relating the frequency band ΔF to the time duration ΔT of the signal.

COLLECTING THE DATA

1. Launch DiSPtool and [Press t o] to go to the [Time domain, One channel] section.
2. [Press g] for signal generation and set
   [0:signal =3],
   [1:c1 =0.0],
   [2:c2 =1.0],
   [3:c3 =50e3],
   [4:c4 =0.0],
   [5:dt =1.0e-6],
   [6:# fft pts =1024].
   [Press q r] to quit the parameters input and view the raw data. Observe that the signal is a sinusoid of many periods.
   [Press f] to see the frequency-domain panels. Observe (by zooming and moving the cursor) that the spectral peak is at 50 kHz, which corresponds to the parameter [3:c3 =50e3].
3. Return to the signal Generation parameters input list and change
   [0:signal =1],
   [1:c1 =200e-6],
   [2:c2 =400e-6],
   leaving the other parameters as before.
   [Press q r] to quit the parameters input and look at the raw data. Observe that the signal is a smoothed triangle of duration $\Delta T = 200\,\mu s$, which corresponds to the difference between the parameters [1:c1] and [2:c2].
   [Press f] to look at the frequency spectrum and observe (by zooming and moving the cursor) where the amplitude first becomes zero.
4. Return to the signal Generation parameters input list and change
   [0:signal =4].

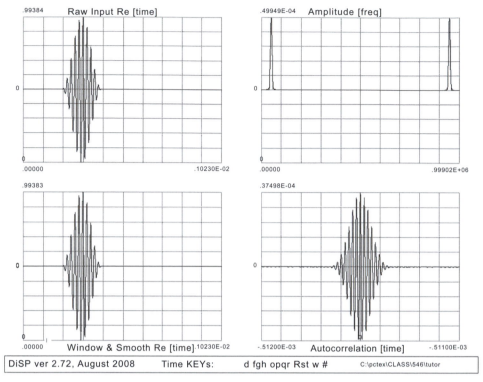

Figure 3.13. A modulated pulse and its amplitude spectrum.

This creates a sinusoid (signal #1) modulated (or windowed) by a smoothed triangle (signal #3).

[Press q r] to quit the parameters input and look at the raw data. Observe that the signal is the sinusoid modulated by the smoothed triangle. It should look like the first panel in Figure 3.13.

[Press f] to look at the frequency spectrum and observe that it has a spectral peak at 50 kHz but it also has a frequency spread. The amplitude spectrum should look like the second panel in Figure 3.13.

Zoom the first panel and estimate the bandwidth of the pulse ΔF (the frequency width of the observable nonzero amplitudes). Record this value and ΔT.

5. Return to the signal Generation input and change the width of the modulating pulse to $\Delta T = 20 \,\mu s$, $40 \,\mu s$, $80 \,\mu s$, $160 \,\mu s$, $320 \,\mu s$. That is, in succession, change

[2:c2 =220e-6],

[2:c2 =240e-6],

[2:c2 =280e-6],

[2:c2 =360e-6],

[2:c2 =520e-6];

each time, view the frequency-domain panels and record the bandwidth of the pulse for each ΔT.

ANALYZING THE DATA

1. Plot the frequency bandwidth ΔF against the reciprocal of pulse duration $1/\Delta T$.
2. Superpose, as continuous plots, the relationships $\Delta F = 1/\Delta T$, $\Delta F = 2/\Delta T$, and $\Delta F = 4/\Delta T$.
3. Which plot best represents the data?
4. Conjecture on the connection between these results and those of the smoothed pulse in Part II.

Part IV. Phase Behavior of Waves

The frequency-domain plots of Figure 3.12 are amplitude spectrums and would look the same irrespective of the location of the pulse along the time axis. The Fourier transform actually produces a complex number with the amplitude and phase related to the real and imaginary parts according to

$$z = a + ib, \qquad A = \sqrt{a^2 + b^2}, \qquad \phi = \tan^{-1}(b/a).$$

It is the phase ϕ that contains the information about the location of the pulse along the time axis.

As developed in Chapter 4, a wave appears as a disturbance shifted in time for each monitored location. That is, there is a definite time–space relationship for the pulse determined by the wave speed. A frequency–domain analysis of waves therefore puts great emphasis on the role played by the phase. The objective of the data collection and analysis in this exploration is to relate the change of phase to the wave speed.

In preparation, use a text editor to create the three files (called <<wave.00>>, <<wave.10>>, <<wave.20>>, respectively) with the two-column [time amplitude] contents of a triangular pulse shifted in time corresponding to the wave being at positions $x = 0$, 10, 20 in:

```
0e-6 0        0e-6 0        0e-6 0
10e-6 0       60e-6 0       110e-6 0
20e-6 1       70e-6 1       120e-6 1
30e-6 0       80e-6 0       130e-6 0
200e-6 0      200e-6 0      200e-6 0
```

The wave travels at a speed of $c_o = 200000$ in/s.

COLLECTING THE DATA

1. Launch DiSPtool from the command line and go to [Time domain, One channel], that is, [press t o].
2. [Press p] to select the parameters input list and change the entries to
   [n:file =wave.00],
   [1:dt =1.0e-6],

```
[3:# fft pts =1024],
[4:MA filt# =1],
[5:MA filt begin =1],
[6:plot#1#2 =1 1024],
[w:window =0].
```
Quit the parameters input by [pressing q].

3. [Press f] to see the frequency-domain data. Note from the [Log Amplitude] panel that there are amplitudes over the full frequency range; we focus on only two frequencies,

```
N=12 (10742 Hz)      and       N=32 (30273 Hz).
```

The third panel contains the phase information. [Press z3] to zoom a full window view and [press E] for a horizontal expansion. Observe that the phase is a series of negatively sloped lines with vertical jumps of 2π. These jumps arise from the ambiguity in the trigonometric arctangent function.

[Press >] to move the cursor to position [N=12]. Record the value of phase. Move the cursor to position [N=32] and record the value of phase. Because there was one jump, the phase to record is the reading minus 2π.

4. [Press q p] to access the parameters list again, and change the file name to [n:file =wave.10].

Repeat the phase readings. Observe that there are more phase jumps over the given frequency range.

5. [Press q p] to access the parameters list again, and change the file name to [n:file =wave.20].

Repeat the phase readings.

ANALYZING THE DATA

1. Plot the phase data against position for the two values of frequency.
   Comment on the plots.
   Are they reasonably represented by straight lines?
2. Estimate the slopes of the lines.
3. The phase of a wave train (single-frequency component) is given by

$$\phi = kx - \omega t \qquad \text{or} \qquad \frac{d\phi}{dx} = k = \text{wave number.}$$

How well do the slopes agree with $k = \omega/c_o$?
4. Conjecture on what the plots would look like if the wave did not travel at a constant speed.

Partial Results

The partial results for the phase behavior in Part IV are shown in Figure 3.14. The higher frequency has the more rapid change of phase.

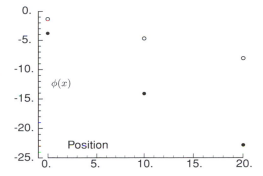

Figure 3.14. Partial results for the phase behavior of waves.

Background Readings

The FFT algorithm is at the core of modern signal analysis. This algorithm is developed in texts such as References [12, 28, 101, 116] with the first and third of these providing source code.

3.5 Effect of Mass and Gravity on Vibrations

The mass has two quite separate effects on the dynamics of a structure. The first is as an inertia, the second is as a preload in the presence of gravity. This section explores these two effects by using the vibration behavior of a slightly curved plate and contrasting it with that of a flat plate (as shown in Figure 3.15).

In preparation, and while **QED** is not running, modify the material file <<qed.mat>> to append the lines

```
        8    ::max # matls
    :
4            ::plastic1
 400000   150000   2.0e-4
5            ::plastic2
 400000   150000   4.0e-4
6            ::plastic3
 400000   150000   8.0e-4
7            ::plastic4
 400000   150000   16.0e-4
8            ::plastic5
 400000   150000   32.0e-4
```

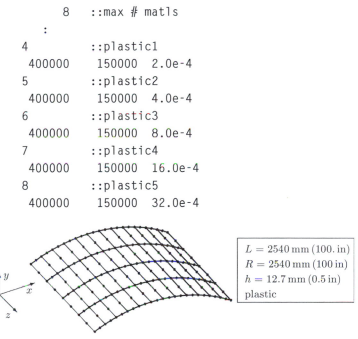

$L = 2540 \, \text{mm} \, (100. \, \text{in})$
$R = 2540 \, \text{mm} \, (100 \, \text{in})$
$h = 12.7 \, \text{mm} \, (0.5 \, \text{in})$
plastic

Figure 3.15. Effect of mass on the vibration of components.

Make sure that the maximum number of materials is specified correctly. This will be the sequence of mass densities studied.

Part I. Eigenanalysis of a Curved Plate under Gravity

We begin with the curved plate shown in Figure 3.15 with gravity load acting downward. The objective of the data collection and analysis is to use a vibration eigenanalysis to determine the effect of the mass and in particular to contrast the effects of gravity versus inertia.

Launch QED from the command line in the working directory.

CREATING THE MODEL

1. Create a solid shell segment of radius 100 and thickness 0.5 using the [Solid:arch] geometry and with properties given by material # 4. That is,
   [0:geometry =5],
   [1:length =100],
   [2:rad_out =100],
   [3:rad_in =99.5],
   [4:angle =60],
   [a:mat# =4],
   [b:constit =1].
   The plasticity yield properties do not play a role in this analysis.

2. Use BCs of [0:type=5(=fixed E/W)] to fix the lateral sides with the other two sides being free. That is,
   [0:type =5].

3. No loads are applied other than gravity. Use Loads of [0:type=1(=near_xyx)] and ensure that all loads are zero. That is,
   [0:type =1],
   and set all loads to zero.

4. Mesh using <12> elements in the hoop direction, <2> through the thickness. and <4> elements through the length. That is,
   [0:Global_ID =111],
   [1:hoop =12],
   [2:thick =2],
   [3:length =4].
   Set the damping and gravity [specials] as
   [a:damping= .0001],
   [by:gravity_Y =-386].
   (Setting the gravity requires [pressing b y] followed by the value.) The gravity acts in the negative y direction.
   Render the mesh by [pressing s]. For clearer viewing, [press o] to set the orientation as $\phi_x = 30$, $\phi_y = 25$, $\phi_z = 0$.

COLLECTING THE DATA

1. Analyze the curved plate for its linear Vibration response. Set
   [0:type =1],

[2:scale[K_G] =1.0],
[8:# modes =16],
[a:algorithm =1].
[Press p] and ensure that all additional loads are zero. [Press q s] to perform the eigenanalysis.

2. View the deformed Shapes.
   For comparative viewing, select the gallery of modes with
   [1:vars =51],
   [2:xMult =5],
   [3:mode =1],
   [5:rate =1].
   Record $\lambda = \omega_1^2$ for the first mode – this is the first number in the upper left of the panel.
   For additional clarity, animate the mode shapes with
   [1:vars =11],
   [2:xMult =5],
   [3:mode =1],
   [5:rate =0],
   [6:size =1],
   [7:max# =4],
   [8:divs =16],
   [9:slow =20].
   Different modes can be observed by changing [3:mode=#].

3. Repeat the Vibration eigenanalysis, changing the scale on the geometric matrix as
   [2:scale[K_G] =2, 4, 8, 16].
   The recorded values of ω_1^2 are DataSet I, which show the effect of gravity on the vibrations.

4. Return to the Geometry section and choose linear material # 5 (plastic2) by
   [a:mat# =5].
   Remesh and redo the Vibration analysis with
   [2:scale[K_G] =1.0].
   View the deformed Shapes and record the first ω_1^2.

5. Repeat this analysis for the other materials so that the total collection has
   [a:mat# =4, 5, 6, 7, 8].
   Each time, remesh and redo the analyses with [2:scale[K_G] =1.0].
   The recorded values of ω_1^2 are DataSet II, which show the effect of mass density on the vibrations.

ANALYZING THE DATA

1. Plot all DataSet I points against the gravity scale factor.
   What is the trend of the data?
2. Plot all DataSet II points against mass-density scale (1, 2, 4, 8, 16).
   What is the trend of the data?

3. Replot all DataSet II points against reciprocal of scale on the density. What is the trend of the data?
4. Because both $[K_G]$ and $[M]$ are proportional to mass density, a simple model for the frequency is

$$\omega^2 = \frac{K_E - \chi_g K_{Go}}{\chi_m M_o},$$

where the subscript nought indicates the reference value.
Does this model explain the data behavior?
5. Do the first two plots cross zero at the same scales?
Conjecture on the implications of your answer.

Part II. Eigenanalysis of a Flat Plate Under Gravity

The objectives here are the same as for Part I, but the plate is made flat. The difference this makes is explored.

Begin by launching QED from the command line in the working directory.

CREATING THE MODEL

1. Create a thick flat plate using the [Solid:block] geometry and with dimensions $L = 100$, width $W = 100$, thickness $h = 0.5$. That is,
   [0:geometry =1],
   [1:X_length =100],
   [2:Y_length =0.5],
   [3:Z_length =100],
   [a:mat# =4],
   [b:constit =1].
   The material corresponds to plastic1 in <<qed.mat>> and the offsets can be set to zero.
2. Use BCs of [0:type=1(=ENWS)] to set fixed BCs on two sides, the other two being free. That is,
   [0:type =1],
   [4:sequence =snew],
   [5:east =000000],
   [6:north =111111],
   [7:west =000000],
   [8:south =111111].
3. Use Loads of [0:type=1(=near_xyx)] and ensure that all loads are zero. That is,
   [0:type =1],
   and set all loads to zero.
4. Mesh using
   [0:Global_ID =111],
   [1:X_dirn =12],
   [2:Y_dirn =2],

```
[3:Z_dirn =4],
[by:gravity_y =-386].
```
The gravity acts in the negative y direction.
Render the mesh by [pressing s]. For clearer viewing, [press o] to set the orientation as $\phi_x = 30$, $\phi_y = 25$, $\phi_z = 0$.

COLLECTING THE DATA

1. Analyze the plate for its linear Vibration response.
   Set
   ```
   [0:type =1],
   [2:scale[K_G] =1.0],
   [8:# modes =16],
   [a:algorithm =1].
   ```
   [Press p] and ensure that all additional loads are zero. [Press q s] to perform the analysis.
2. View the deformed Shapes.
   For comparative viewing, select the gallery of modes with
   ```
   [1:vars =51],
   [2:xMult =20],
   [3:mode =1],
   [5:rate=0].
   ```
   Record the lowest ω_1^2, which is the first number in the upper left.
   For additional clarity, animate the mode shapes with
   ```
   [1:vars =11],
   [2:xMult =10],
   [3:mode =1],
   [5:rate =0],
   [6:size =1],
   [7:max# =4],
   [8:divs =16],
   [9:slow =20].
   ```
   Different modes can be observed by changing [3:mode=#].
3. Repeat the analysis, changing the scale on the geometric matrix as
   ```
   [2:scale[K_G] =2, 4, 8, 16].
   ```
   The recorded values of ω_1^2 are DataSet I.
4. Return to the Geometry section and choose linear material # 5 (plastic2) by
   ```
   [a:mat# =5].
   ```
   Remesh and redo the Vibration analysis with
   ```
   [2:scale[K_G] =1.0].
   ```
   View the deformed Shapes and record ω_1^2.
5. Repeat the analysis for the other materials
   ```
   [a:mat# =6, 7, 8],
   ```
   ensuring each time to remesh.
   The recorded values of ω_1^2 are DataSet II.

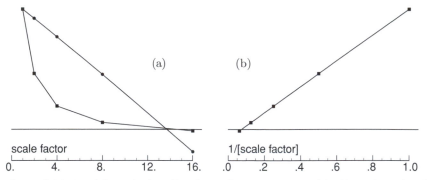

Figure 3.16. Partial results for vibrations of a curved plate: (a) effect of gravity, (b) effect of mass density.

ANALYZING THE DATA

1. Plot all DataSet I points against scale factor.
   What is the trend of the data?
2. Plot all DataSet II points against mass-density scale (1, 2, 4, 8, 16).
   What is the trend of the data?
3. Replot all DataSet II points against reciprocal of scale on density.
   What is the trend of the data?
4. In what way is the flat-plate behavior significantly different than that of the curved plate?
   Conjecture why they are different in this respect.

Partial Results

Partial results for the curved plate of Part I are shown in Figure 3.16. The mesh is shown in Figure 3.15.

Background Readings

Gravity effects are typically neglected for bulky structures but are significant for buildings, bridges, and flexible structures. Reference [134] has a documented example of a tapered column under its own weight.

3.6 Vibration of Shells

This section explores the nature of vibrations in shells. Of particular interest is the transition from 3D solids modeling of the shell to a thin-walled theory.

Part I. 3D Solid Modeling

We begin with a solid Hex20 element modeling of the shell as shown in Figure 3.17. What is not clear in this figure is that there are two elements (and hence five nodes)

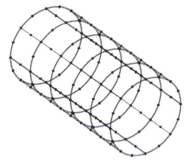

$L = 508\,\text{mm}\,(20.\,\text{in})$
$R = 127\,\text{mm}\,(5.0\,\text{in})$
$h = 20.3\,\text{mm}\,(0.8\,\text{in})$
aluminum

Figure 3.17. Solid modeling of a cylindrical shell. The shown mesh is coarse so as to emphasize the curved nature of the Hex20 element.

through the thickness. The objective of the data collection and analysis is to assess the performance of the Hex20 element in the limit of thin elements.

Launch **QED** from the command line in the working directory.

CREATING THE MODEL

1. Create a solid aluminum cylindrical shell of radius $R = 5$, length $L = 20$, and thickness $h = 0.8$ using the [Solid:cylinder] geometry. That is,
   [0:geometry =4],
   [1:length =20],
   [2:rad_out =5.4],
   [3:rad_in =4.6],
   [a:mat# =1],
   [b:constit s/e =1].

2. Use BCs of [0:type=7(=ring B/F)] to set simply supported BCs on each end. That is,
   [0:type =7],
   [1:radius =5.0],
   [5:front =001000],
   [7:back =001000].
   Only those nodes on the given ring circumference have the set BCs; because only displacements are constrained, this gives a simply supported condition. Note that the shell is not constrained in the axial direction.

3. No loads are to be applied. Use Loads of [0:type=1(=near_xyx)] and ensure that all loads are zero. That is,
   [0:type =1],
   and set all loads to zero.

4. Mesh using <16> elements in the hoop direction, <2> through the thickness, and <10> through the length. That is,
   [0:Global_ID =111],
   [1:hoop =16],
   [2:thick =2],
   [3:length =10].

Set the reduced integration [specials] as

[d:red_Intn =2]

with others zero.

[Press s] to render the mesh. For clearer viewing, [press o] to set the orientation as $\phi_x = 25$, $\phi_y = 35$, $\phi_z = 0$. The mesh may look coarse in the hoop direction but that is only because QED does not plot the middle node of the Hex20 element.

COLLECTING THE DATA

1. Analyze the cylinder for its Vibration response using the subspace iteration algorithm. Set

   [0:type =1],

   [2:scale[K_G] =0.0],

   [8:# modes =24],

   [a:algorithm =1].

   [Press s] to perform the eigenanalysis.

   Before exiting the analysis section, record the eigenvalues (frequencies) from the <<stadyn.out>> file.

2. View the deformed Shapes.

   For comparative viewing, select the gallery of modes with

   [1:vars =51],

   [2:xMult =0.5],

   [3:mode =1],

   [5:rate =0],

   and ensure all tagged surfaces are toggled "on." Observe that the first mode is a rigid-body mode.

3. Return to the Mesh section and change the reduced integration [specials] to

   [d:red_Intn =3].

   Remesh and redo the analysis. Observe that there is only a slight difference between the full integration and the reduced integration results.

4. Return to the Mesh section and change both the reduced integration [specials] and the number of elements as

   [2:thick =1],

   [d:red_Intn =2].

   Remesh and redo the analysis. Observe that the difference in frequency between one and two elements through the thickness is only slight.

5. Return to the Geometry section and change the thickness to $h = 0.4$ with

   [2:rad_out =5.2],

   [3:rad_in =4.8].

   Remesh and reanalyze corresponding to the three preceding cases.

6. Return to the Geometry section and change the thickness to two more cases:

   [2:rad_out =5.1],

   [3:rad_in =4.9], $(h = 0.2)$

   and

[2:rad_out =5.05],

[3:rad_in =4.95]. ($h = 0.1$)

Remesh and reanalyze corresponding to the three preceding cases.

ANALYZING THE DATA. The objective of the data analysis is to assess the performance of the Hex20 element in the limit of thin elements.

1. Using just the first two distinct nonrigid-body modes, use a distinguishing symbol to identify each case and plot all frequency data against shell thickness. Connect those data associated with a particular case.
2. What conclusions can be drawn about the role of reduced integration?
3. What conclusions can be drawn about the number of elements through the thickness?

Part II. Thin-Walled Shell Modeling

The cylinder is now modeled by using shell elements. These elements have just one node through the thickness and therefore are computationally more efficient than the solid elements. Their drawback, however, is that it is not always clear a priori as to the maximum thickness they can adequately model. The objective of the data collection and analysis is to assess the adequacy of the thin-walled shell elements for the same cases as in Part I.

Begin by launching **QED** from the command line in the working directory.

CREATING THE MODEL

1. Create an aluminum thin-walled cylindrical shell of radius $R = 5$, length $L = 20$, and thickness $h = 0.8$ using the [Closed:circular] geometry. That is,

   [0:geometry =1],

   [1:length =20],

   [2:radius =5.0],

   [a:mat# =1],

   [b:thick_s =0.8],

   [d:area =.0001],

   [e:end plate =0].

   The small stringer area [4:area=.0001] means that the shell will not have any reinforcements.
2. Use BCs of [0:type=1(L/R plane)] to set simply supported BCs on each end. That is,

   [0:type =1],

   [8:L_bc =001111],

   [9:R_bc =001111].
3. No loads are to be applied. Use Loads of [0:type=1(=near_xyx)] and ensure that all loads are zero. That is,

   [0:type =1],

   and set all loads to zero.

4. Mesh using $<16>$ modules through the length and $<36>$ in the hoop direction; this gives a model with about the same number of DoFs as the solid cylinder with one element through the thickness. That is,
```
[0:Global_ID =32],
[1:length =16],
[2:hoop =36],
[3:bias V/H =1],
[4:Z_rate =16];
```
the [specials] can be set to zero.
[Press s] to render the mesh. If necessary, [press o] to set the orientation as $\phi_x = 25, \phi_y = 35, \phi_z = 0$.

COLLECTING THE DATA

1. Analyze the cylinder for its linear Vibration response using the subspace iteration algorithm. Set
```
[0:type =1],
[2:scale[K_G] =0.0],
[8:# modes =24],
[a:algorithm =1].
```
[Press p] and ensure that all additional loads are zero. [Press q s] to perform the eigenanalysis.
Before exiting the analysis section, record the eigenvalues (frequencies) from the <<stadyn.out>> file.

2. View the deformed Shapes.
For comparative viewing, select the gallery of modes with
```
[1:vars =51],
[2:xMult =0.5],
[3:mode =1],
[5:rate =0].
```
Observe that the first mode is a rigid-body mode.
For additional clarity, animate the mode shapes with
```
[1:vars =11],
[2:xMult =0.5],
[3:mode =2],
[5:rate =0],
[6:size =1.0],
[7:max# =4],
[8:divs =16],
[9:slow =40].
```
Different modes can be observed by changing [3:mode=#]. Observe that the fourth mode is a global mode.
Record a note about the mode shapes, e.g., how many half-waves in the hoop and length directions.

3. Return to the Geometry section and, in turn, change the thickness to the additional cases
   [2:thick_s =0.4, 0.2, 0.1].
   Remesh and reanalyze for each case.

ANALYZING THE DATA
1. Using just the first two distinct nonrigid-body modes, plot all frequency data against shell thickness. Connect those data associated with a particular mode.
2. How do these values compare with the case of 3D solid modeling?
3. For thin shells, the modes are dominated by the local transverse displacement. Assume a Ritz function where $w(s, z) = w_o \sin(ns/2R) \sin(m\pi z/L)$; then an approximation for the energies is

$$\mathcal{U}_E = \tfrac{1}{2} \int_A D[\nabla^2 w]^2 ds\, dz, \qquad T = \tfrac{1}{2} \int_A \rho h [\dot{w}]^2 ds\, dz, \qquad D = \frac{Eh^3}{12(1 - v^2)}.$$

This gives an effective frequency estimate as

$$\omega = \sqrt{\frac{D}{\rho h} \left[(\frac{n}{2R})^2 + (\frac{m\pi}{L})^2 \right]}.$$

How well does this simple model agree with actual data over the full range of thicknesses and mode shape?

Part III. Global Modes

One of the interesting features of shells and similar structures is that they simultaneously exhibit local mode shapes in the form of wavy patterns and global mode shapes that resemble those of rods and beams. This exploration investigates the occurrence of global modes in a cylindrical shell.

The objective of the data collection and analysis is to construct a simple model for the global modes. Begin by launching **QED** from the command line in the working directory.

CREATING THE MODEL
1. Create an aluminum cylindrical shell of radius $R = 5$, length $L = 40$, and thickness $h = 0.8$ using the [Solid:cylinder] geometry. That is,
   [0:geometry =4],
   [1:length =40],
   [2:rad_out =5.4],
   [3:rad_in =4.6],
   [a:mat# =1],
   [b:constit s/e =1].
2. Use BCs of [0:type=7(=ring B/F)] to set simply supported BCs on each end. That is,
   [0:type =7],
   [1:radius =5.0],

[5:front =001000],
[7:back =001000].

3. No loads are to be applied. Use Loads of [0:type=1(=near_xyx)] and ensure that all loads are zero. That is,
[0:type =1],
and set all loads to zero.

4. Mesh using < 16 > elements in the hoop direction, < 1 > through the thickness, and < 16 > through the length. That is,
[0:Global_ID =111],
[1:hoop =16],
[2:thick =1],
[3:length =16].
Set the reduced integration [specials] as
[d:red_intn =2].
[Press s] to render the mesh. If necessary, [press o] to set the orientation as $\phi_x = 25, \phi_y = 35, \phi_z = 0$.

COLLECTING THE DATA

1. Analyze the cylinder for its Vibration response using the subspace iteration algorithm. Set
[0:type =1],
[2:scale[K_G] =0.0],
[8:# modes =24],
[a:algorithm =1].
[Press s] to perform the eigenanalysis.

2. View the deformed Shapes.
For comparative viewing, select the gallery of modes with
[1:vars =51],
[2:xMult =0.5],
[3:mode =1],
[5:rate =0].
For additional clarity, animate the mode shapes with
[1:vars =11],
[2:xMult =1],
[3:mode =2],
[5:rate =0],
[6:size =1],
[7:max# =4],
[8:divs =16],
[9:slow =25].
Different modes can be observed by changing [3:mode=#].
Identify (and record) the global bending modes, that is, those modes that have a significant overall lateral displacement.

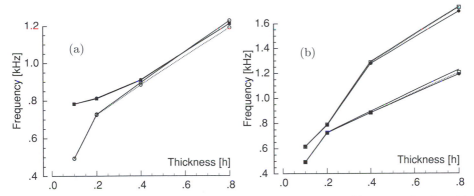

Figure 3.18. Partial results for vibrations of a circular cylinder: (a) effect of reduced integration and number of elements through the thickness, (b) comparison of thin-walled shell modeling for two natural frequencies.

Identify (and record) the torsional mode. The deformed shape for the torsional mode is deceptive in that the rotation appears as an expansion of the diameter – this is because the linear solution is exaggerated.

3. Return to the Geometry section, and, in succession, change the length to [1:length =80, 160].
   Remesh and redo the Vibration analysis. View the deformed Shapes and identify (and record) the bending and torsional modes.

ANALYZING THE DATA

1. Using a distinguishing symbol to identify each mode, plot all frequency data against shell length. Connect those data associated with a particular mode.
2. For the bending modes, assume a Ritz function where $w(z) = w_o \sin(m\pi z/L)$; then an approximation for the energies is

$$ \mathcal{U}_E = \frac{1}{2} \int_L EI \left[\frac{\partial^2 w}{\partial z^2} \right]^2 dz, \quad \mathcal{T} = \frac{1}{2} \int_L \rho A [\dot{w}]^2 ds \, dz, \quad I = \frac{1}{2} \pi R^3 h, \quad A = 2\pi R h. $$

This gives a frequency estimate of

$$ \omega = \frac{m^2 \pi^2}{L^2} \sqrt{\frac{EI}{\rho A}}. $$

How well does this simple model agree with actual data over the full range of lengths?

3. Conjecture about the construction of a simple model for the torsional modes.

Partial Results

The partial results for the effect of shell thickness in Parts I and II are shown in Figure 3.18.

Background Readings

Analytical treatments of the vibrations of plates and shells are given in References [78, 79, 115]. The analysis is complicated by the boundary conditions, and typically approximate methods must be used. Reference [83] does an excellent job in assessing the different shell theories used in dynamics; his conclusion is that "all you need is Love" – a pun on one of the shell theories.

4 Wave Propagation

Wave propagation is the transport of energy in space and time. That is, the essence of wave propagation is the space–time localization of energy that moves with definite speed and amplitude characteristics. This contrasts with vibrations that set each point in the structure in motion simultaneously. Figure 4.1 illustrates some characteristic wave behaviors. It shows the velocity response of a semi-infinite two-material rod free at one end and impacted at the junction. The pulse in the lower semi-infinite part travels at a constant speed, conducting energy away from the joint. Observe how the pulse is initially trapped in the upper material (resulting in multiple reflections) but eventually leaks away after the multiple reflections.

The general wave in a structure is *dispersive*; that is, it changes its shape as it propagates, and so identifying the appropriate propagating entities is quite difficult. For example, Figure 4.2 shows an example of the deflected shapes of a plate transversely impacted; observe that, although "something" is propagating out from the impacted region, it is not obvious how to characterize it.

The collection of explorations in this chapter considers waves in extended media as well as in particular types of waveguides with an emphasis on understanding dispersive behavior. The first exploration uses a pretensioned cable to introduce the fundamental ideas in wave propagation; namely, the speed with which entities propagate in space and time and their amplitude variation. The second exploration looks at different wave types in extended media and their associated speeds. The third exploration investigates the concept of *dispersion*; this requires the analytical tool of *spectral analysis* whereby the waves are analyzed in the frequency domain. The type of wave generated depends on the nearness of neighboring free boundaries, and the fourth exploration uses deep *waveguides* (long members with free lateral surfaces) to determine the effect of these boundaries. There is an underlying connection between vibrations and wave propagation, and the fifth exploration uses spectral analysis to show this connection. Transient loads cause transient stresses, and their concentrations can be quite different from those of their static counterparts; the final exploration considers some aspects of dynamic stress concentrations.

The transient dynamics of a linear structure is described by

$$[M]\{\ddot{u}\} + [C]\{\dot{u}\} + [K_E]\{u\} = \chi_1\{P\} f_1(t) + \chi_2\{G\} f_2(t),$$

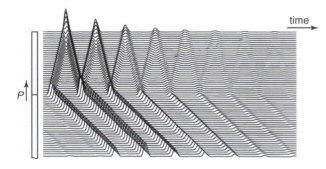

Figure 4.1. Wave propagation in a bi-material rod.

where $[M]$ is the mass matrix, $[C]$ is the damping matrix, $[K_E]$ is the elastic stiffness, $\{P\}$ is the vector of applied loads, and $\{G\}$ is the vector of gravity loads. Each load regime can have a different history specified through $f_i(t)$. The effect of prestress is estimated by solving the problem

$$[M]\{\ddot{u}\} + [C]\{\dot{u}\} + [[K_E] + \chi[K_G]]\{u\} = \{\delta P\}f(t),$$

where $[K_G]$ is the geometric stiffness matrix computed from the preexisting static loads. The current applied load $\{\delta P\}f(t)$ is specified as part of this type of analysis.

The spectral form of these equations is described by

$$[[K_E] + i\omega[C] - \omega^2[M]]\{\hat{u}\} = \{\hat{P}\} \qquad \text{or} \qquad [K_D]\{\hat{u}\} = \{\hat{P}\},$$

where $[K_D]$ is called the dynamic stiffness matrix. Because wave-propagation, forced-frequency, and free-vibration problems all deal with the same matrices, then, not surprising, all three problems are interrelated. The spectral analysis of waves shows this interrelation.

The key to the construction of simple models is first to identify appropriate quantities that propagate. Energy balance can then often be used to estimate the speed and amplitude decay of these quantities.

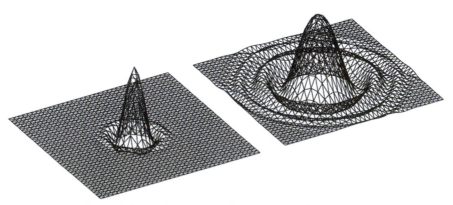

Figure 4.2. Deformed shapes (exaggerated) of an isotropic plate at two times after a central impact.

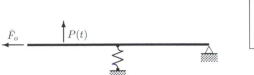

$$L = 5080 \, \text{mm} \, (200. \, \text{in})$$
$$b = 0.85 \, \text{mm} \, (0.0333 \, \text{in})$$
$$h = 5.08 \, \text{mm} \, (0.2000 \, \text{in})$$
steel

Figure 4.3. Wave propagation in a prestressed cable: (a) very thin pretensioned beam, (b) properties.

4.1 Introduction to Wave Propagation

This section explores some basic aspects of free-wave propagation. A pretensioned ribbon cable as shown in Figure 4.3 is used to illustrate the concept of wave-propagation speed and reflections at boundaries.

The cross section of the cable will be made so thin that the dominant transverse stiffness will arise from the pretension associated with the axial load $\bar{F}_o$.

In preparation, create two force files; call them <<force.41a>> and <<force.41b>>, respectively, with the two-column [time force] contents:

```
0.0       0        0.0       0
500e-6    1        200E-6    1
1000E-6   0        4800E-6   1
1500E-6  -1        5000E-6   0
2000E-6   0        1.0       0
1.0       0
```

The first is a *ping* load used to set the cable in motion in such a way that the disturbance is initially localized in space and time; the second is a long-duration pulse of magnitude $P_o \Delta T$ that simulates a step loading. Additionally, and while QED is not running, modify the material file <<qed.mat>> to append the lines

```
        7    ::max # matls
    :
7            ::massless spring
30   12   7.5e-8
```

Make sure that the maximum number of materials is specified correctly. This material will be used to implement an elastic boundary.

Part I. Uniform Pretension

We begin by observing nondispersive (no change of shape) wave behavior. The objective of the data collection and analysis is to determine the effect of the prestress level on the propagation speed.

Launch QED from the command line in the working directory.

CREATING THE MODEL

1. Create an axially loaded very slender steel beam of dimensions [200 × 0.0333 × 0.20] using the [Frame:1D_beam] geometry. That is,
   [0:geometry =1],
   [1:X_len =200],
   [4:X_span =1],
   [a:mat # =2],
   [b:thick_1 =0.0333],
   [c:depth_1 =0.2000].

2. Use BCs of [0:type=1(=AmB)] to fix one end. That is,
   [0:type =1],
   [5:end_A =111111],
   [6:others =111111],
   [7:end_B =000000].
   Note that end A is unconstrained.

3. Use Loads of [0:type=1(=near xyz)] to put the beam in axial tension. Set
   [0:type =1],
   [1:X_pos =0.],
   [4:X_load =-500.].
   All other loads are set to zero.
   This axial load, $\bar{F}_o$, will be treated as a pretension and the stiffness of the beam will be dominated by the associated geometric stiffness. In this way, the very slender beam will behave essentially as a pretensioned cable.

4. Mesh using <200> elements through the length and restrict the motion to being planar. That is,
   [0:Global_ID =22],
   [1:X_elems =200].
   Set the damping and all gravity components as zero.
   [a:damping =.000],
   [b?:gravity =0.0].
   (The gravity special requires [pressing b] followed by [?] which is to be [x y z] as appropriate.)
   Render the mesh by [pressing s]. Set the orientation of the beam to $\phi_x = 30$, $\phi_y = 35$, $\phi_z = 0$, by [pressing o].

COLLECTING THE DATA

1. Analyze the problem as linear Transient motion with prestress. That is, set
   [n:P(t)name =force.41a],
   [0:type =2],
   [2:scale[K_G] =1],
   [5:time inc =25e-6],
   [6:#steps =2000],
   [7:snap@ =2],
   [9:node =1].

2. The prestresses arise from the loads specified in <<qed.sdf>> and gravity. The transient disturbance is specified by [pressing p] to access its submenu. Set
   [1:X_pos =100],
   [5:Y_load =2000],
   with all other loads being zero.
   This ping load is sufficient to cause a geometrically nonlinear problem; however, we will nonetheless do a linear analysis and use this value of load as a scale so the deflections will be visually large.
   [Press q s] to quit the submenu and perform the time integration.

3. View a movie of the deformed Shapes with
   [1:vars =111],
   [2:xMult =1],
   [3:snapshot =1],
   [5:rate =0],
   [6:size =1.5],
   [7:max# =1111],
   [8:divs =2],
   [9:slow =20].
   (Note that [8:divs =?], and [9:slow =?] may need to be adjusted for particular computers.) [Press s] to process the snapshots and [press s] again to show the movie.
   The movie can be replayed by double [pressing s].
   Observe that two disturbances propagate from the impact site and both are compact in space. On reaching the boundary, the disturbance at the free end reflects with the same sign whereas the disturbance at the fixed end flips sign.

4. Return to the Transient Analysis section and change the location of the ping load to the free end of the cable. That is,
   [press p] to access the submenu and set
   [1:X_pos =0],
   [5:Y_load =2000].
   Quit and change the number of time steps to
   [6:#steps =2000].
   [Press s] to rerun the analysis.

5. View a movie of the deformed Shapes and observe that the behavior is essentially the same as for the central impact.

6. View time Traces of the responses. For displacements at the pinged location, set
   [0:style type =1],
   [1:vars =1],
   [2:xMult =1],
   [5:rate =0],
   [6:node# =1],
   [7:elem# =1],
   [8:elem type =3].

[Press s] to show the results; individual panels can be zoomed by pressing [z #], where the panels are numbered 1 through 6, top row first. (This can also be done while in zoom mode.) Note that the response is predominantly a transverse deflection.

Store the arrival time and amplitude of the first peak transverse (v, [Var:V]) response. The simplest way to do this is to zoom on the appropriate panel; this will automatically store the data in a file called <<qed.dyn>>. Then launch DiSPtool and go to [View]. [Press p] for the parameters menu and specify
[n:file =qed.dyn],
[1:plot_1 =1 2],
[6:line#1#2 =1 1001].
[Press q] to see the same trace as shown by QED. [Press z 1] in DiSPtool to zoom the panel and then move the cursor by [pressing >]. (The cursor speed can be adjusted up/down by [pressing F/f].)

7. Return to QED and repeat the measurement for locations
   [6:node# =21, 41, 61, ... 161, 181].
8. Return to the Loads section and change the prestress load to
   [1:X_pos =0.],
   [4:X_load =-1000.].
   Remesh and rerun the Transient analysis, and view the Traces.
   Store the arrival time and amplitude of the peak transverse (v, [Var:V]) response at locations
   [6:node# =1, 21, 41, 61, ... 161, 181].

ANALYZING THE DATA

1. On a single graph, plot both sets of amplitude data against arrival time.
   What conclusions can be drawn?
2. On a single graph, plot both sets of location data against arrival time.
   What conclusions can be drawn?
   Estimate the speed of wave propagation from the slope of the plot.
3. A simple model for estimating wave speed assumes that the main energies in the cable are

$$\mathcal{U} = \tfrac{1}{2} \int \bar{F}_o [\frac{\partial v}{\partial x}]^2 \, dx \,, \qquad \mathcal{T} = \tfrac{1}{2} \int \rho A \dot{v}^2 \, dx.$$

The observed responses are localized and described by $v(x,t) = v_o f(ct - x)$, where c is the speed. The space and time derivatives are given by, respectively,

$$\frac{\partial v}{\partial x} = \frac{\partial v}{\partial(ct - x)} \frac{\partial(ct - x)}{\partial x} = -f' \,, \qquad \frac{\partial v}{\partial t} = \frac{\partial v}{\partial(ct - x)} \frac{\partial(ct - x)}{\partial t} = cf',$$

where the prime indicates differentiation with respect to the argument $(ct - x)$. The energies then become

$$\mathcal{U} = \tfrac{1}{2} \bar{F}_o v_o^2 \int f'^2 dx = \tfrac{1}{2} \bar{F}_o v_o^2 \beta \,, \qquad \mathcal{T} = \tfrac{1}{2} \rho A v_o^2 c^2 \int f'^2 dx = \tfrac{1}{2} \rho A v_o^2 c^2 \beta,$$

where the integration is over the space span of the response. Assume that the total energy is constant and partitioned as $\mathcal{U} = \alpha\mathcal{E}$, $\mathcal{T} = (1 - \alpha)\mathcal{E}$ so that

$$\mathcal{E} = \tfrac{1}{2}\bar{F}_o A v_o^2 f'^2 \beta/\alpha = \tfrac{1}{2}\rho A v_o^2 c^2 f'^2 \beta/(1 - \alpha) \qquad \text{or} \qquad c = \sqrt{\frac{\bar{F}_o}{\rho A}}\sqrt{\frac{1 - \alpha}{\alpha}}.$$

Estimate a value of α that best represents the data.
Conjecture on the meaning of this value of α.
Replace α with the nearest simple fraction. Conjecture on the meaning of this fraction.

Part II. Nonuniform Pretension

The previous exploration showed that the wave speed depends on the level of pre-stress; here we explore what happens as the wave propagates into a region of varying prestress. The objective of the data collection and analysis is to determine the effect of a varying prestress on the propagation behavior.

The same model as that of Part I is used. The varying prestress is implemented by the gravity mechanism. If **QED** is not already running, begin by launching it from the command line in the working directory.

COLLECTING THE DATA

1. Return to the Loads section and change the prestress load back to
   `[1:X_pos =0.0]`,
   `[4:X_load =-500.]`.
   In the Mesh section change the gravity loading to
   `[b?:gravity_X =-1e6]`.
   This uses the mechanism of gravity to apply an additional prestress loading that varies from zero at the free end to $\rho g L A$ at the fixed end. The numbers are such that $F(x) = \bar{F}_o[1 + \beta x/L]$, $\beta = 2$.
   Remesh and reanalyze as Transient with
   `[n:P(t)name =force.41a]`,
   `[0:type =2]`,
   `[2:scale[K_G] =1]`,
   `[5:time inc =25e-6]`,
   `[6:#steps =2000]`,
   `[7:snap@ =2]`,
   `[9:node =1]`.
   `[Press p]` to access the submenu and set
   `[1:X_pos =0]`,
   `[5:Y_load =2000]`.
   `[Press q s]` to quit the submenu and rerun the analysis.
2. View a movie of the deformed Shapes with
   `[1:vars =111]`,
   `[2:xMult =1]`,

```
[3:snapshot =1],
[5:rate =0],
[6:size =1.5],
[7:max# =1111],
[8:divs =2],
[9:slow =2].
```
Observe that, as the wave propagates, the pulse decreases amplitude and broadens its span. Also observe that this is recovered on reflection.

3. View time Traces of the displacements with
```
[0:style type =1],
[1:vars =1],
[2:xMult =1],
[5:rate =0],
[6:node# =1],
[7:elem# =1],
[8:elem type =3].
```
Store the arrival time, peak amplitude, and duration of the pulse at locations
```
[6:node# =1, 21, 41, 61, .... 161].
```

4. Return to the Mesh section and change the gravity loading to
```
[b?:gravity_X =-0.5e6].
```
This gives $F(x) = \bar{F}_o[1 + \beta x/L]$, $\beta = 1$.
Remesh and reanalyze.

5. View time Traces of the displacements and store the arrival time, peak amplitude, and duration of the pulse at locations
```
[6:node# =1, 21, 41, 61, .... 161].
```

ANALYZING THE DATA

1. On a single graph, plot for both data sets the product [peak amplitude × duration] against arrival time.
   What conclusions can be drawn?

2. On a single graph, plot for both data sets the location against arrival time.
   What conclusions can be drawn?

3. A simple propagation model assumes that the instantaneous speed depends on the current stress level. That is,

$$c(x) = \frac{dx}{dt} = \sqrt{\frac{F(x)}{\rho A}} = \sqrt{\frac{\bar{F}_o}{\rho A}}\sqrt{1 + \beta x/L} \quad \text{or} \quad dt = \frac{dx}{c_o\sqrt{1 + \beta x/L}},$$

where $c_o = \sqrt{\bar{F}_o/\rho A}$ is the initial speed. Integrate this to get

$$t = t_o + \frac{2L}{c_o\beta}\left[\sqrt{1 + \beta x/L} - 1\right],$$

where t_o is a reference time. Superpose this result on the preceding arrival plots. How well do the data and the model match?

Part III. Reflection and Transmission at an Elastic Boundary

All waves eventually meet some boundary where there are reflections and transmissions. Part I showed the effect of a free boundary and a fixed boundary. Here we explore the effect of an elastic boundary, and the objective of the data collection and analysis is to construct a simple model to explain the difference between the reflected and the transmitted waves.

Begin by launching **QED** from the command line in the working directory.

CREATING THE MODEL

1. Create an axially loaded, very slender steel beam of dimensions [200 × 0.0333 × 0.20] using the [Frame:1Dw/el_sup] geometry to implement an intermediate elastic support. That is,
   [0:geometry =7],
   [1:X_len =200],
   [8:X_pos =100],
   [9:Y_len =0.2],
   [a:mat #1 =2],
   [b:thick1 =0.0333],
   [c:depth1 =0.2000],
   [d:mat #2 =7] (material number for massless spring),
   [e:thick2 =1.0],
   [f:depth2 =1.0].
   The stiffness of the spring is given by $K = E_2 A_2 / L_2$; [9:Y_len =?] can therefore be used to conveniently adjust it.

2. Use BCs of [0:type=1(=AmB)] to fix the right end and the spring support. That is,
   [0:type =1],
   [5:end_A =111111],
   [6:others =000000],
   [7:end_B =000000].

3. Use Loads of [0:type=1(=near xyz)] to put the beam in axial tension. Set,
   [0:type =1],
   [1:X_pos =0.],
   [4:X_load =-500.].
   All other loads are set to zero.

4. Mesh using <200> elements through the length and restrict the motion to being planar. That is,
   [0:Global_ID =22],
   [1:X_elems =200],

```
[a:damping =.000],
[b?:gravity =0.0].
```
Render the mesh by [pressing s].

COLLECTING THE DATA

1. Analyze the problem as linear Transient motion with prestress using the ping loading. That is, set
   ```
   [n:P(t)name =force.41a],
   [0:type =2],
   [2:scale[K_G] =1],
   [5:time inc =25e-6],
   [6:#steps =1000],
   [7:snap@ =2],
   [9:node =1].
   ```
   The prestresses arise from the loads specified in <<qed.sdf>> and gravity. The transient disturbance is specified by [pressing p] to access its submenu. Set
   ```
   [1:X_pos =0],
   [5:Y_load =200],
   ```
   with all other loads being zero. This will be P_o.

   This step loading is sufficient to cause a geometrically nonlinear problem; however, we nonetheless do a linear analysis and use this value of load as a scale so the deflections will be visually large.

   Quit the submenu and [press s] to perform the time integration.

2. View a movie of the deformed Shapes with
   ```
   [1:vars =111],
   [2:xMult =10],
   [3:snapshot =1],
   [5:rate =0],
   [6:size =1.5],
   [7:max# =1111],
   [8:divs =1],
   [9:slow =20].
   ```
   [Press s] to process the snapshots, and [press s] again to show the movie. Observe that the transmitted pulse does not resemble (in shape) that of the incident pulse.

3. Redo the Transient analysis using the step loading. That is, change
   ```
   [n:P(t)name =force.41b].
   ```
   [Press s] to perform the time integration and View a movie of the deformed Shapes with
   ```
   [1:vars =111],
   [2:xMult =1].
   ```

Observe that the step loading causes a linear displacement profile for the duration of the step. Observe that the wave, on reaching the elastic support, reflects
an unloading wave that brings the cable back to the zero-deflection position.
Also observe that there is a transmitted pulse that is flat; that is, the linear incident profile causes a flat transmitted profile. The multiplication factor can be
changed to [2:xMult=4] so as to better observe this.

4. View time Traces of the responses. For displacements of the incident and reflected waves set

[0:style type =1],
[1:vars =1],
[2:xMult =1],
[5:rate =0],
[6:node# =51] (position $x = 50$),
[7:elem# =1],
[8:elem type =3].

For the transverse deflection $v(t)$ ([Var:V]), observe the linear rise up to a constant value followed by an almost-linear unloading.

Record the maximum deflection (from the numbers displayed above the plot)
and call it v_{im}.

View the velocity traces by changing [5:rate =1]. Observe the almost-
rectangular incident profile whereas the reflection shows a smoothed rise to the
constant value.

Record the maximum incident velocity and call it $\dot{v}_{im}$; record the maximum reflected velocity and call it $\dot{v}_{rm}$.

View the transmitted displacement trace with

[6:node# =122], (position $x = 120$)
[5:rate =0].

Observe the smoothed rise to a maximum plateau.

Record the maximum deflection and call it v_{tm}.

ANALYZING THE DATA

1. Consider a small free body taken from the $x = 0$ end of the cable. Vertical equilibrium gives

$$P + V = 0 \qquad \text{or} \qquad P + \bar{F}_o \frac{\partial v}{\partial x} = 0.$$

Let the generated wave have the form $v(x, t) = f(ct - x)$; then, as shown in
Part I, the space and time derivatives are given by, respectively,

$$\frac{\partial v}{\partial x} = -f', \qquad \frac{\partial v}{\partial t} = cf' \qquad \Rightarrow \qquad \frac{\partial v}{\partial x} = -\frac{1}{c}\frac{\partial v}{\partial t}.$$

The equilibrium equation can therefore be written as

$$P - \frac{\bar{F}_o}{c}\frac{\partial v}{\partial t} = 0 \qquad \text{or} \qquad \sqrt{\rho A \bar{F}_o}\,\dot{v} = P.$$

Interestingly, this is a special form of $M\ddot{v} + C\dot{v} + Kv = P$, where there is only a loaded viscous damper with damping $C = \sqrt{\rho A \bar{F}_o}$. Thus, when a wave propagates energy out of a localized region, the region can be viewed as losing energy, analogous to damping.

Treat the step loading as a pulse $P_o \Delta T$, where ΔT is the duration of the rectangular pulse. The simple model then predicts for the generated incident wave

$$\dot{v}_{im} = P_o / \sqrt{\rho A \bar{F}_o}, \qquad v_{im} = P_o \Delta T / \sqrt{\rho A \bar{F}_o}.$$

How well do these agree with the recorded values?

2. Consider the small region surrounding the elastic support. The transmitted and reflected waves can be thought of as being due to viscous dampers; the incident wave can be thought of as being due to a negative viscous damper. Continuity and equilibrium then give, respectively,

$$v_i + v_r = v_t = v_j, \qquad -Kv_j + C\dot{v}_i - C\dot{v}_r - C\dot{v}_t = 0,$$

where v_j is the displacement of the joint. These can be rewritten as

$$2C\dot{v}_r + Kv_r = -Kv_i, \qquad 2C\dot{v}_t + Kv_t = 2C\dot{v}_i.$$

The reflection is driven by the incident displacement, whereas the transmission is driven by the incident velocity. These inhomogeneous differential equations can be integrated to give [during the $P(t) = P_o = $ constant stage]

$$\dot{v}_r(t) = -\frac{v_{im}}{\Delta T}\left[1 - e^{-Kt/2C}\right], \qquad v_t(t) = \frac{2Cv_{im}}{K}\left[1 - e^{-Kt/2C}\right].$$

Does the exponential decay explain the observed smoothed behavior of the reflected and transmitted pulses?

Does the peak reflected velocity and transmitted deflection agree with the measured values?

3. Suppose the cable has a concentrated mass instead of an attached spring; conjecture in what way(s) the simple model must be adjusted.

Conjecture how a concentrated mass would be implemented using the [Frame:1Dw/el_sup] geometry.

Partial Results

The partial results for the wave behavior of Part II are shown in Figure 4.4.

Background Readings

General aspects of waves are covered in References [36, 48, 67, 81], frequency-domain methods are covered in Reference [28], and asymptotic methods are covered in Reference [1]. Analytical treatments of dynamic stress concentrations are given in References [97, 98].

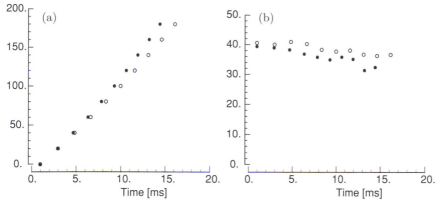

Figure 4.4. Partial results for waves in a cable with nonuniform prestress: (a) speed behavior, (b) amplitude behavior.

4.2 General Exploration of Wave Speeds

We explore some basic wave-propagation behaviors. In particular, we consider the effect of geometry (mass and stiffness distribution) on the types of waves produced and their speeds.

Waves are profoundly affected by the presence of boundaries. The general problem is shown in Figure 4.5, which has multiple types of boundaries. The sequence of simpler geometries shown in Figure 4.6 is designed to elucidate the evolution of different types of waves.

In preparation, create a force-history file; (call it <<force.30>>) that has a pulse of about 30-μs duration with the following [time, force] data:

```
0        0
15e-6    1.0
30e-6    0
1000e-6  0
```

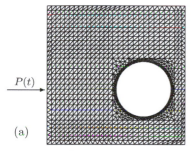

$P(t)$

(a)

(b)

Figure 4.5. Stress waves in a plate with a hole: (a) mesh and load point, (b) contours of von Mises stress at time $t = 35\,\mu$s.

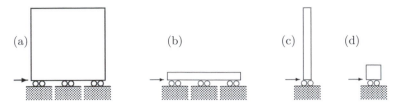

Figure 4.6. Sequence of geometries. The lower edge is a line of symmetry. (a) Extended body, (b) rod, (c) beam, (d) concentrated mass.

This will be interpolated as needed. A better pulse is given by the formula

$$t < t_1, \; t_2 < t : \qquad P(t) = 0,$$

$$t_1 < t < t_2 : \qquad P(t) = [\sin(\frac{t - t_1}{t_2 - t_1}\pi)]^2,$$

with $t_2 - t_1$ being the duration of the pulse. This gives a smoothed pulse with a well-defined frequency content and very small side lobes. This pulse can be produced using DiSPtool as documented in Section 3.4, Part III or Subsection 6.5.1.

Part I. General Problem

First we analyze the evolution of the waves in the model of Figure 4.5 with the data analysis objective of identifying the number of waves generated. This task is not nearly as simple as it might first seem.

Begin by launching **QED** from the command line in the working directory.

CREATING THE MODEL

1. Create a $[10 \times 10 \times 0.10]$ aluminum plate with a hole radius of 2.0 using the [Hole:circle] geometry. That is,
   [0:geometry =1],
   [1:inner_R =2],
   [3:X_dim =10],
   [4:Y_dim =10],
   [5:X_pos =7],
   [6:Y_pos =4],
   [a:mat# =1],
   [b:thick =0.1].

2. Use BCs of [0:type=1(=ENWS)] to set all the boundaries as free [111111]. That is,
   [0:type =1],
   [4:sequence =enws],
   [5:east =111111],
   [6:north =111111],
   [7:west =111111],
   [8:south =111111].

These BCs would normally cause an instability for static problems, but it is not the case for transient problems.

3. Use Loads of `[0:type=1(near xyz)]` to apply a horizontal force near the middle of the west edge. That is,

```
[0:type =1],
[1:X_pos =0.0],
[2:Y_pos =4.0],
[4:X_load =1.0],
[5:Y_load =0],
```

with all other loads being zero.

4. Mesh using

```
[0:Global_ID =22],
[1:X_mods =30],
[2:Y_mods =30],
[3:Inner divs =60],
[4:Inset divs =2],
[7:# smooth =0],
[8:max elems =9000],
[9:mesh style =2];
```

set all `[specials]` as zero.

`[Press s]` to render the mesh. If necessary, reorient the mesh (`[press o]`) to $\phi_x = 0$, $\phi_y = 0$, $\phi_z = 0$. The mesh should look like Figure 4.5(a).

COLLECTING THE DATA

1. Analyze the plate for its linear Transient response. Set

```
[n:P(t)name =force.30],
[0:type =1],
[1:scale{P} =100],
[2:scale{Grav} =0],
[5:time inc =1e-6],
[6:#steps =200],
[7:snaps@ =1],
[9:node =1].
```

The snapshots are stored at every time increment. `[Press s]` to perform the time integration.

2. View the Contours of von Mises stress with fixed scales. That is, set

```
[0:type =2],
[1:vars =51],
[2:xMult =1.0],
[3:snapshot =11],
[5:rate =0],
[6:var_max =200],
[7:var_min =0],
[8:fsig/h =20].
```

This displays snapshot #11, which corresponds to time $10\,\mu s$ when the force is still increasing. (Note that the peak stress contours are missing because fixed scales are used.) Individual panels can be zoomed by pressing [z #], where the panels are numbered 1 through 6, top row first. (This can also be done while in zoom mode.)

3. Look at [3:snapshot=16], which corresponds to time $15\,\mu s$ just when the force reaches its peak.

   Increase the snapshot to [3:snapshot=26], which corresponds to time $25\,\mu s$ when the stress wave is interacting with the hole.

   Look at [3:snapshot=31, 36]; the former corresponds to time $30\,\mu s$ just when the force goes to zero. Observe that, even though the force is now zero (there are currently no applied loads), there is a good deal of stress activity in the body. Also observe that only a portion of the body is stressed; this is because it takes a finite time for the wave energy to move or propagate a distance.

4. Complete viewing the snapshots as [3:snapshot=41, 51, ...] until the wave has totally dispersed. Observe how the maximum stress propagates along the free surfaces and how a stress concentration occurs at the back of the hole.

5. View the stress contour sequence having set the axes scales as fixed with
   [0:type =2],
   [1:vars =21],
   [5:rate =0],
   [6:var_max =200],
   [7:var_min =-200].
   Concentrate on the snapshots up to about $60\,\mu s$ and in the region immediately ahead of the applied load point. By snapshot [41] $(40\,\mu s)$, it seems that the σ_{xx} ([Var:Smxx]) component of the pulse has disappeared. However, it reappears again at about snapshot [46], but the central region is tensile. Observe that the σ_{yy} ([Var:Smyy]) stress becomes zero in the central region but has opposite signs on the free boundaries.

ANALYZING THE DATA

1. Given that the essence of wave propagation is the space–time localization of energy that moves with definite speed characteristics, how many waves can be identified that were generated by this single impact?

Part II. Body and Surface Waves

A characteristic of waves is that their energy is transported with a definite space–time relationship. For extended media, there are two body waves. The P wave travels with speed

$$c_P = \sqrt{\frac{E}{\rho(1 - v^2)}} \approx 1.03\sqrt{\frac{E}{\rho}} \approx 206,000\,\text{in/s} \approx 5000\,\text{m/s}.$$

The S wave travels with speed

$$c_S = \sqrt{\frac{G}{\rho}} \approx 126000 \,\text{in/s} \approx 3000 \,\text{m/s}.$$

Stress waves can also propagate along a free surface; these are called Rayleigh waves and have a speed $c_R \approx c_S$.

In this exploration, the model of Figure 4.5 is simplified by removing the hole so that the data collection and analysis can focus on the body and surface waves, in particular, to determine their characteristic behaviors of speed and amplitude decay.

If QED is not already running, begin by launching it from the command line in the working directory.

CREATING THE MODEL

1. Create a $[10 \times 9 \times 0.1]$ aluminum plate using the [Plate:quadrilateral] geometry. That is,
   [0:geometry =1],
   [1:X_2 =10],
   [2:Y_2 =0],
   [3:X_3 =10],
   [4:Y_3 =9],
   [5:X_4 =0],
   [6:Y_4 =9],
   [a:mat# =1],
   [b:thick =0.1].

2. Use BCs of [0:type =1(=ENWS)] to set the south side as symmetric with all others being free. That is,
   [0:type =1],
   [4:sequence =ewns],
   [5:east =111111],
   [6:north =111111],
   [7:west =111111],
   [8:south =100000].

3. Use Loads of [0:type =1(=near xyz)] to apply a horizontal point load $P_x = 1.0$ at the origin. That is,
   [0:type =1],
   [1:X_pos =0],
   [2:Y_pos =0],
   [4:X_load =1];
   set all other loads to zero.

4. Mesh using $<40>$ modules through the length and $<36>$ transverse modules. That is,
   [0:Global_ID =22],
   [1:X_mods =40],

[2:Y_mods =36];

all [specials] can be set to zero.

[Press s] to render the mesh. If necessary, reorient the mesh by [pressing o] and setting $\phi_x = 0$, $\phi_y = 0$, $\phi_z = 0$.

COLLECTING THE DATA

1. Analyze the plate for its linear Transient response. Set

   [n:P(t)name =force.30],
   [0:type =1],
   [1:scale{P} =100],
   [2:scale{Grav} =0],
   [5:time inc =1e-6],
   [6:#steps =200],
   [7:snaps@ =1],
   [9:node =1].

   [Press s] to perform the time integration.

2. View the Contours of stress with autoscaling on and no multiplication. That is,

   [0:type =1],
   [1:vars =21],
   [2:xMult =1],
   [3:snapshot =31],
   [5:rate=0].

   This displays snapshot #31, which corresponds to time $30\,\mu s$ just when the force goes to zero. Individual panels can be zoomed by [pressing z #], where the panels are numbered 1 through 6, top row first. (This can also be done while in zoom mode.)

3. Record the time, $(x, y = 0)$ position and value of the peak (blue) σ_{xx} ([Var:Smxx]) stress. Record the time, $(x \approx 0, y)$ position and value of the peak (blue) σ_{xy} ([Var:Smxy]) stress.

4. Increment the snapshots by $5\,\mu s$ by choosing [3:snapshot=36]. Record the time, position, and value of the peak (blue) σ_{xx} and σ_{xy} stresses.

5. Repeat the measurements for subsequent snapshots at 5-μs intervals up to a time of $t = 60\,\mu s$. At this time, the σ_{xx} pulse is lost in the reflections.

ANALYZING THE DATA

1. Plot the position and magnitude of the σ_{xx} stress against time.
   Does the pulse travel into the body with a constant speed and diminishing amplitude of

   $$c \approx c_P, \qquad \sigma_{xx} \approx \frac{A}{x}, \qquad A = \text{constant}.$$

2. Plot the position and magnitude of the σ_{xy} stress against time.
   Does the pulse travel along the free surface with a constant speed of

   $$c \approx c_R \approx c_S, \qquad \sigma_{xy} \approx A, \qquad A = \text{constant}.$$

3. Conjecture why the surface wave behavior seems different from the body wave behavior.

Part III. Longitudinal Waves

A body that extends in one direction with free lateral surfaces is called a *waveguide*. Waveguides have wave speeds depending on the type of wave they carry. A longitudinal wave (where the displacements are only along the axis of the member) travels with speed

$$c_o = \sqrt{\frac{E}{\rho}} \approx 200000 \text{ in/s} \approx 5000 \text{ m/s}.$$

This is close to the speed of the P body wave.

In this exploration, we reduce the lateral dimension so that the body is the waveguide of Figure 4.6(b). We assume that QED is already running so that creating the model is simply a matter of changing the lateral dimension. The data collection and analysis focus on determining the speed and amplitude behavior of the longitudinal wave.

COLLECTING THE DATA

1. Return to the Geometry section and change the dimensions of the plate to $[10 \times 1 \times 0.1]$. That is,
   `[4:Y_3 =1]`,
   `[6:Y_4 =1]`.
   Remesh after changing
   `[2:Y_mods =4]`.
2. Redo the linear Transient analysis.
3. View the Contours of stress with
   `[0:type =1]`,
   `[1:vars =31]`,
   `[2:xMult =1]`,
   `[3:snapshot =31]`,
   `[5:rate=0]`.
   Record the time, $(x, y = 0)$ position, and value of the peak (blue) σ_{xx} (`[Var:Smxx]`) stress at 5-μs intervals from time $t = 30 \, \mu$s up to a time of $t = 60 \, \mu$s.

ANALYZING THE DATA

1. Plot the position and magnitude of the σ_{xx} stress against time. Does the wave travel with the characteristics

$$c \approx c_o, \qquad \sigma_{max} \approx \frac{P_{max}}{\text{area}}?$$

2. Conjecture why this wave, in contrast to the P wave of the previous exploration, does not seem to have an amplitude decay, although it does have a somewhat similar wave speed.

Part IV. Flexural Waves

In this exploration, the orientation of the waveguide is changed to that of Figure 4.6(c) so that the transverse loading causes flexural waves. We assume that QED is already running so that creating the model is simply a matter of changing the dimensions of the earlier models.

The data collection and analysis try to use time-domain parameters (speed and amplitude) to characterize the wave properties. However, it is not nearly as successful as for the longitudinal wave primarily because the "shape" of the wave seems to change as it propagates. A wave that changes its shape during propagation is called a *dispersive* wave, and a flexural wave is an example of such. There is something unique about dispersive waves that needs to be more fully explored, which is done in the next two sections.

COLLECTING THE DATA

1. Return to the Geometry section and change the dimensions of the plate to $[1 \times 10 \times 0.1]$. That is,
   [1:X_2 =1],
   [2:Y_2 =0],
   [3:X_3 =1],
   [4:Y_3 =10],
   [5:X_4 =0],
   [6:Y_4 =10].
   Remesh, having changed the modules to
   [1:X_mods =8],
   [2:Y_mods =40].
2. Redo the linear Transient analysis.
3. View the Contours of stress.
   Observe that the peak shear σ_{xy} ([Var:Smxy]) stresses occur along the midline in between the peak bending σ_{yy} ([Var:Smyy]) stresses that occur on the boundaries.
4. Beginning at time $t = 30\,\mu$s, record the time, $(x = h/2, y)$, position, and value of the peak σ_{xy} stress at time intervals of $\Delta t = 5\,\mu$s. Note that some shear maxima are visible at only the larger times.
5. Flexural behavior is best visualized by means of a movie. To that end, the transient analysis will be rerun with an exaggerated force so as to enhance the total displacement.
   Redo the linear Transient analysis, having changed
   [1:scale{P} =100e3],
   [7:snaps@ =5].

View a movie of displaced Shapes with
```
[1:vars =111],
[2:xMult =1],
[3:snapshot =1],
[5:rate =0],
[6:size =1],
[7:max# =44],
[8:divs =4],
[9:slow =30].
```
[Press s] to process the movie frames and [press s] to show the movie. Observe the "wiggling" motion as the wave extends along the beam.

Also observe that there is not an obvious "thing" that propagates along the beam because (at a later time) the whole beam seems to be in motion at the same time.

ANALYZING THE DATA

1. Plot the position of the shear-stress maximums σ_{xy} against time.
2. Plot the magnitude of the bending-stress maximums σ_{yy} against time.
3. Do any of these data seem to have a unique speed characteristic?

Part V. Rigid-Body Dynamics

For a given input pulse width, wave-propagation effects are noticeable only if the body is sufficiently extended. If multiple reflections occur within the loading period, then only an overall motion dominated by the rigid-body motions will be observed. The motion of a rigid body is governed by

$$M\ddot{u} = P(t), \qquad \dot{u} = \frac{1}{M} \int P \, dt,$$

where M is the total mass. This shows that a pulse of finite duration leads to a constant (particle) speed, i.e., the body as a whole moves with a constant speed.

In this exploration, the dimensions of the body are reduced to those of Figure 4.6(d), so that within the time periods considered earlier there are a great number of reflections. The data collection and analysis try to make this connection between wave propagation and rigid-body dynamics.

As before, we assume that QED is already running so that creating the model is simply a matter of changing the dimensions of the earlier models.

COLLECTING THE DATA

1. Return to the Geometry section and change the dimensions of the plate to $[1 \times 0.9 \times 0.1]$. That is,
```
[1:X_2 =1.0],
[2:Y_2 =0.0],
[3:X_3 =1.0],
```

        [4:Y_3 =0.9],
        [5:X_4 =0.0],
        [6:Y_4 =0.9].
        Remesh, having changed the modules to
        [1:X_mods =10],
        [2:Y_mods =9].
  2. Redo the linear Transient analysis with
        [1:scale{P} =1000],
        [6:#steps =1000],
        [7:snaps@ =10].
  3. View the gallery of displaced Shapes with
        [1:vars =51],
        [2:xMult =1],
        [3:snapshot =1].
        The panel begins at [3:snapshot=#], and the time is given immediately after
        [mode_#].
  4. Record the position of the impact site (by estimating from the grid) at a number
     of times over the complete time range.

ANALYZING THE DATA
  1. Plot the position of the plate against time.
  2. Do the position data agree with

$$u(t) = \frac{P_{\max}\Delta T}{2M}\, t,$$

     where ΔT is the pulse width of $30\,\mu s$?
  3. Conjecture about the velocity of any particular point on the (small) plate.
     Would it follow that

$$\dot{u}(t) = \frac{P_{\max}\Delta T}{2M},$$

     or would there be some wave-propagation residual effects?

Partial Results

The partial results for the body and surface wave behaviors of Part II are shown in
Figure 4.7. Both seem to have a constant wave speed, but their amplitude behaviors
are different.

    The partial results for the flexural wave behavior of Part III are shown in Fig-
ure 4.8. It could be argued that the identified peaks in stress travel at a constant
speed but, because the amplitudes change, it is not obvious as to "what" is actually
propagating.

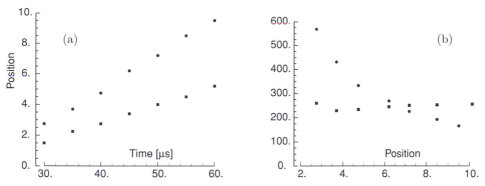

Figure 4.7. Partial results for body and surface wave behaviors: (a) propagation behavior, (b) amplitude behavior.

Background Readings

Analytical treatments of body waves in extended media are given in References [1, 28, 48, 126]. These references also consider the reflection of body waves. The classic solution of a point load on a half-plane is given in Reference [71], and additional solutions for waves in layered media are given in Reference [39].

4.3 Dispersion of Waves

Part IV of the previous exploration showed that it was difficult to use time-domain methods to analyze flexural waves. This is because flexural waves are *dispersive*, that is, they change their shape as they propagate. Frequency-domain methods are essential for analyzing dispersive systems. The source of dispersion is usually associated with some type of coupling. It could be physical coupling as in a beam on an elastic foundation (Figure 4.9) or two parallel beams attached by springs, or mode coupling as in a curved beam that couples the longitudinal and flexural waves. This section explores the dispersive nature of waves.

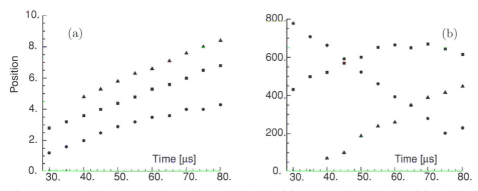

Figure 4.8. Partial results for flexural wave behavior: (a) propagation behavior, (b) amplitude behavior.

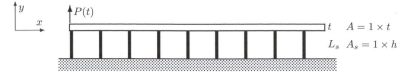

Figure 4.9. Beam on an elastic foundation.

The systems are interrogated with two types of pulses, so, in preparation, use DiSPtool (as described in Section 3.4) to create a sine-squared pulse of duration $\Delta T = 30\,\mu s$ (name it <<force.30s>> and store it to disk) and a narrow-banded pulse of central frequency 20 kHz, sine-squared modulated for a pulse of duration $\Delta T = 600\,\mu s$ (name this <<force.20k>> and store it to disk). At the time of creating these pulses, also store their amplitude spectrums.

Additionally, edit the file <<qed.mat>> (while QED is not running) to add a fourth material with properties similar to those of aluminum, but change the mass density from [2.50E-04] to [2.50E-08]; this material will be used to implement the massless springs of the foundation and the attachment.

Part I. Free-Beam Response

We begin with a simple free-standing beam impacted transversely. A beam has two modes of deformation, the transverse deflection and the rotation of the cross section. The governing equations [see, for example, Equation (7.24)] are a set of differential equations coupling these modes, which leads to the dispersion.

The objective of the data collection and analysis is to characterize the wave response in terms of its speed and amplitude behaviors. Begin by launching QED from the command line in the working directory.

CREATING THE MODEL

1. Create a beam on an elastic foundation using the [Frame:2D_frame] geometry. In this modeling, the walls will be the distributed elastic support and the free-standing beam will be obtained by releasing the ground attachments. Set the geometry as
[0:geometry =2],
[1:X_len =0.25],
[2:Y_hgt =4.00],
[4:# X_bays =400],
[5:# Y_flrs =1],
[7:brace =0].
For the walls/springs set
[a:mat#1 =4],
[b:thick1 =0.0512],
[c:depth1 =1.0000].

For the floors/beam set
[d:mat#2 =1],
[e:thick2 =0.2000],
[f:depth2 =1.0000].

2. Use BCs of type [0:type=1(=AmB)] to set all boundaries to free [111111]; this will essentially nullify the effect of the springs because they are massless. Set
[0:type =1],
[5:corners_A =111111],
[6:others =111111],
[7:corners_B =111111].

3. Use Loads of type [0:type=1(=near xyz)] to apply a vertical point load $P_y = 1.0$ at the origin. That is, set
[0:type =1],
[1:X_pos =0.0],
[2:Y_pos =4.0],
[5:Y_load =1];
all other loads and positions being zero.

4. Mesh using <1> element for each member and add heavy damping. That is,
[0:Global_ID =22],
[1:X_elems =1],
[2:Y_elems =1],
[a:damping =0.1].
[Press s] to render the mesh. (Note that, for this type of mesh, there will be an extra amount of scrolling in the GenMesh command window.) If necessary, orient the mesh ([press o]) to $\phi_x = 0$, $\phi_y = 0$, $\phi_z = 0$.

COLLECTING THE DATA

1. Analyze the problem as linear Transient using the narrow-banded force history. That is,
[n:P(t)name =force.20k],
[0:type =1],
[1:scale{P} =100],
[2:scale{Grav} =0.0],
[5:time inc =2e-6],
[6:#steps =1500],
[7:snaps@ =1],
[9:node =2].
[Press s] to perform the time integration.

2. View time Traces of the responses. For velocities (first time derivative of displacements) set
[0:style type =1],
[1:vars =1],

[2:xMult =1],
[5:rate =1],
[6:node# =2],
[7:elem# =1],
[8:elem type =3].
[Press s] to show the results; individual panels can be zoomed by pressing [z #], where the panels are numbered 1 through 6, top row first. (This can also be done while in zoom mode.)

3. Record the peak value and arrival time for the $\dot{v}$ ([Var:V]) response at every $\Delta x = 10$, that is, in succession, change
[6:node#=82, 162, 242, 322, 402, 482].
These data are referred to as the speed data. The best way to interrogate these responses is to zoom on the appropriate panel (this will automatically store the data in a file called <<qed.dyn>>), launch DiSPtool, and View the file. For subsequent responses, it will be just a matter of [pressing R] to refresh the new file into DiSPtool.

4. Return to the Mesh section and change the damping to
[a:damping=0.01].
Remesh and redo the Transient analysis, and record the same response information. Observe the exaggerated reflections.

5. Record the full transverse velocity response (by zooming and copying <<qed.dyn>>) at nodes
[6:node# =82, 90],
and call them <<vel.82>>, <<vel.90>>, respectively. These are a distance $\Delta x = 1.0$ apart.

ANALYZING THE DATA

1. Plot the speed data as position versus arrival time.
Does the pulse have a constant speed?
Does damping have an effect on the speed?
Estimate the propagation speed for the smaller damped case. We refer to this as the *group speed*.

2. Launch DiSPtool and go to [Time Two channels] by [pressing t t]. [Press p] for the parameters list and set
[a:chan_AA =vel.82],
[b:chan_BB =vel.90],
[1:dt =0.5e-6],
[3:#fft pts =2048],
[6:plot#1#2 =1 2048].
[Press q x] to quit and view the cross-correlation results. Zoom on the third panel [X correlation [time] Re] and use the cursor keys (</>) to locate the first positive peak with a time lag Δt (i.e., negative time). This is the time phase difference between the predominant sinusoids at the two locations; the

phase speed is $c = \Delta x / \Delta t$. The cursor speed can be changed faster/slower by [pressing F/f].

3. A simple model for estimating the phase wave speed assumes that the main energies in the beam are

$$\mathcal{U} = \tfrac{1}{2} \int EI[\frac{\partial^2 v}{\partial x^2}]^2 \, dx, \qquad \mathcal{T} = \tfrac{1}{2} \int \rho A[\dot{v}]^2 \, dx.$$

The observed sinusoids are described by $v(x, t) = v_o \sin(\omega t - kx)$, where k is the *wave number*. The energies become (on integrating over a representative number of cycles)

$$\mathcal{U} = \tfrac{1}{2} EI v_o^2 k^4 \beta, \qquad \mathcal{T} = \tfrac{1}{2} \rho A v_o^2 \omega^2 \beta.$$

Assume that the total energy is constant and partitioned equally between $\mathcal{U}$ and $\mathcal{T}$ so that

$$\tfrac{1}{2} EI v_o^2 k^4 \beta = \tfrac{1}{2} \rho A v_o^2 \omega^2 \beta \qquad \text{or} \qquad k = \sqrt{\omega} \left[\frac{\rho A}{EI} \right]^{1/4}.$$

This gives the phase speed as

$$c = \frac{\omega}{k} = \sqrt{\omega} \left[\frac{EI}{\rho A} \right]^{1/4} = \sqrt{2\pi f} \left[\frac{EI}{\rho A} \right]^{1/4}.$$

This important result shows that the phase speed is frequency dependent, a result characteristic of all dispersive systems.

The disturbance propagating in the beam is actually a collection of slightly different frequencies – this is called a *wave packet* or *wave group*. These sinusoids superpose constructively when their phases are not too different, so that the maximum is observed when

$$\frac{\partial}{\partial \omega} [\omega t - kx] = 0 \qquad \text{or} \qquad \frac{x}{t} = \frac{\partial \omega}{\partial k} = c_g.$$

This is called the *group speed* and is given by

$$c_g = \frac{\partial \omega}{\partial k} = 2\sqrt{\omega} \left[\frac{EI}{\rho A} \right]^{1/4} = 2\sqrt{2\pi f} \left[\frac{EI}{\rho A} \right]^{1/4}.$$

4. Make plots against frequency (in hertz) for the phase and group speeds from the simple model. Superpose on this figure the amplitude spectrum (appropriately scaled) of the narrow-banded pulse.

   Superpose on this figure the group speed and phase speed from the data.

   How well do they match the modeling?

5. Normalize the amplitude data by dividing by the first recorded value (at $x = 10$). Plot the natural log of the normalized amplitudes against position. Determine how well the slopes agree with

$$\text{slope} = \frac{\eta}{2 c_g \rho_o}, \qquad \eta = \text{FE damping}, \qquad \rho_o = \text{average density} = \tfrac{1}{2} \rho.$$

The half in the density occurs because the springs are massless.

Part II. Beam on an Elastic Foundation

The same model as in Part I is used here, except that the springs are attached to the ground. The objective of the data collection and analysis is to establish the effect of the elastic foundation on the wave-propagation characteristics.

If QED is not already running, begin by launching it from the command line in the working directory.

COLLECTING THE DATA

1. Return to the boundary conditions and fix ([000000]) the springs. That is, set
    [0:type =1],
    [5:corners_A =000000],
    [6:others =000000],
    [7:corners_B =000000].
    Remesh after resetting the damping to [a:damping =0.1].

2. Analyze the problem as linear Transient with the broad-banded force history. That is,
    [n:P(t)name =force.30s],
    [0:type =1],
    [1:scale{P} =100],
    [2:scale{Grav} =0.0],
    [5:time inc =2e-6],
    [6:#steps =1500],
    [7:snaps@ =1],
    [9:node =2].
    [Press s] to perform the time integration.

3. View the time Traces with
    [0:style type =1],
    [1:vars =1],
    [2:xMult =1],
    [5:rate =1],
    [6:node# =#],
    [7:elem# =],
    [8:elem type =3].
    Monitor the transverse velocity at locations
    [6:node# =2, 12, 22, 32, 42]
    and observe the evolution of the response from a few zero crossings to many zero crossings; this is the hallmark of a dispersive response in the time domain – the excitation with no zero crossings has caused a response with many crossings. These types of responses are best analyzed in the frequency domain.

4. Monitor the transverse velocity at location
    [6:node# =242].
    Zoom on the second panel [this will automatically store the $\dot{v}(t)$ trace data in the file <<qed.dyn>>], launch DiSPtool, and go to [Time:One] to interrogate the file. [Press p] for the parameters menu and set

```
[n:file =qed.dyn]),
[1:dt =2.0e-6],
[3:# fft pts =4096],
[4:MA filt# =1],
[5:MA filt begin =1],
[6:plot#1#2 =1 4096],
[w:window =0].
```

The spectra are then obtained by [Pressing qf]. For subsequent responses, it will be just a matter of [pressing R] to refresh the file in DiSPtool and immediately see its transform.

The *cutoff frequency* (in this case) is the frequency below which the amplitude spectrum is essentially zero. Record the cutoff frequency by zooming on the panel and moving the cursor with [>/<].

5. Return to the Geometry section and change the area of the spring to
   ```
   [b:thick_1 =0.2048],
   [c:depth_1 =1.00].
   ```
   Remesh and redo the Transient analysis, view the time Traces, and zoom on the second panel.
   Switch to DiSPtool and [press R] to refresh the file. Record the cutoff frequency.
6. Collect the cutoff-frequency information for the additional spring areas:
   ```
   [b:thick_1 =0.0128, 0.0032, 0.0008, 0.0002],
   [c:depth_1 =1.00].
   ```

ANALYZING THE DATA

1. Plot the cutoff-frequency data versus the square root of the spring thickness ($\sqrt{h}$). What can be said about the data trend?
2. A simple model assumes that the sinusoids are described by $v(x, t) = v_o \sin(\omega t - kx)$, leading to the energies (on integrating over a representative number of cycles)

$$\mathcal{U}_b = \tfrac{1}{2} E I v_o^2 k^4 \beta, \qquad \mathcal{T} = \tfrac{1}{2} \rho A v_o^2 \omega^2 \beta, \qquad \mathcal{U}_s = \tfrac{1}{2} K_s v_o^2 \beta,$$

where K_s is the foundation spring stiffness per unit length (of beam). Assume that the total energy is partitioned equally between $\mathcal{U} = \mathcal{U}_b + \mathcal{U}_s$ and $\mathcal{T}$ so that

$$E I k^4 + K_s = \rho A \omega^2.$$

This leads to an estimate of the group speed given by

$$c_g = \frac{\partial \omega}{\partial k} = 2\sqrt{2\pi f} \left[\frac{EI}{\rho A}\right]^{1/4} [1 - f_c^2/f^2]^{3/4}, \qquad f_c = \frac{1}{2\pi} \sqrt{\frac{K_s}{\rho A}}.$$

3. Make a plot against frequency (in hertz) for the group speed for values of h in the range 0.0 to 0.2024, where the spring stiffness is computed from

$$K_s = 4 \times \frac{E_s A_s}{L_s}, \qquad A_s = 1.0 \times h, \qquad L_s = 4.0.$$

Superpose on this figure the amplitude spectrum (appropriately scaled) of the broad-banded pulse.
What conclusions about the wave response can be drawn from this figure?

4. Superpose on the cutoff-frequency plot the relation $f_c = \sqrt{4E_s h/\rho A L_s}\,/2\pi$. How well does this simple model relation agree with the cutoff-frequency data?

Part III. Elastically Coupled Beams
The ground of the model of Part II is now replaced with another beam. The coupled beams add another source of dispersion. The objective of the data collection and analysis is to determine the wave-propagation characteristics in a couple system.

Begin by launching QED from the command line in the working directory.

CREATING THE MODEL
1. Create a coupled beam system using the [Frame:2D_frame] geometry. Set the geometry as
   [0:geometry =2],
   [1:X_len =0.25],
   [2:Y_hgt =8.00],
   [4:# X_bays =400],
   [5:# Y_flrs =2],
   [7:brace =0].
   For the walls/springs set
   [a:mat#1 =4],
   [b:thick1 =0.0512],
   [c:depth1 =1.0000].
   For the floors/beam set
   [d:mat#2 =1],
   [e:thick2 =0.2000],
   [f:depth2 =1.0000].
2. Use BCs of type [0:type=1(=AmB)] to set all boundaries to free [111111]. Set
   [0:type =1],
   [5:corners_A =111111],
   [6:others =111111],
   [7:corners_B =111111].
   This effectively gives two beams, a distance $L_s = 8$ apart, coupled by massless springs.
3. Use Loads of type [0:type=1(=near xyz)] to apply a vertical load at the uppermost left position. That is,
   [0:type =1],
   [1:X_pos =0.0],

```
[2:Y_pos =16.0],
[5:Y_load =1.0].
```
4. Mesh using <1> element for each member and add light damping. That is,
```
[0:Global_ID =22],
[1:X_elems =1],
[2:Y_elems =1],
[a:damping =0.01].
[Press s] to render the mesh.
```

COLLECTING THE DATA

1. Analyze the problem as linear Transient using the narrow-banded force history. That is,
```
[n:P(t)name =force.20k],
[0:type =1],
[1:scale{P} =100],
[2:scale{Grav} =0.0],
[5:time inc =2e-6],
[6:#steps =1500],
[7:snaps@ =1],
[9:node =2].
[Press s] to perform the time integration.
```
2. View the time Traces and record the multiple peak values and arrival times for the $\dot{v}$ response at locations
```
[6:node# =123, 243, 363, 483, 603, 723].
```

ANALYZING THE DATA

1. Plot the position against arrival time for the peak data. Determine the wave speeds.

   Conjecture why there are two waves propagating.

2. Normalize the amplitude data by dividing by the first recorded value (at $x = 10$, [node=123]). Plot the natural log of the normalized amplitudes against position. Compare the loss of amplitude with the situation in Part I; comment.

   Each mode seems to have a different loss factor; conjecture why.

3. A simple model views the coupled beam system as having two modes: a symmetric mode in which the beam is on an elastic foundation of length $L_s/2$, and an antisymmetric mode in which both beams move in unison and the springs have no effect. This leads to estimates of the group speeds as

$$c_{g1} = 2\sqrt{2\pi f}\left[\frac{EI}{\rho A}\right]^{1/4}[1 - f_c^2/f^2]^{3/4}, \qquad f_c = \frac{1}{2\pi}\sqrt{\frac{K_s}{\rho A}} = \frac{1}{2\pi}\sqrt{\frac{E_s h}{\rho A}},$$

$$c_{g2} = 2\sqrt{2\pi f}\left[\frac{EI}{\rho A}\right]^{1/4}.$$

How well do these speeds agree with the data?

Part IV. Longitudinal Wave Propagation in a Curved Rod

A small axial load applied to a straight rod causes just an axial displacement. The same load applied to a curved rod (or beam) causes both an axial and a transverse displacement. Not surprising then that a longitudinal wave propagating in a curved rod also causes both displacements and this coupling effect leads to dispersion of the wave. The dispersion effect is similar to that for a straight rod encased in a surrounding elastic material [28]; the main characteristic is that it exhibits a cutoff frequency.

The objective of the data collection and analysis is to infer the origin of the cutoff-frequency effect. Begin by launching **QED** from the command line in the working directory.

CREATING THE MODEL

1. Create a curved aluminum rod using the [Frame:2D_arch] geometry. The thickness will be made relatively small so that the responses are dominated by the longitudinal behaviors. Set the geometry as
   [0:geometry =6],
   [1:radius =8.0],
   [2:angle =359] (the arch is almost a complete ring),
   [a:mat# =1],
   [b:thick =0.2],
   [c:depth =1.0].

2. Use BCs of type [0:type=1(=AmB)] to set all boundaries to free [111111]; that is, set
   [0:type =1],
   [5:end_A =111111],
   [6:others =111111],
   [7:end_B =111111].

3. Use Loads of type [0:type=1(=near xyz)] to apply a horizontal point load $P_x = 1.0$ at the origin. That is, set
   [0:type =1],
   [1:X_pos =0.1] (this distinguishes the end points),
   [2:Y_pos =0.0],
   [4:X_load =1],
   [5:Y_load =0];
   all other loads and positions being zero.

4. Mesh using <128> elements and add light damping. That is,
   [0:Global_ID =22],
   [1:X_elems =128],
   [a:damping =0.04].
   [Press s] to render the mesh. If necessary, orient the mesh ([press o]) to $\phi_x = 0$, $\phi_y = 0$, $\phi_z = 0$.

COLLECTING THE DATA

1. Analyze the problem as linear Transient using the broad-banded force history. That is,

   `[n:P(t)name =force.30s]`,
   `[0:type =1]`,
   `[1:scale{P} =100]`,
   `[2:scale{Grav} =0.0]`,
   `[5:time inc =1e-6]`,
   `[6:#steps =1000]`,
   `[7:snaps@ =1]`,
   `[9:node =66]`.

   `[Press s]` to perform the time integration.

2. View time Traces of the responses.

   (Before viewing the traces, confirm the node numbers by launching the utility PlotMesh and viewing the structure data file `<<qed.sdf>>`. The quarter points on the ring are expected to have the node numbers `[6:node#=65,129,66,1, 64]`.)

   For velocities (first time derivative of displacement) set

   `[0:style type =1]`,
   `[1:vars =1]`,
   `[2:xMult =1]`,
   `[5:rate =1]`,
   `[6:node# =65]`,
   `[7:elem# =1]`,
   `[8:elem type =3]`.

   `[Press s]` to show the results.

   Observe that there is a substantial transverse ($v(t)$, `[Var:V]`) component of velocity; this is an indication that the axial displacements and transverse displacements are tightly coupled.

   Observe that the axial ($u(t)$, `[Var:U]`) component of velocity seems to be divided into three portions: an initial positive response that mimics the shape of the input force history; this is followed by an almost-constant-velocity stage that is about half of the maximum velocity; finally, midway through the trace there is an abrupt change in response that is tentatively associated with reflections.

3. The first goal of the data collection is to establish the wave speed characteristics. From the axial velocity trace, record the times of the beginning of the first pulse and the beginning of the suspected reflections. The best way to interrogate these responses is to zoom on the appropriate panel (this will automatically store the data in a file called `<<qed.dyn>>`), launch DiSPtool, and View the file; moving the cursor will then display the times. For subsequent responses, it will be just a matter of `[pressing R]` to refresh the new file into DiSPtool.

   Repeat the measurement for the other quarter point nodes at
   `[6:node#=65, 129, 66, 1, 64]`.

Note that the axial and transverse velocities switch designations at the east and west points versus the south and north points.

Observe that the two times converge at the most distant point; this confirms that there is a wave-propagation phenomenon occurring and that the second disturbances are indeed reflections.

4. Zoom on the axial velocity at
   [6:node# =64].
   Launch DiSPtool and go to [Time:One] to interrogate the file. [Press p] for the parameters menu and set
   [n:file =qed.dyn],
   [1:dt =1.0e-6],
   [3:#fft pts =2048],
   [4:MA filt# =1111],
   [5:MA filt begin =500],
   [6:plot#1#2 =1 2048],
   [w:window =0].
   The very large moving average filter is set to remove the reflections; at present, this is more convenient than the alternative of using the windowing options in DiSPtool. [Press qr] to see the filtered file.
   [Press f] to obtain the spectra. Zoom on the first panel and [press E] a few times to expand the low-frequency range. Observe that most of the low frequencies have been filtered; that is, there appears to be a cutoff frequency effect.
   Record the cutoff frequency by moving the cursor with [>/<] to a point somewhere around half the rising portion of the spectrum.

5. Switch back to QED and View the time Traces at
   [5:rate =1],
   [6:node#=66].
   Zoom on the axial velocity and copy <<qed.dyn>> to <<dyn.uu1>>.
   View the accelerations with
   [5:rate =2],
   [6:node#=66].
   Zoom on the transverse acceleration and copy <<qed.dyn>> to <<dyn.vv2>>.

ANALYZING THE DATA

1. Reference [28] shows that the pertinent geometric behaviors for a curved beam of constant radius R are given by

$$\epsilon_{ss} = \frac{\partial u}{\partial s} - \frac{v}{R} - y\left[\frac{1}{R}\frac{\partial u}{\partial s} + \frac{\partial^2 v}{\partial s^2}\right], \qquad \phi = \frac{u}{R} + \frac{\partial v}{\partial s},$$

where s is the distance around the curve and ϕ is the total rotation of the cross section. To construct a simple model for the longitudinal behavior, we assume that the thickness is small so that the flexural strains (the preceding y

dependence) can be neglected. We also assume that the contributions to the rotation can be parameterized as

$$\frac{u}{R} = \alpha\phi, \qquad \frac{\partial v}{\partial s} = [1 - \alpha]\phi \qquad \text{or} \qquad u = \frac{\alpha}{1 - \alpha}R\frac{\partial v}{\partial s},$$

with α ranging over $0 \leq \alpha \leq 1$. Estimate the energies in the rod as

$$\mathcal{U} = \tfrac{1}{2}\int EA\epsilon_{ss}^2\,ds = \tfrac{1}{2}\int EA\Big[\frac{\alpha}{1-\alpha}R\frac{\partial^2 v}{\partial s^2} - \frac{v}{R}\Big]^2\,ds,$$

$$\mathcal{T} = \tfrac{1}{2}\int \rho A[\dot{u}^2 + \dot{v}^2]\,ds = \tfrac{1}{2}\int \rho A\Big[\big(\frac{\alpha}{1-\alpha}R\frac{\partial \dot{v}}{\partial s}\big)^2 + \dot{v}^2\Big]\,ds.$$

As done earlier, assume that the individual sinusoids are described by $v(s, t) = v_o \sin(\omega t - ks)$ where k is the wave number; this allows the energies to be integrated explicitly over a representative number of cycles. Also, assume that the total energy is constant and partitioned equally between $\mathcal{U}$ and $\mathcal{T}$; this leads to

$$k^2 = \Big[\frac{\rho A\omega^2 - EA/R^2}{EA}\Big]\frac{\alpha}{1-\alpha}.$$

This core result shows that the wave response has a cutoff frequency at

$$\omega_c = 2\pi f_c = \frac{c_o}{R}, \qquad c_o \equiv \sqrt{\frac{EA}{\rho A}}.$$

What is significant is that the prediction is independent of the parameterization α; we will set $\alpha = 0.5$.

The group speed is given by

$$c_g = \frac{\partial \omega}{\partial k} = \sqrt{\frac{EA}{\rho A}}\sqrt{1 - \frac{EA}{R^2\rho A\omega^2}} = c_o\sqrt{1 - \frac{EA}{R^2\rho A\omega^2}}.$$

This indicates that the high-frequency components asymptote to the longitudinal wave speed c_o for a straight rod.

2. Plot the propagation data as position ($s = n2\pi R/4$, where n is the number of the quarter point) versus arrival time.

   Does the head of the pulse have a constant speed?

   Do the reflections have the same constant speed?

   How well do these speeds agree with $c_o = \sqrt{E/\rho} = 200000$?

3. For propagation behavior in which $v(s, t) = v(s - c_o t)$, it can reasonably be assumed that

$$\frac{\partial v}{\partial s} = \frac{\partial v}{\partial t}/(-c_o).$$

Consequently, the rotation partition assumption (with $\alpha = 0.5$) reduces to

$$c_o\dot{u} = -R\ddot{v}.$$

Make a plot of <<dyn.vv2>> against time. Superpose on this a plot of <<dyn.uu1>> $\times(-c_o/R)$.

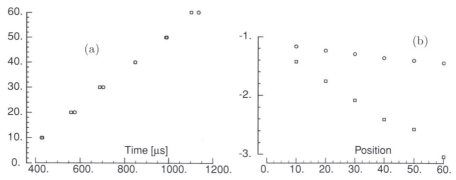

Figure 4.10. Partial results for the free-beam wave behavior: (a) propagation behavior, (b) scaled amplitude behavior for different amounts of damping.

How well do the early time responses match?
Conjecture why the later time data do not match.

4. Conjecture on the connection between the curved-rod problem and the beam on an elastic foundation. In particular, conjecture on the origin of the cutoff frequency for the curved rod.

Partial Results

The partial results for the wave behavior in Part I are shown in Figure 4.10. The partial results for the wave behaviors in Parts II and III are shown in Figure 4.11.

Background Readings

Analytical treatments of the coupled-beam problem are given in References [28, 111]. Experimental studies of wave propagation in curved beams are given in References [23, 54]. This problem is part of the more general problem of wave propagation in coupled systems, a particularly interesting example being that of

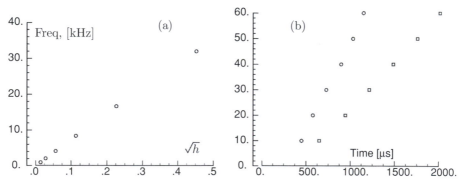

Figure 4.11. Partial results for the elastically constrained beams: (a) cutoff frequency for beam on an elastic foundation, (b) propagation behavior for two coupled beams.

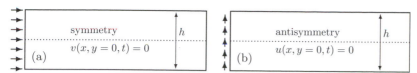

Figure 4.12. Deep waveguides; only above the centerline is modeled: (a) symmetric modes, (b) antisymmetric modes.

structure–fluid interaction [41, 59, 91]. The finer points of structure–fluid interaction are covered in the excellent summary paper of Reference [20].

4.4 Deep Waveguides

The nearness (or remoteness) of lateral boundaries profoundly affect the wave-propagation behaviors in waveguides. The measure of nearness is related to the wavelength (and hence frequency) content of the propagating pulse. When the wavelengths are long, the waveguide is referred to as being slender; when the wavelengths are short it is referred to as being deep. This section explores the wave-propagation behaviors of a disturbance traveling in a deep waveguide; of particular interest is the transition to slender waveguide modeling.

The systems are interrogated by using a sine-squared pulse of duration $\Delta T = 30\,\mu s$, so in preparation, use DiSPtool (as described in Section 3.4) to create this pulse and name it <<force.30s>>. Note that this pulse was also created in Section 4.3.

Part I. Symmetric Waves in a Deep Waveguide

We begin with a long rectangular plate excited by a uniform normal traction along one side. This puts the whole cross section in motion simultaneously such that the centerline is a line of symmetry, as indicated in Figure 4.12(a). The objective of the data collection and analysis is to identify the type of wave propagating with the most energy.

Launch QED from the command line in the working directory.

CREATING THE MODEL

1. Create a $[20 \times 8 \times 0.2]$ aluminum plate using the [Plate:quadrilateral] geometry. That is, set
   [0:geometry =1],
   [1:X_2 =20],
   [2:Y_2 =0],
   [3:X_3 =20],
   [4:Y_3 =8],
   [5:X_4 =0],
   [6:Y_4 =8],

```
[a:mat# =1],
[b:thick =0.2].
```

2. Use BCs of [0:type=1(=ENWS)] to set the south side as symmetric with all others being free. That is,
```
[0:type =1],
[4:sequence =enws],
[5:east =111111],
[6:north =111111],
[7:west =111111],
[8:south =100000].
```

3. Use Loads of type [0:type=2(=traction)] to apply a uniformly distributed traction $t_x = 1.0$ on the west side. That is,
```
[0:type =2],
[1:side# =4],
[4:X_trac =1.0],
[5:Y_trac =0.0];
```
all other loads being zero.

4. Mesh using $<80>$ modules through the length and $<32>$ transverse modules. That is, set
```
[0:Global_ID =22],
[1:X_mods =80],
[2:Y_mods =32];
```
all [specials] can be set as zero.

Render the mesh by [pressing s].

(Note that this is a relatively dense mesh, 5120 elements and 2673 nodes, so it may be necessary to adjust the memory allocations in <<genmesh.cfg>>.)

By [pressing o], ensure the orientation is set as $\phi_x = 0, \phi_y = 0, \phi_z = 0$.

COLLECTING THE DATA

1. Analyze the model for its linear Transient response. Set
```
[n:P(t)name =force.30s],
[0:type =1],
[1:scale{P} =1000],
[2:scale{Grav} =0.0],
[5:time inc =1e-6],
[6:#steps =200],
[7:snaps@ =1],
[9:node =1].
```
[Press s] to perform the time integration.

(Again, because this is a relatively dense mesh, it may be necessary to adjust the memory allocations in <<stadyn.cfg>>.)

2. View the Contours of strain energy with
```
[0:type =1],
[1:vars =81],
```

```
[2:xMult =1],
[3:snapshot =31],
[5:rate =0].
```

The selection [3:snapshot=31] corresponds to time $t = 30\,\mu s$ just when the traction history goes to zero. Individual panels can be zoomed by [pressing z #], where the panels are numbered 1 through 6, top row first. (This can also be done while in zoom mode, and the six panels can be refreshed by [pressing o].)

Observe (by zooming to see the legend) that the $\sigma_{xx}\epsilon_{xx}$ strain energy is dominant and essentially uniform through the depth.

Record the $(x, y = 0)$ and $(x, y = h/2)$ positions and values of the peak $\sigma_{xx}\epsilon_{xx}$ strain energy,

View the contours of particle velocity by changing

```
[1:vars =1],
[5:rate =1].
```

Observe that the $\dot{u}$ distribution coincides with that of $\sigma_{xx}\epsilon_{xx}$. Observe that $\dot{v}$ is localized to the surface and trailing at about 45°.

Record the $(x, y = 0)$ and $(x, y = h/2)$ peak values of $\dot{u}$. Record the $(x, y = h/2)$ peak value of $\dot{v}$.

3. Increment the snapshots by $10\,\mu s$ by choosing

```
[3:snapshot =41].
```

Repeat the preceding energy and velocity measurements.

Repeat the measurements for subsequent snapshots at 10-μs intervals up to a time of $t = 90\,\mu s$.

4. Observe at the maximum elapsed time that there is a significant trailing behavior to the pulse.

5. Return to the Geometry section and change the dimensions of the plate to $[20 \times 4 \times 0.2]$. That is, change

```
[4:Y_3 =4],
[6:Y_4 =4].
```

Mesh using $<80>$ modules through the length and $<16>$ modules through the height, that is, with

```
[0:Global_ID =22],
[1:X_mods =80],
[2:Y_mods =16];
```

this keeps the same density of elements.

Redo the Transient analysis.

View the progression of energies (without recording measurements). Observe that the peak $\sigma_{xx}\epsilon_{xx}$ energy has hardly changed.

6. Change the dimensions of the plate to

$[20 \times 2 \times 0.2]$,
$[20 \times 1 \times 0.2]$,
$[20 \times .5 \times 0.2]$,

and remesh in each case. Repeat the observations without measurements.

ANALYZING THE DATA

1. On a single graph, plot the position data against arrival time.

   Does the wave propagate as a plane wave?

   Estimate the wave speed.

   Which of the following speeds is the estimate closest to?

$$c_o = \sqrt{\frac{E}{\rho}}\,, \qquad c_P = 1.2\,c_o\,, \qquad c_S = \sqrt{\frac{G}{\rho}} = 0.61\,c_o\,, \qquad c_R = 0.93\,c_R = 0.57\,c_o.$$

2. On a single graph, plot the energy and velocity amplitude data against position.
   Comment on the dispersion of the wave.

Part II. Partition of Symmetric Mode Energies

Following on from Part I, this exploration investigates the partitioning of the strain and kinetic energies. The objective of the data collection and analysis is to identify the significant energy contributions that would then be suitable for inclusion in a waveguide theory.

If QED is not already running, begin by launching it from the command line in the working directory.

COLLECTING THE DATA

1. Return to the Geometry section and change the dimensions of the plate to [40 × 4 × 0.2]. That is,

   [1:X_2 =40],

   [2:Y_2 =0],

   [3:X_3 =40],

   [4:Y_3 =4],

   [5:X_4 =0],

   [6:Y_4 =4].

   Mesh using <160> modules through the length and <16> modules through the height with

   [0:Global_ID =22],

   [1:X_mods =160],

   [2:Y_mods =16].

   Redo the Transient analysis, having changed the number of time steps to

   [6:# steps =300].

2. View the time Traces. To look at particle velocities set

   [0:style type =1],

   [1:vars =1],

   [2:xMult =1],

   [5:rate =1],

   [6:node# =1369],

   [7:elem# =1],

   [8:elem type =4].

The selected node corresponds to the center of the modeled plate.
Record the maximum magnitudes of $\dot{u}$ and $\dot{v}$.
3. View strain energy Traces by changing
   [1:vars =81],
   [5:rate =0].
   Record the maximum magnitudes of the $\sigma_{xx}\epsilon_{xx}$ and $\sigma_{xy}\gamma_{xy}$ energy densities.
4. Return to the Geometry section and change the dimensions of the plate to [40 × 2 × 0.2].
   Mesh using <160> modules through the length and <8> modules through the height.
   Redo the Transient analysis and View the time Traces.
   At node [3:node=725], record the maximum magnitudes of $\dot{u}$ and $\dot{v}$ and $\sigma_{xx}\epsilon_{xx}$, $\sigma_{xy}\gamma_{xy}$.
5. In succession, change the dimensions of the plate to
   [40 × 1 × 0.2],
   [40 × 0.5 × 0.2],
   [40 × 0.25 × 0.2],
   [40 × 0.125 × 0.2],
   [40 × 0.0625 × 0.2].
   Each time, remesh using <160> modules through the length and <8> modules through the height and redo the Transient analysis.
   View the Time traces at node [3:node=725] (it may be necessary each time to use PlotMesh to confirm the node number of the center of the modeled plate) and record the maximum magnitudes of $\dot{u}$ and $\dot{v}$ and $\sigma_{xx}\epsilon_{xx}$, $\sigma_{xy}\gamma_{xy}$.

ANALYZING THE DATA

1. On a single graph, plot $\sqrt{\sigma_{xx}\epsilon_{xx}}$, $\sqrt{\sigma_{xy}\gamma_{xy}}$, $\dot{u}$, $\dot{v}$ (all suitably scaled) against $\log(h)$.
2. Identify which energy terms become dominant in the limit of a slender waveguide, that is, when $h \to 0$.
3. Conjecture about constructing a waveguide theory. In particular, conjecture about which energy terms would be included.

Part III. Antisymmetric Waves in a Deep Waveguide

This is the antisymmetric counterpart of Part I: The long rectangular plate is excited by a uniform shearing traction such that the centerline is a line of antisymmetry as indicated in Figure 4.12(b). As in Part I, the objective of the data collection and analysis is to identify the type of wave propagating with the most energy.
   Launch QED from the command line in the working directory.

CREATING THE MODEL

1. Create a [20 × 8 × 0.2] aluminum plate using the [Plate:quadrilateral] geometry. That is, set
   [0:geometry =1],
   [1:X_2 =20],

```
    [2:Y_2 =0],
    [3:X_3 =20],
    [4:Y_3 =8],
    [5:X_4 =0],
    [6:Y_4 =8].
    [a:mat# =1],
    [b:thick =0.2].
```

2. Use BCs of [0:type=1(=ENWS)] to set the south side as antisymmetric with all others being free. That is,

```
    [0:type =1],
    [4:sequence =enws],
    [5:east =111111],
    [6:north =111111],
    [7:west =111111],
    [8:south =011111].
```

3. Use Loads of type [0:type=2(=traction)] to apply a uniformly distributed shear traction $t_y = 1.0$ on the west side. That is,

```
    [0:type =2],
    [1:side# =4],
    [4:X_trac =0.0],
    [5:Y_trac =1.0];
```

all other loads being zero.

4. Mesh using $<80>$ modules through the length and $<32>$ transverse modules. That is, set

```
    [0:Global_ID =22],
    [1:X_mods =80],
    [2:Y_mods =32];
```

all [specials] can be set as zero.

Render the mesh by [pressing s].

COLLECTING THE DATA

1. Analyze the model for its linear Transient response. Set

```
    [n:P(t)name =force.30s],
    [0:type =1],
    [1:scale{P} =1000],
    [2:scale{Grav} =0.0],
    [5:time inc =1e-6],
    [6:#steps =200],
    [7:snaps@ =1],
    [9:node =1].
```

[Press s] to perform the time integration.

2. View the Contours of strain energy density with

```
    [0:vars =1],
    [1:vars =81],
```

```
[2:xMult =1],
[3:snapshot =31],
[5:rate =0].
```
The selection [3:snapshot=31] corresponds to time $t = 30\,\mu s$ just when the traction history goes to zero. Individual panels can be zoomed by [pressing z #], where the panels are numbered 1 through 6, top row first.

Observe that the $\sigma_{xy}\gamma_{xy}$ strain energy term is dominant and essentially uniform, except at the top boundary where it goes to zero.

Observe that the other two energy terms are localized to surfaces and zero along the line of antisymmetry.

Record the $(x, y = h/2)$ position and value of the peak $\sigma_{xx}\epsilon_{xx}$ energy. Record the $(x, y = 0)$ position and value of the peak $\sigma_{xy}\gamma_{xy}$ energy.

View the contours of particle velocity with
```
[1:vars =1],
[5:rate =1].
```
Observe that the $\dot{v}$ distribution coincides (except at the free surface) with that of $\sigma_{xy}\gamma_{xy}$ and that the localized $\dot{u}$ has about the same magnitude.

Record the $(x, y = h/2)$ position and value of the peak $\sigma_{xx}\epsilon_{xx}$ energy. Record the $(x, y = 0)$ position and value of the peak $\sigma_{xy}\gamma_{xy}$ energy.

3. Increment the snapshots by $10\,\mu s$ by choosing
```
[3:snapshot =41].
```
Record the $(x, y = h/2)$ position and value of the peak $\sigma_{xx}\epsilon_{xx}$ energy. Record the $(x, y = 0)$ position and value of the peak $\sigma_{xy}\gamma_{xy}$ energy.

4. Repeat the measurements for subsequent snapshots at 10-μs intervals up to a time of $t = 110\,\mu s$.

   Observe at the maximum time that the peak $\sigma_{xy}\gamma_{xy}$ is no longer at $(x, y = 0)$ but close to the top free surface. Observe the emergence of a new peak $\sigma_{xy}\gamma_{xy}$ ahead of the one being tracked. Also observe that there is a significant trailing behavior to the pulse.

5. Change the dimensions of the plate to $[20 \times 4 \times 0.2]$. Mesh using $<80>$ modules through the length and $<16>$ modules through the height.

   Redo the Transient analysis.

   View the progression of energies (without recording measurements). Observe that the peak $\sigma_{xx}\epsilon_{xx}$ energy hardly changes. Observe that the emergence of the new peak $\sigma_{xy}\gamma_{xy}$ occurs at a shorter distance than in the previous case.

6. Change the dimensions of the plate to
   $[20 \times 2.0 \times 0.2]$,
   $[20 \times 1.0 \times 0.2]$,
   $[20 \times 0.5 \times 0.2]$,
   and remesh after each change. Repeat the observations without measurements.

ANALYZING THE DATA

1. On a single graph, plot the position data against arrival time. Estimate the wave speeds.

Which of the following speeds is the estimate closest to?

$$c_o = \sqrt{\frac{E}{\rho}}, \qquad c_P = 1.2\,c_o, \qquad c_S = \sqrt{\frac{G}{\rho}} = 0.61\,c_o, \qquad c_R = 0.93\,c_S = 0.57\,c_o.$$

2. On a single graph, plot the energy magnitudes against position. Comment on the dispersion of the waves.

Part IV. Partition of Antisymmetric Mode Energies

Following on from Part III, this exploration investigates the partitioning of the strain and kinetic energies. As in Part II, the objective of the data collection and analysis is to identify the significant energy contributions that would then be suitable for inclusion in a waveguide theory.

If QED is not already running, begin by launching it from the command line in the working directory.

COLLECTING THE DATA

1. Return to the Geometry section and change the dimensions of the plate to [40 × 4 × 0.2].
   Mesh using <160> modules through the length and <16> modules through the height.
   Redo the Transient analysis, having changed the number of time steps to [6:#steps =400].
2. View the time Traces. To look at particle velocities set
   [0:style type =1],
   [1:vars =1],
   [2:xMult =1],
   [5:rate =1],
   [6:node# =1369],
   [7:elem# =1],
   [8:elem type =4].
   Record the maximum magnitudes of $\dot{u}$ and $\dot{v}$.
   View the strain energy histories by changing
   [1:vars =81],
   [5:rate =0].
   Record the maximum magnitudes of the $\sigma_{xx}\epsilon_{xx}$ and $\sigma_{xy}\gamma_{xy}$ energy densities.
3. Change the dimensions of the plate to [40 × 2 × 0.2].
   Mesh using <160> modules through the length and <8> modules through the height.
   Redo the Transient analysis and View the time Traces.
   At node
   [3:node=725],
   record the maximum magnitudes of $\dot{u}$ and $\dot{v}$ and $\sigma_{xx}\epsilon_{xx}$, $\sigma_{xy}\gamma_{xy}$.

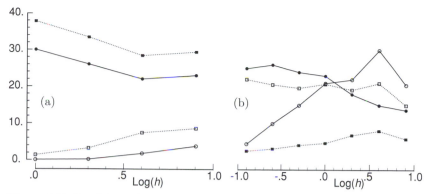

Figure 4.13. Partial results for the partitioning of energies in a waveguide: (a) symmetric modes, (b) antisymmetric modes.

4. In succession, change the dimensions of the plate to
   $[40 \times 1 \times 0.2]$,
   $[40 \times 0.5 \times 0.2]$,
   $[40 \times 0.25 \times 0.2]$,
   $[40 \times 0.125 \times 0.2]$,
   $[40 \times 0.0625 \times 0.2]$.
   Each time remesh using $<160>$ modules through the length and $<8>$ modules through the height and redo the Transient analysis.
   View the time Traces at node [3:node=725] (it may be necessary each time to use PlotMesh to confirm the node number of the center of the modeled plate) and record the maximum magnitudes of $\dot{u}$ and $\dot{v}$ and $\sigma_{xx}\epsilon_{xx}$, $\sigma_{xy}\gamma_{xy}$.

ANALYZING THE DATA

1. On a single graph, plot $\sqrt{\sigma_{xx}\epsilon_{xx}}$, $\sqrt{\sigma_{xy}\gamma_{xx}}$, $\dot{u}$, $\dot{v}$ (all suitably scaled) against $\log(h)$.
2. Identify which energy terms become dominant in the limit of a slender waveguide, that is, when $h \to 0$.
3. Conjecture about constructing a waveguide theory. In particular, conjecture about which energy terms would be included.

Partial Results

The partial results for the partitioning of energies in Parts II and IV are shown in Figure 4.13. The circles are strain energy and squares the kinetic energy.

Background Readings

Waveguides are treated extensively in References [28, 48, 103]. The main emphasis is on establishing a reduced model that captures the main propagation effects. Invariably, however, there are coupling effects; for example, waves in rings couple

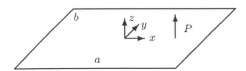
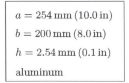

$a = 254 \, \text{mm} \, (10.0 \, \text{in})$

$b = 200 \, \text{mm} \, (8.0 \, \text{in})$

$h = 2.54 \, \text{mm} \, (0.1 \, \text{in})$

aluminum

Figure 4.14. Clamped plate with transverse impact loading.

longitudinal behavior with flexural behavior, resulting in dispersive systems; this is considered in Reference [51].

4.5 Relation Between Waves and Vibrations

The governing equations for wave propagation and vibrations are the same, so it is not surprising that both of these phenomena have many similar underlying features. This section explores the connection between wave-propagation and vibration behaviors. The key is to view the responses in the frequency domain.

A rectangular plate, excited transversely at different points, is the test model (see Figure 4.14). In preparation, use DiSPtool [Time:One] to create a sine-squared pulse of duration $\Delta T = 100 \, \mu s$ – see Section 3.4, Part III, entry 2, for details on running DiSPtool. Call the file <<force.100>>. Store it with a time resolution of $dt = 1.0 \, \mu s$.

Part I. Transient Analyses

The archetypal wave is a localized disturbance traveling in an infinite uniform body. For real structures, however, the wave encounters multiple obstructions and multiple reflections and transmissions are thereby generated. Over time, parts of the original wave then occupy the complete structure simultaneously. That is, the complete structure is simultaneously in motion and exhibits vibrational behavior.

The objective of the data collection and analysis is to determine the effect of the multiple reflections on the responses. Begin by launching QED from the command line in the working directory.

CREATING THE MODEL

1. Create an aluminum plate of size $[10 \times 8 \times 0.1]$ using the [Plate: quadrilateral] geometry. That is,

   [0:geometry =1],
   [1:X_2 =10],
   [2:Y_2 =0],
   [3:X_3 =10],
   [4:Y_3 =8],
   [5:X_4 =0],
   [6:Y_4 =8],
   [a:mat# =1],
   [b:thick =0.1].

2. Use BCs of [0:type=1(=ENWS)] to set fixed BCs on all sides as
   [0:type =1],
   [4:sequence =senw],
   [5:east =000000],
   [6:north =000000],
   [7:west =000000],
   [8:south =000000].
3. Use Loads of type [0:type=1(=near xyz)] to apply a transverse load $P_z = 1.0$ at location $x = 5$, $y = 4$. That is,
   [0:type =1],
   [1:X_pos =5.0],
   [2:Y_pos =4.0],
   [6:Z_load =1.0];
   all other loads and positions being zero.
4. Mesh using $<40>$ modules through the length and $<32>$ transverse modules. That is, set
   [0:Global_ID =23],
   [1:X_mods =40],
   [2:Y_mods =32],
   [a:damping =.01],
   with all other specials as zero. Note that the [0:Global_ID =23] means that the only free DoFs are w, ϕ_x, ϕ_y.
   Render the mesh by [pressing s].
   For clearer viewing, [press o] to set the orientation as $\phi_x = -55$, $\phi_y = 0$, $\phi_z = -20$.

COLLECTING THE DATA

1. Analyze the plate for its linear Transient response. Set
   [n:P(t)name =force.100],
   [0:type =1],
   [1:scale{P} =1000],
   [2:scale{Grav} =0.0],
   [5:time inc =10e-6],
   [6:#steps =1000],
   [7:snaps@ =1],
   [9:node = 668].
   [Press s] to perform the time integration. (Because this is a relatively dense mesh, it may be necessary to adjust the memory allocations in <<stadyn.cfg>>.)
2. View the gallery of displaced Shapes starting at time $100\,\mu$s. That is,
   [1:vars =51],
   [2:xMult =50],
   [3:snapshot =11],
   [5:rate =0].

Zoom and observe the undulation. (If the whole mesh does not appear, then increase the number of lines requested in <<qed.cfg>> – do this while QED is not running.)

3. View the Contours of displacement at the same times. That is,
   [0:type =1],
   [1:vars =1],
   [2:xMult =1],
   [3:snapshot =11],
   [5:rate =0].
   View the contours of velocity using
   [5:rate =1].
   Observe the alternating sign and note that the maximum velocity is no longer at the center.

4. View the contours of displacement and velocity at
   [3:snapshot =21, 31, 41, ...].
   Observe how the displacement pattern is always simpler than that of the velocity. Observe how the double symmetry is always maintained.

5. View Traces of the velocity responses at the center of the plate. That is,
   [6:style type =1],
   [1:vars =1],
   [2:xMult =1],
   [5:rate =1],
   [6:node# =668],
   [7:elem# =1],
   [8:elem type =4].
   [Press s] to show the results; individual panels can be zoomed by pressing [z #], where the panels are numbered 1 through 6, top row first. (This can also be done while in zoom mode.)
   (It may be necessary to confirm the node numbers by launching the utility PlotMesh and viewing the SDF <<qed.sdf>>.)

6. Store the out-of-plane response. The simplest way to store the response is to zoom on the appropriate panel because this will automatically store the data in a file called <<qed.dyn>>, which then can be copied.

7. Launch DiSPtool and go to [Time:One]. [Press p] for the parameters menu and specify
   [n:file =qed.dyn],
   [1:dt =10.0e-6],
   [3:# fft pts =4096],
   [4:MA filt# =1],
   [5:MA filt begin =1],
   [6:plot#1#2 =1 4096],
   [w:window =0].
   [Press q r] to compare the raw file with the windowed and smoothed versions.

8. We now use an exponential decay filter to isolate the effect of the various re-
    flections. [Press w] for the windowing menu and specify
    [0:type =e],
    [1:n1 =1],
    [2:n2 =2],
    [3:n3 =11],
    [4:n4 =21].
    [Press q r] to compare the raw file with the windowed and smoothed versions.
    Observe by zooming panel 3 and expanding ([pressing E] multiple times) that
    only the initial impact pulse has been retained. Observe in panel 2 that the fre-
    quency spectrum is that of a smoothed triangular pulse.
    Change the windowing parameters to
    [3:n3 =21],
    [4:n4 =41],
    and observe how the spectrum is no longer smooth.
    Change the windowing parameters to include some more reflections,
    [3:n3 =41],
    [4:n4 =81],
    and observe how peaks appear in the spectrum.
    In succession, change
    [3:n3 =81], [4:n4 =161],
    [3:n3 =161], [4:n4 =321],
    [3:n3 =321], [4:n4 =641],
    [3:n3 =641], [4:n4 =1281],
    and observe how the peaks in the spectrum become sharper.
    For the final window, store the amplitude spectrum by [pressing s] and
    setting
    [n:filename =amptt.1a],
    [1:1st col =11],
    [2:2nd col =14],
    [3:3rd col =0],
    [4:# pts =250],
    [5:# thin =1].
    Write to disk by [pressing s]. This stores about 6 kHz of the spectrum.
9. In QED, choose the velocity response at
    [3:node =933],
    which should correspond to position $x = 7$, $y = 4$.
    [Press s] then [press z 3] to store the response in <<qed.dyn>>.
    In DiSPtool, [press R] to refresh with the new <<qed.dyn>>. Store its ampli-
    tude spectrum as <<amptt.1b>>.
10. In QED, write the velocity response at
    [3:node=955],
    which should correspond to position $x = 7$, $y = 5.5$. [Press s] then [press
    z 3] to store the response in <<qed.dyn>>.

In DiSPtool, [press R] to refresh with the new <<qed.dyn>>). Store its amplitude spectrum as <<amptt.1c>>.

11. In QED, return to the Loads section and change the load position to
[1:X_pos =7.0],
[2:Y_pos =4.0].
Remesh and redo the linear Transient analysis, and observe by Viewing the Contours that some of the symmetry in the response is lost.
Follow the preceding procedure using DiSPtool to store the amplitude spectrums (for the unwindowed responses) at
[3:node =668, 933, 955]
as <<amptt.2a,2b,2c>>, respectively.

12. Change the load position to
[1:X_pos =7.0],
[2:Y_pos =5.5].
Remesh and redo the linear Transient analysis, and observe by Viewing the Contours that all of the symmetry is lost. Store the amplitude spectrums at
[3:node =668, 933, 955]
as <<amptt.3a,3b,3c>>, respectively.

ANALYZING THE DATA

1. On a single figure, plot the amplitude spectrums <<amptt.1a,1b,1c>> shifted vertically. Observe that the spectral peaks coincide in frequency but not amplitude. Conjecture why.

2. On a single figure, plot the amplitude spectrums <<amptt.2a,2b,2c>> shifted vertically. Observe that some of the spectral peaks do not coincide in frequency with those of the first figure. Conjecture why.

3. On a single figure, plot the amplitude spectrums <<amptt.3a,3b,3c>> shifted vertically. Observe that all nine spectrums of the three figures form a $[3 \times 3]$ symmetric matrix. Conjecture why.

4. Do the corresponding time traces also show a similar symmetry? If yes, conjecture why; if no, conjecture why not.

Part II. Frequency-Domain Analyses

Part I explored the evolution of spectral peaks over time. This information can be obtained directly by doing a forced-frequency analysis. The objective of the data collection and analysis is to relate the forced-frequency response to a vibration eigenanalysis and thus make the connection between Part I and vibrations.

The same model as in Part I is used. If QED is not already running, begin by launching it from the command line in the working directory.

COLLECTING THE DATA

1. Return to the Loads section and place the load at location
[1:X_pos =5.0],
[2:Y_pos =4.0].
Remesh.

2. Analyze the plate for its (forced) Frequency response. Set
   `[0:type=1]`,
   `[1:scale{P} =1000]`,
   `[4:freq_0 =1.0e-3]`,
   `[5:freq_inc =50.]`,
   `[6:#steps =40]`,
   `[7:snaps@ =1]`.
   This will give a frequency range of 2000 Hz. [`Press s`] to perform the frequency scan.

3. View the Traces of the responses against frequency.
   To monitor the displacement amplitude at the center, set
   `[0:style type =1]`,
   `[1:vars =1]`,
   `[2:xMult =1]`,
   `[5:rate =0]`,
   `[6:node# =668]`,
   `[7:elem# = 1]`,
   `[8:elem type =4]`.
   [`Press s`] to show the results; individual panels can be zoomed by pressing [`z #`], where the panels are numbered 1 through 6, top row first. (This can also be done while in zoom mode.) Also look at the log trace with [`0:style=3`]; observe both the sharp spectral peaks and sharp negative peaks called *antiresonances*.

4. Store the out-of-plane response at
   `[3:node=668, 933, 955]`.
   The simplest way to store the response is to zoom on the appropriate panel (this will automatically store the data in a file called <<qed.dyn>>), then copy it to another file name, <<ampff.1a>>, say.

5. View the Contours of displacements with
   `[0:type =1]`,
   `[1:vars =1]`,
   `[2:xMult =1]`,
   `[3:snapshot =1]`,
   `[5:rate =0]`.
   Increment through the snapshots and especially observe the shapes corresponding to frequencies for which the response shows spectral peaks. Make a note of these shapes.

6. Place the load at location
   `[1:X_pos =7.0]`,
   `[2:Y_pos =4.0]`
   and remesh.
   Repeat the frequency analysis and record the same data and call them <<ampff.2a,2b,2c>>, say.

7. Place the load at location
   `[1:X_pos =7.0]`,

[2:Y_pos =5.5]

and remesh.

Repeat the frequency analysis and record the same data and call them <<ampff.3a,3b,3c>>, say.

8. Analyze the plate for its Vibration response using the subspace iteration algorithm. Set

[0:type =1],

[2:scale[K_G] =0.0],

[8:# modes =16],

[a:algorithm =1].

Perform the analysis by [pressing s].

Before exiting the vibration analysis section, record the modal frequencies stored in the file <<stadyn.out>>.

9. View the Contours of displacements with

[0:type =1],

[1:vars =1],

[2:xMult =1],

[3:mode =#],

[5:rate =0],

and confirm that the displacements are only out of the plane of the plate.

10. View the deformed Shapes. For comparative viewing, select the gallery of modes with

[1:vars =51],

[2:xMult =0.1],

[3:mode =1],

[5:rate =0].

For additional clarity, animate the mode shapes using

[1:vars =11],

[2:xMult =0.1],

[3:mode =1],

[5:rate =0],

[6:size =1.5],

[7:max# =4],

[8:divs =16],

[9:slow =50].

ANALYZING THE DATA

1. On a single figure, plot the amplitude spectrums <<ampff.1a,1b,1c>> shifted vertically. Superpose the vibration eigenanalysis modal frequencies as sharp spectral lines (vertical lines at the frequency value). Observe that only some spectral lines coincide with the spectral peaks. Conjecture why.

2. On a single figure, plot the amplitude spectrums <<ampff.2a,2b,2c>> shifted vertically. Superpose the vibration eigenanalysis modal frequencies as sharp spectral lines. Observe that only some spectral lines coincide with the spectral

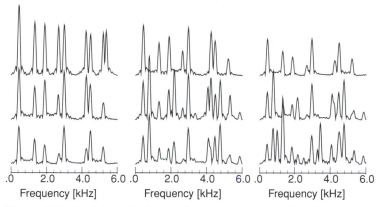

Figure 4.15. Partial results for the frequency-response functions.

peaks and that they are different than those of the first figure. Conjecture why.

3. On a single figure, plot the amplitude spectrums <<ampff.3a,3b,3c>> shifted vertically. Superpose the vibration eigenanalysis modal frequencies as sharp spectral lines. Again, observe that only some spectral lines coincide with the spectral peaks.

4. Observe that all nine spectrums of the three figures form a $[3 \times 3]$ symmetric matrix. Conjecture why.

    Observe that, taken all together, the peaks of the nine spectrums coincide with the spectral lines. Conjecture why.

5. Conjecture on the relation between the analyses of Part I and Part II.

Partial Results

Partial results for the amplitude spectrums of Part I are shown in Figure 4.15. The nine responses form a $[3 \times 3]$ symmetric matrix.

Background Readings

The problem of a point loaded plate is covered analytically in References [28, 61, 80, 86]. These references also consider the reflection of flexural waves at various boundaries.

4.6 Dynamic Stress Concentrations

This section explores the idea of stress concentrations under dynamic loading. In particular, it considers the ways in which the dynamic situation is different from the static counterpart. The general problem to be understood is that of a circular hole in a bar (as shown in Figure 4.16), and by analyzing a series of simpler geometries the essential contributing factors are to be identified.

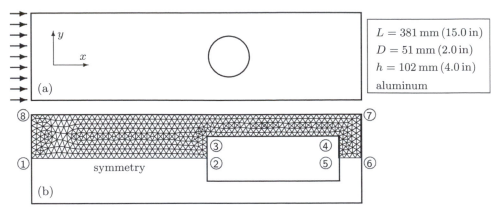

Figure 4.16. Model geometries.

The loading to be used is a smoothed step function so that the long-term behavior is analogous to a static loading. In preparation, use DiSPtool (as described in Section 3.4) to create a smoothed ramp with a rise time of duration $\Delta T = 10\,\mu s$. Name it <<force.10s>> and store it to disk. This gives a loading with a well-defined frequency content and very small side lobes.

Part I. General Problem
We begin with the general problem of a hole in a bar, as shown in Figure 4.16(a). The objective of the data collection and analysis is to identify those features of the response that seem different from the equivalent static problem.

Launch QED from the command line in the working directory.

CREATING THE MODEL
1. Create a $[15 \times 4 \times 0.25]$ aluminum plate with a hole radius of 1.0 using the [Hole:circle] geometry. That is,
   [0:geometry =1],
   [1:inner_R =1],
   [3:X_dim =15],
   [4:Y_dim =4],
   [5:X_pos =9],
   [6:Y_pos =2],
   [a:mat# =1],
   [b:thick =0.25].
2. Use BCs of [0:type=1(=ENWS)] to set all the boundaries as free [111111] except the east side, which is fixed. That is,
   [0:type =1],
   [4:sequence =nsew],
   [5:east =000000],
   [6:north =111111],

      [7:west =111111],
      [8:south =111111].
3. Use Loads of [0:type=2(=traction)] to apply a horizontal uniform load on
   the west side. That is,
      [0:type =2],
      [1:side# =4],
      [4:X_trac =1.0],
   with all other loads being zero.
4. Mesh using
      [0:Global_ID =22],
      [1:X_mods =60],
      [2:Y_mods =16],
      [3:Inner divs =32],
      [4:Inset divs =2],
      [7:# smooth =0],
      [8:max elems =9000],
      [9:mesh style =2];
   set all [specials] as zero.
   [Press s] to render the mesh. If necessary, reorient the mesh ([press o]) to
   $\phi_x = 0$, $\phi_y = 0$, $\phi_z = 0$.

COLLECTING THE DATA
1. Analyze the plate for its linear Static response. Set
      [0:type =1],
      [1:scale{P} =100],
      [2:scale{Grav} =0.0].
   Perform the analysis by [pressing s].
2. View the Contours of von Mises stress with
      [0:type =1],
      [1:vars =51], (specials)
      [2:xMult =1],
      [5:rate =0],
   and confirm that the von Mises contours are in agreement with expectation.
   That is, the largest stresses are on the north and south tips of the hole. Individual
   panels can be zoomed by [pressing z #], where the panels are numbered 1
   through 6, top row first.
   Record the maximum von Mises stress. The applied traction is 100, so the ratio
   of the maximum to this is the static stress concentration factor.
3. Analyze the plate for its linear Transient response. Set
      [n:P(t)name =force.10s],
      [0:type =1],
      [1:scale{P} =100],
      [2:scale{Grav} =0],

```
[5:time inc =0.5e-6],
[6:#steps =400],
[7:snaps@ =1],
[9:node =647].
```
The snapshots are stored at every time increment. [Press s] to perform the time integration.

4. View the Contours of von Mises stress with fixed scales. That is, set
```
[0:type =2],
[1:vars =51],
[2:xMult =1.0],
[3:snapshot =81],
[5:rate =0],
[6:var_max =250],
[7:var_min =0],
[8:fsig/h =20].
```
This displays snapshot #81, which corresponds to time $40\,\mu s$ when the stress wavefront reaches the hole. Observe that the wavefront is nearly straight and the ramp occupies about one unit of length. Observe that the trailing portion of the wave is nearly constant at the applied level.

Increment the snapshot to [3:snapshot=101], which corresponds to when the peak reaches the north tip of the hole. Observe that the maximum stress occurs before the north tip of the hole. Also observe an unloading wave at the west side of the hole.

Increase the snapshot to [3:snapshot=131] when the stress ramp has passed the hole. Observe that the maximum stress has risen considerably.

Complete viewing the snapshots and observe that the maximum stress continues to increase. If desired, change the fixed scales to a larger range.

5. View time Traces of the responses with
```
[0:style type =1],
[1:vars =51],
[2:xMult =1],
[5:rate =0],
[6:node# =647],
[7:elem# =1],
[8:elem type =4].
```
The monitored node corresponds to the north tip of the hole. [Press s] to show the results; individual panels can be zoomed by [pressing z #], where the panels are numbered 1 through 6, top row first.

Observe that the maximum stress peaks at a value about 10 times that of the nominal, and it does this at a time of about $150\,\mu s$. Also observe that the stress plateaus at a time of about $80\,\mu s$.

Store the stress response. The simplest way to do this is to zoom on the appropriate panel; this will automatically store the data in a file called <<qed.dyn>>, which can then be renamed as appropriate. Call the file <<stress.1a>>.

ANALYZING THE DATA

1. In the Static case, the average stress on the narrowest cross section is $\sigma = \sigma_o 4/(4-2) = 2\sigma_o$. Conjecture why the maximum stress is significantly larger than this.

2. Conjecture why the Transient stress concentration is about twice that of the Static value.

3. Conjecture about the origin of the plateau and why the time for it to form does not seem to bear any connection with the rise time of the stress ramp.

Part II. Plate Without a Hole

The first task is to ascertain the dynamics of the plate without the hole. This will also make clear the role played by the overall boundaries of the plate separate from the boundaries of the hole. This is the objective of the data collection and analysis.

Begin by launching QED from the command line in the working directory.

CREATING THE MODEL

1. Create a $[15 \times 4 \times 0.25]$ aluminum plate using the [Plate:quadrilateral] geometry. That is,

   [0:geometry =1],
   [1:X_2 =15],
   [2:Y_2 =0],
   [3:X_3 =15],
   [4:Y_3 =4],
   [5:X_4 =0],
   [6:Y_4 =4],
   [a:mat# =1],
   [b:thick =0.25].

2. Use BCs of [0:type =1(=ENWS)] to set the east side as fixed with all others being free. That is,

   [0:type =1],
   [4:sequence =nsew],
   [5:east =000000],
   [6:north =111111],
   [7:west =111111],
   [8:south =111111].

3. Use Loads of [0:type =2(=traction)] to apply a uniform traction on the west side acting in the x direction. That is,

   [0:type =2],
   [1:side# =4],
   [4:X_trac =1.0],

   with all other loads being zero.

4. Mesh using <60> modules through the length and <16> transverse modules. That is,

   [0:Global_ID =22],

[1:X_mods =60],
[2:Y_mods =16];
all [specials] can be set to zero.
[Press s] to render the mesh. If necessary, reorient the mesh by [pressing o]
and setting $\phi_x = 0$, $\phi_y = 0$, $\phi_z = 0$.

COLLECTING THE DATA

1. Analyze the plate for its linear Transient response. Set
   [n:P(t)name =force.10s],
   [0:type =1],
   [1:scale{P} =100],
   [2:scale{Grav} =0],
   [5:time inc =0.5e-6],
   [6:#steps =400],
   [7:snaps@ =1],
   [9:node =621].
   [Press s] to perform the time integration.

2. View the Contours of von Mises stress with fixed scales. That is, set
   [0:type =2],
   [1:vars =51],
   [2:xMult =1.0],
   [3:snapshot =81],
   [5:rate =0],
   [6:var_max =250],
   [7:var_min =0],
   [8:fsig/h =20].
   Observe that the wavefront is essentially uniform and that the trailing stress
   level is nearly uniform.
   Increment the snapshot to [3:snapshot=151], which corresponds to time 85 μs
   when the stress wavefront reaches the east side. Observe that the whole plate is
   now (nearly) uniformly loaded.
   Increment the snapshot to [3:snapshot=221], which corresponds to time
   110 μs when the reflected stress wave reaches the position of the hole in Part I.
   Observe the maximum stress is about double the nominal value.

3. View time Traces of the responses with
   [0:style type =1],
   [1:vars =51],
   [2:xMult =1],
   [5:rate =0],
   [6:node# =621],
   [7:elem# =1],
   [8:elem type =4].
   The monitored node corresponds to the hole position of Part I.

Observe that the maximum stress peaks at a value about double that of the nominal, and it does this at a time of about 140 μs.

Store the stress response as described earlier and call it <<stress.1b>>.

4. Return to the Geometry section to make a tapered plate. That is, change
    [2:Y_2 =1],
    [4:Y_3 =3].
    Remesh and rerun the Transient analysis.
5. View time Traces of the responses with
    [6:node# =621].
    Store the stress response and call it <<stress.2b>>.
    Change the monitored node to
    [6:node# =689, 757],
    which is one length unit away from the previous position. Store the stress responses and call them <<stress.3b>> and <<stress.4b>>.

ANALYZING THE DATA
1. Plot all files on a single graph with <<stress.#>> scaled to 25%.
    Observe that the early time history of <<stress.1a>> is different from the others.
2. Conjecture why the histories <<stress.#b>> show oscillations after the first peak.
3. Conjecture why the taper affects the stress level but not the early time history.

Part III. Plate With a Rectangular Cutout

This part focuses on the effect of the cutout. In particular, the objective of the data collection and analysis is to understand the nature of the reverberations in the material surrounding the cutout.

The model used here is not one of QED's preparameterized ones but uses the [General:] forms motif; Section 2.6 gives a tutorial for this type of model creation that has additional details left out here. Begin by launching QED from the command line in the working directory.

CREATING THE MODEL
1. Create a model similar to Figure 4.16(b) by using the [General:arbitrary] geometry forms. The paradigm is that of forms, and the tabs are shown on the top ribbon. Access the form for constructing arbitrary 2D shapes by [pressing fa]. This allows construction of the model geometry; BCs, Loads, material properties are specified through the SDF form.
2. All entries are changed by pressing two characters: the first letter of the box followed by the first letter of the line. This will invoke a dialog box for data input; the data entry is completed by [pressing ENTER]. For example, specify the element type by [pressing et], then typing [4] in the dialog box, followed

by [ENTER]. Choose material [m_tag=1] (aluminum) and [name=qed.msh]. A summary of these instructions is

[et 4],

[em 1],

[en qed.msh].

A colon will be used to separate the command from the input.

3. The geometry is constructed by specifying a collection of control points ([Points of control:]) and connecting them ([Connect pts:]) in various ways. The model of Figure 2.17 has eight control points. The locations of these points are specified by

[p1: =0.0 0.0],

[p2: =8.0 0.0],

[p3: =8.0 1.0],

[p4: =14. 1.0],

[p5: =14. 0.0],

[p6: =15. 0.0],

[p7: =15. 2.0],

[p8: =0.0 2.0].

Ensure remaining entries are zero.

These points need not be specified consecutively.

Segments are specified through connecting two control points and specifying a line style. The six available line styles are given in the information box in the lower right-hand corner; in the present case, all segments are straight. Connect the eight segments as

[c1: =1 2 1],

[c2: =2 3 1],

[c3: =3 4 1],

[c4: =4 5 1],

[c5: =5 6 1],

[c6: =6 7 1],

[c7: =7 8 1],

[c8: =8 1 1].

Ensure remaining entries are zero.

These segments need not be specified consecutively.

Finally, we need to divide each segment into a number of elements. Divide as

[ds:smooth =10],

[d1: =2 4 4],

[d2: =6 8 8],

[d3: =1 1 32],

[d4: =3 3 24],

[d5: =5 5 4],

[d6: =7 7 60].

Ensure remaining entries are zero.

Note that ranges of segments are specified and these can overlap (with the later specification taking precedence). Smoothness of the created mesh is achieved by trying to equalize the areas of neighboring elements.

4. The mesh is rendered by [pressing s] and can be viewed in PlotMesh as file <<qed.msh>>. It should be similar to Figure 4.16(b).

5. Access the SDF form by [pressing fs]. This allows the specification of BCs, Loads, material Properties and so on. In the [Make SDFile:] block, specify

[m1:in mesh =qed.msh],
[m2:out SDF =qed.sdf],
[m3:Global =22],
[m4:BW_red =1].

The last of these reduces the bandwidth of the stiffness matrix. In the [Properties:] block, specify

[p1:tag_1 =1],
[p2:h_1 =0.25].

The BCs are imposed by isolating the appropriate nodes inside a cylinder. We impose the conditions that the bottom is a line of symmetry and the east side is fixed:

[b1:R =.01],
[b2:X1 =-0.1 0.0 0.0],
[b3:X2 =15.1 0.0 0.0],
[b4:D =100000],
[b5:R =.01],
[b6:X1 =15.0 -0.1 0.0],
[b7:X2 =15.0 2.1 0.0],
[b8:D =000000].

Ensure that the radii for the other cylinders are zero.

Loads can be specified either through a cylinder or as a point load at the nearest node. We impose the former along the west side:

[l1:R =.01],
[l2:X1 =0.0 -0.1 0.0],
[l3:X2 =0.0 2.1 0.0],
[l4:P =0.5 0.0 0.0],
[l5:T =0.0 0.0 0.0],
[l6:R =.01],
[l7:X1 =0.0 0.1 0.0],
[l8:X2 =0.0 1.9 0.0],
[l9:P =0.5 0.0 0.0],
[la:T =0.0 0.0 0.0].

These are overlapping cylinders arranged to put unity load on the interior nodes and half that on the corner nodes.

Ensure that all other loads are zero.

6. The structure data file is rendered by [pressing s] and can be viewed in PlotMesh as file <<qed.sdf>>. The model is now ready for analysis. If necessary, reorient the mesh by [pressing o] and changing $\phi_x = 0$, $\phi_y = 0$, $\phi_z = 0$.

COLLECTING THE DATA

1. Analyze the plate for its linear Static response. Set
   [0:type =1],
   [1:scale{P} =6.25],
   [2:scale{Grav} =0.0].
   The load scale is such that it produces a traction of 100 similar to the other parts of this exploration. Perform the analysis by [pressing s].

2. View the Contours of von Mises stress with fixed scales:
   [0:type =2],
   [1:vars =51],
   [2:xMult =1],
   [5:rate =0],
   [6:var_max =250],
   [7:var_min =0].
   There are stress singularities at the interior corners but they are not of immediate concern here. Observe that the stress is nearly uniform in the narrow section at a level about twice that of the applied traction. Observe that the stress is not uniform in the wide section.
   Why is there a significant stress on the west vertical side of the cutout?

3. Analyze the plate for its linear Transient response. Set
   [n:P(t)name =force.10s],
   [0:type =1],
   [1:scale{P} =6.25],
   [2:scale{Grav} =0],
   [5:time inc =0.5e-6],
   [6:#steps =400],
   [7:snaps@ =1],
   [9:node =386].
   [Press s] to perform the time integration.

4. View the Contours of von Mises stress with fixed scales. That is, set
   [0:type =2],
   [1:vars =51],
   [2:xMult =1.0],
   [3:snapshot =81],
   [5:rate =0],
   [6:var_max =250],
   [7:var_min =0],
   [8:fsig/h =20].
   This displays the snapshot at time $40\,\mu s$ when the wavefront just reaches the cutout.

Observe that the ramp portion of the wave occupies about one-half the length of the wide section. Also observe that the trailing end of the wave is essentially at the applied traction of 100.

Increment the snapshot to [3:snapshot=101] and observe that the top portion of the wave has penetrated the narrow section unaffected. Observe an unloading reflection occurring in the bottom of the wide section.

Increment the snapshot to [3:snapshot=121] and observe how the stress singularity at the corner has begun to distort the contours in the narrow section.

Increment the snapshot to [3:snapshot=141], which corresponds to time $80\,\mu s$ when the stress wavefront reaches the east side of the cutout. Observe that the trailing portion of the wave (in the narrow section) is at an elevated stress level and is not uniform. Indeed the pattern of the nonuniformity is analogous to the superposition of a flexural wave with alternating positive and negative stresses. Conjecture why there would be a flexural wave in the narrow section.

Observe that the stress in the wide section has established a significant value along the west vertical side of the cutout.

5. View time Traces of the responses with

    [0:style type =1],
    [1:vars =51],
    [2:xMult =1],
    [5:rate =0],
    [6:node# =386],
    [7:elem# =1],
    [8:elem type =4].

The monitored node corresponds to one length unit into the narrow section and is nearly at the midposition. (In the subsequent monitorings, it may be necessary to use the utility PlotMesh for viewing <<qed.sdf>> to verify the node numberings – to find the node nearest to any point [press nx] and input the coordinates.)

[Press s] to show the trace results.

Observe that the stress initially plateaus but with an oscillation, and the oscillation is about a mean nearly 4/3 times that of the nominal. The plateau is followed by a very large increase in stress.

Why is there an oscillation in the response? Conjecture why the plateau is followed by a very large increase in stress.

Store the stress response and call the file <<stress.1c>>.

6. Return to the [General:arbitrary] geometry forms and change the positions of the cutout to

    [p4: =12. 1.0],
    [p5: =12. 0.0].

It is also necessary to change the element density with

    [d4: =3 3 16],
    [d5: =5 5 12].

Render the mesh by [pressing s].

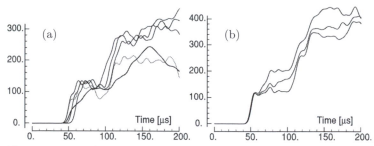

Figure 4.17. Partial results for the stress responses: (a) straight and tapered plates compared with the plate with a hole, (b) plate with a rectangular cutout.

Access the SDF form by [pressing fs]. No changes are necessary here but the SDF needs to be updated by [pressing s].

7. Redo the linear Transient analysis.

8. View time Traces of the responses with
   [0:style type =1],
   [1:vars =51],
   [2:xMult =1],
   [5:rate =0],
   [6:node# =391],
   [7:elem# =1],
   [8:elem type =4].
   Store the stress response and call the file <<stress.2c>>.

9. Return to the [General:arbitrary] geometry forms and change the positions of the cutout and mesh density to
   [p4: =10. 1.0],
   [p5: =10. 0.0].
   [d4: =3 3 8],
   [d5: =5 5 20].
   Render the mesh by [pressing s]. Access the SDF form by [pressing fs] and update the file by [pressing s].
   Redo the linear Transient analysis and View the stress response at
   [6:node# =385].
   Rename the file <<stress.3c>>.

ANALYZING THE DATA

1. Plot all <<stress.#c>> files on a single graph.
   Observe that the early time history of each is the same, but the time when the plateau ends is different.

2. Conjecture on the connection between the cutout size and the ending of the plateau.

3. A proposed conjecture is that the slow rising stress in file <<stress.1a>> is due to the multiple reflections in the narrow section above the hole.
   Do the plotted data support this conjecture?

Partial Results

The partial results for the stress histories of Part II are shown in Figure 4.17(a), and those for Part III are shown in Figure 4.17(b).

Background Readings

Reference [97] is the classic work on the analysis of dynamic stress concentrations. Experimental studies are given in References [64, 69, 70, 84, 137].

5 Nonlinear Structural Mechanics

There are two main sources of nonlinearities in the mechanics of solids and structures. The first is geometric in nature and arises from large deflections, large rotations, and large strains. The second arises from the material behavior and is typified by elastic-plastic and rubberlike materials. Contact problems are also nonlinear even when the contacting bodies themselves remain linear elastic. Figure 5.1 is an example of a long, thin cantilevered plate undergoing large deflections and rotations because of an end moment.

The explorations in this chapter look at aspects of each of these nonlinear behaviors. The first exploration considers the concept of stress and strain under large deformation conditions because they need to be refined relative to the ideas explored in Chapter 2. The second and third explorations consider nonlinear material (constitutive) behavior in the form of elastic-plastic and rubber elasticity, respectively. The presence of a nonlinearity can affect the fundamental behavior of phenomena; the fourth exploration shows the rich, complex behaviors of nonlinear vibrating systems. Within a nonlinear context, applied loads affect the stresses, which in turn affect the stiffness and thus the deformations; this in turn affects the stresses, leading to a complicated connection between the applied loads and the responses. The fifth exploration uses vibrations under gravity to illustrate this point. The final exploration considers the contact that is due to impact of various shaped bodies and the resulting contact force histories.

The dynamics of a general nonlinear system is described by

$$[M]\{\ddot{u}\} + [C]\{\dot{u}\} = \chi_1\{P\}f_1(t) + \chi_2\{G\}f_2(t) - \{F\},$$

where $[M]$ is the mass matrix, $[C]$ is the damping matrix, $\{F\}$ is the vector of element nodal forces, and $\{P\}$, $\{G\}$ are the applied and gravity loads, respectively, with individually specified histories. The nonlinearity arises through $\{F\}$ being a function of the deformation, and therefore these problems must be solved with an incremental–iterative strategy.

The strategy for constructing simple models often entails reducing the size of the problem down to the essential DoF that still exhibits the nonlinearity. This invariably results in a nonlinear ODE, which is why QED has a special analysis section devoted to solving these nonlinear ODEs.

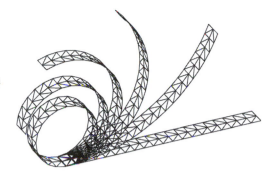

Figure 5.1. Deformation of a cantilevered plate with an end moment.

5.1 Nonlinear Geometric Behavior of Structures

This collection of explorations investigates the nonlinear geometric behavior of structures and structural members. They consider the large-deflection, large-rotation, and large-strain behaviors.

5.1.1 Large Deformations and Strains

The typical nonlinear problem in solid mechanics is that in which the load is specified and the resulting deformation is to be calculated. The situations explored here are those in which the deformation is specified and aspects of the solution (such as stress and strain) are interrogated. The three implemented plane strain ($x_3 = x_3^o$) cases are

$$\text{shear:} \quad x_1 = \lambda_1 x_1^o + k_1 x_2^o, \qquad\qquad x_2 = k_2 x_1^o + \lambda_2 x_2^o;$$

$$\text{bending:} \quad x_1 = [R - x_2^o]\sin(x_1^o/R), \qquad x_2 = R - [R - x_2^o]\cos(x_1^o/R);$$

$$\text{general:} \quad x_1 = (1 + E_1)\cos\phi_1 x_1^o \qquad x_2 = (1 + E_1)\sin\phi_1 x_1^o$$
$$+ (1 + E_2)\sin\phi_2 x_2^o, \qquad\qquad + (1 + E_2)\cos\phi_2 x_2^o,$$

where x_i^o are the original positions. The associated geometries are shown in Figure 5.2.

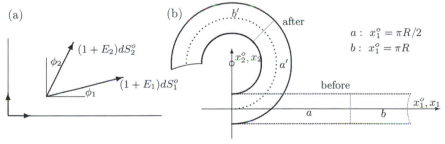

Figure 5.2. Large strains and rotations: (a) generalized homogeneous deformation, (b) shape before and after a deformation that resembles a beam in pure bending.

The material is described by Hooke's law in the form of Equation (7.8). Although the stress–strain relationship itself is linear, it is a nonlinear constitutive relation because the measures of stress and strain used are nonlinear and appropriate for large-strain, large-rotation problems. The next two sections illustrate other nonlinear constitutive relations.

QED gives control over four parameters that are inputted as c_1, c_2, c_3, c_4. These have the following meanings:

$$\text{shear:} \qquad k_1 = c_1 f(t), \quad k_2 = c_3, \quad \lambda_1 = 1, \quad \lambda_2 = c_4;$$

$$\text{bending:} \qquad 1/R = c_1 f(t);$$

$$\text{general:} \qquad \phi_1 = c_1 f(t), \quad \phi_2 = c_2 f(t), \quad E_1 = c_3, \quad E_2 = c_4,$$

where $f(t)$ is a supplied time history. In preparation, create this history with the time–load columns

```
0      0      0
1      0      0
11     10     0
101    100    0
```

and call it <<force.51a>>.

Part I. Strain and Stress Under Rigid-Body Rotation

Rotations play a very important role in the behavior of stress and strain under large deformations, so we begin with the case of a stressed block that rotates but otherwise there are no additional deformations. The objective of the data collection and analysis is to show how the different definitions of stress and strain depend on the orientation of material lines and faces. Of particular interest is the contrast between the behaviors of the Kirchhoff and Cauchy stresses.

Begin by launching QED from the command line in the working directory.

CREATING THE MODEL

1. Create a $[2 \times 2 \times 0.5]$ aluminum block using the [Solid:block] geometry. That is,
   [0:geometry =1],
   [1:X_length =2],
   [2:Y_length =2],
   [3:Z_length =0.5],
   [a:mat# =1],
   [b:constit s/e =1];
   the yield properties are not relevant to this analysis. Position the offsets at zero.
2. Use BCs of [0:type=1(=ENWS)] to set all boundaries as free. That is,
   [4:sequence =snew],
   [5:east =111111],
   [6:north =111111],

[7:east =111111],
[8:south =111111].
Because the deformation will be specified, the BCs do not actually play a role.
3. Use Loads of [0:type=1(=near xyz)] to apply a load of $P_y = 1$ at position $x = 0$, $y = 0$. (This will have no effect but it makes the SDF complete and makes QED fully specified.) That is, set
[0:type =1],
[1:X_pos =0],
[2:Y_pos =0],
[3:Z_pos =0],
[5:Y_load =1];
make all other loads zero.
4. Mesh using <2> elements through the length, <2> through the height, and <1> through the thickness direction. That is, set
[0:Global_ID =111],
[1:X_dirn =2],
[2:Y_dirn =2],
[3:Z_dirn =1];
all [specials] can be set to zero except
[d:red_intn =3].
[Press s] to render the mesh.
For clearer viewing, [press o] to set the orientation as $\phi_x = 30$, $\phi_y = 25$, $\phi_z = 0$. (If the whole mesh does not display, ensure that all tags are set to 1; that is, [press t # 1].)

COLLECTING THE DATA
1. Analyze the problem using the nonlinear Implicit integration scheme, the opening screen should look like Figure 5.3. Set
[n:P(t)name =force.51a],
[0:type =4],
[3:geom =3],
[5:time inc =1],
[6:#steps =21],
[7:snaps@ =1],
[9:node =11].
To achieve a stretching followed by a rigid-body rotation, set the parameters to
[a:c1 =0.157],
[b:c2 =-0.157],
[c:c3 =1],
[d:c4 =0].
[Press s] to create the time snapshots.
2. View the deformed Shapes. Set the movie parameters as
[1:vars =111],
[2:xMult =1],
[3:snapshot =1],

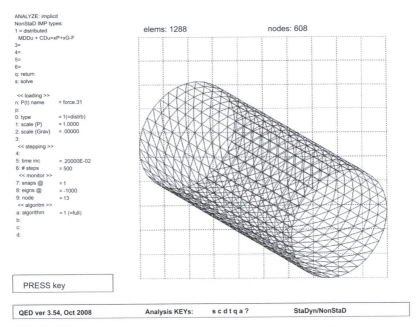

ANALYZE: implicit
NonStaD IMP types:
1 = distributed
  MDDu + CDu=xP+xG-F
3=
4=
5=
6=
q: return
s: solve

  << loading >>
n: P(t) name   = force.31
p:
0: type   = 1(=distrb)
1: scale {P}   = 1.0000
2: scale {Grav} = .00000
3:
  << stepping >>
4:
5: time inc   = .20000E-02
6: # steps   = 500
  << monitor >>
7: snaps @   = 1
8: eigns @   = -1000
9: node   = 13
  << algoritm >>
a: algorithm   = 1 (=full)
b:
c:
d:

elems: 1288   nodes: 608

PRESS key

QED ver 3.54, Oct 2008   Analysis KEYs: s c d t q a ?   StaDyn/NonStaD

Figure 5.3. Opening screen for QED implicit nonlinear analysis.

> [5:rate =0],
> [6:size =1],
> [7:max# =44],
> [8:divs =16],
> [9:slow =50].
> [Press s] to process the snapshots, and [press s] again to run the movie.
> Confirm that the deformation is in agreement with expectation.

3. View the history Traces. Select the displacement traces for Node 39 (position $x = 2$, $y = 0$, $z = 0$, other node positions can be obtained using the PlotMesh utility) with
> [0:style type =1],
> [1:vars =1],
> [2:xMult =1],
> [5:rate =0],
> [6:node# =39],
> [7:elem# =1].
> [Press s] to show the panel of responses; only the first three panels are relevant for solids.
> The numbers for the traces can be found in the file <<stadyn.dyn>>; this will need to be copied if the results are to be saved.

4. Select the nodal average strain history with
> [1:vars=14].
> (A complete list of viewable variables can be seen by [pressing h] for the help menu.)

Observe that the only significant strain is E_{xx}, and this remains constant during the rotation.

Select the nodal average Kirchhoff stress history with
[1:vars=24].

Observe that there are three significant normal stresses, and they remain constant during the rotation.

Select the nodal average Cauchy stress history with
[1:vars=34].

Observe that there are significant normal and shear stresses, but only σ_{zz} remains constant during the rotation.

5. Verify that each node has the same stress and strain.

ANALYZING THE DATA

1. Describe the major characteristics of the stress and strain histories.
2. Using the discussions of stress and strain in Section 7.1 as background, conjecture why the Cauchy stresses oscillate whereas the Kirchhoff stresses remain constant.

Part II. Strain and Stress Under Simple Shear Deformation

In a plane rigid-body rotation, all lines rotate by the same amount; by contrast, in a simple shear deformation, each line has a different rotation, some being positive and others negative. The objective of the data collection and analysis is similar to that of Part I, but in addition, a quantitative assessment of the strain data will be performed.

We use the same model as in Part I, so if QED is still running, then the analysis can be done directly. Otherwise, relaunch QED from the command line in the working directory and ensure that the current model is that of Part I.

COLLECTING THE DATA

1. Use the same model as in Part I. Analyze the problem using the nonlinear Implicit integration scheme with
[n:P(t)name =force.51a],
[0:type =4)],
[3:geom =1],
[5:time inc =1],
[6:#steps =11],
[7:snaps@ =1],
[9:node =11].
To achieve a simple shear deformation, set the parameters to
[a:c1 =0.15],
[b:c2 =0.0],
[c:c3 =0.0],
[d:c4 =1.0].
[Press s] to perform the time integration.

2. View a movie of the deforming Shape with
   [1:vars =111],
   [2:xMult =1],
   [3:snapshot =1],
   [5:rate =0],
   [6:size =1.5],
   [7:max# =44],
   [8:divs =16],
   [9:slow =50].
   [Press s] to process the snapshots and [press s] again to run the movie.
   Confirm that the deformation is in agreement with expectation.
3. View the history Traces. Select the displacement traces for Node 49 (position
   $x = 2$, $y = 2$, $z = 0$) with
   [0:style type =1],
   [1:vars =1],
   [2:xMult =1],
   [5:rate =0],
   [6:node/IP =49],
   [7:elem# =1].
   [Press s] to show the panel of responses; only the first three panels are relevant for solids.
   Observe that there is only a horizontal displacement.
   Select the nodal average strain history with
   [1:vars =14].
   Observe that the significant strains are E_{yy} and E_{xy} but they have different functional forms. Record this strain history by copying the file <<stadyn.dyn>>.
   Select the nodal average Kirchhoff stress history with
   [1:vars =24].
   Observe that all in-plane stresses are significant but only σ_{xy}^K varies linearly.
   Select the nodal average Cauchy stress history with
   [1:vars =34].
   Observe that all in-plane stresses are significant and all seem to have the same
   functional forms.
4. Verify that each node has the same stress and strain.

ANALYZING THE DATA

1. Describe the major characteristics of the stress and strain histories.
2. Plot the strain histories against time.
3. Superpose, as a continuous line, the theoretical Lagrangian strains given by

$$[E_{ij}] = \frac{1}{2}\begin{bmatrix} 0 & k \\ k & k^2 \end{bmatrix}, \qquad k = k_1 = c_1 f(t).$$

How well does this agree with the data?

Part III. Strain and Stress Under Bending Deformation

In Parts I and II, the deformations are homogeneous, that is, the strains and stresses are the same at all points. Here we explore inhomogeneous deformations in a block undergoing bending-type loading as in Figure 5.2(b). The objective of the data collection and analysis is similar to that of the earlier parts, but in addition, this exploration also investigates the performance of the Hex20 element under large strain conditions.

Begin by launching QED from the command line in the working directory.

CREATING THE MODEL

1. Create a slender aluminum block of size $[10 \times 1 \times 0.5]$ using the [Solid:block] geometry. Set
   [0:geometry =1],
   [1:X_length =10],
   [2:Y_length =1],
   [3:Z_length =0.5],
   [6:X_offset =0],
   [7:Y_offset =-0.5],
   [8:Z_offset =0],
   [a:mat# =1],
   [b:constit s/e =1],
   The [off-sets] position the centerline of the block to be along $y^o = 0$.
2. Use BCs of [0:type=1(=ENWS)] to set all boundaries as free. That is,
   [4:sequence =snew],
   [5:east =111111],
   [6:north =111111],
   [7:east =111111],
   [8:south =111111].
   Because the deformation will be specified, the BCs do not actually play a role.
3. Use Loads of [0:type=1(=near xyz)] to apply a load of $P_y = 1$ at position $x = 0$, $y = 0$. (This will have no effect but it makes the SDF complete and makes QED fully specified.) That is, set
   [0:type =1],
   [1:X_pos =0],
   [2:Y_pos =0],
   [3:Z_pos =0],
   [5:Y_load =1];
   make all other loads zero.
4. Mesh using
   [0:Global_ID =111],
   [1:X_dirn =10],
   [2:Y_dirn =2],
   [3:Z_dirn =1].
   Render the mesh by [pressing s].

COLLECTING THE DATA

1. Analyze the problem using the nonlinear Implicit integration scheme with
   [n:P(t)name =force.51a],
   [0:type =4],
   [3:geom =2],
   [5:time inc =0.5],
   [6:# steps =21],
   [7:snaps@ =1],
   [9:node =196].
   To achieve a bending-type deformation, set the parameters as
   [a:c1 =0.062],
   [b:c2 =0.0],
   [c:c3 =0.0],
   [d:c4 =0.0].
   [Press s] to perform the time integration.

2. View the deforming Shape movie with
   [1:vars =111],
   [2:xMult =1],
   [3:snapshot =1],
   [5:rate =0],
   [6:size =1.4],
   [7:max# =44],
   [8:divs =16],
   [9:slow =50].
   [Press s] to process the snapshots, and [press s] again to run the movie.
   Confirm that the deformation is in agreement with expectation.

3. View the history Traces. Select the displacement traces for Node 196 (position $x^o = 10$, $y^o = 0.5$, $z^o = 0$) with
   [0:style type =1],
   [1:vars =1],
   [2:xMult =1],
   [5:rate =0],
   [6:node/IP =196],
   [7:elem# =1].
   [Press s] to show the panel of responses. Observe that both displacements achieve peak values.
   Select the nodal average strain history with
   [1:vars =14].
   Observe that the only significant strain is E_{xx}, and it achieves a value of the order of 25%.
   Record the strain history for this node by copying the file <<stadyn.dyn>>.
   Record the strain history for the corresponding node on the lower side and at the center with [6:node/IP =197] and [6:node/IP =191], respectively.
   Select the nodal average Kirchhoff stress history with

[1:vars =24].

Observe that there are three significant normal stresses, and all increase mono-tonically.

Select the nodal average Cauchy stress history with

[1:vars =34].

Observe that all stresses are significant and appear to behave somewhat errati-cally in time.

4. Check to see if each node on the same radius has the same strain and stress.
5. Return to the Mesh section and refine the mesh with twice as many elements in the x and y directions. That is, set

[0:Global_ID =111],

[1:X_dirn =20],

[2:Y_dirn =4],

[3:Z_dirn =1].

[Press s] to render the mesh.

6. Repeat the analysis and view the behavior at Node 675.

Record the strain histories for nodes

[6:node/IP =675, 663, 671].

ANALYZING THE DATA

1. Describe the major characteristics of the stress and strain histories.
2. Plot the strain histories against time.
3. Superpose, as a continuous line, the theoretical in-plane Lagrangian strain given by

$$E_{xx} = -\frac{y^o}{R} + \frac{1}{2}\left(\frac{y^o}{R}\right)^2.$$

How well does this agree with the data?

What is the effect of mesh refinement?

Partial Results

The partial results for the bending-type strains are shown in Figure 5.4 for two mesh densities. All strains at a given radius should be the same; it is clear that this is achieved with the finer mesh. Elementary beam theory indicates that the top and bottom strains should be the same; this in not true in the nonlinear case.

5.1.2 Large Deflections of Plates

Thin-walled flexible structures derive a significant amount of their (transverse) stiff-ness from membrane stresses. This is through the same mechanism as for cables; the difference is that the stresses are generated as part of the deformation itself and not preset as for the cable. Furthermore, these stresses are generated only when the

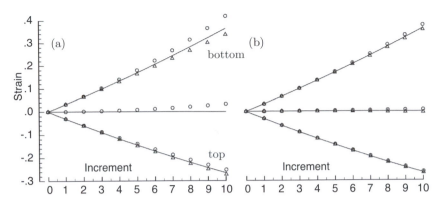

Figure 5.4. Partial results for strains under bending deformation: (a) coarse mesh, (b) refined mesh.

structures are constrained in some way. The explorations in this section focus on the effect of constraints on the deformation of thin-walled components.

The load is applied over time although slow enough that no inertia effects are generated. In preparation, create a load-history file called <<force.51b>> with the three-column [time force gravity] contents:

```
0.00   0.0   0.0
10.0   1.0   0.0
20.0   2.0   0.0
100.   2.0   0.0
```

The load history will be automatically interpolated by StaDyn/NonStaD as needed.

Part I. Plate Constrained on One Edge
We begin with the narrow cantilevered plate of Figure 5.5(a) with a uniform transverse load at the end. This has a minimum of constraints, and so a very large deflection must occur before significant stiffness can be generated. The objective of the data collection and analysis is to investigate the stiffening effect.

Launch QED from the command line in the working directory.

CREATING THE MODEL
1. Create a [10 × 1 × 0.1] aluminum plate using the [Plate:quadrilateral] geometry. That is,
   [0:geometry =1],
   [1:X_2 =10],
   [2:Y_2 =0],
   [3:X_3 =10],
   [4:Y_3 =1],
   [5:X_4 =0],
   [6:Y_4 =1],

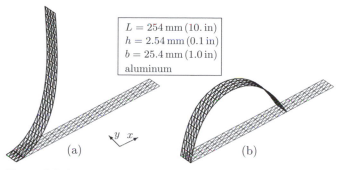

Figure 5.5. Two narrow plates with different constraints: (a) one end completely constrained, (b) both ends simply supported.

        `[a:mat# =1]`,
        `[b:thick =0.1]`.
2. Use BCs of `[0:type=1(=ENWS)]` to simply support [000010] the west edge with all others being free [111111]. That is,
        `[0:type =1]`,
        `[4:sequence =snew]`,
        `[5:east =111111]`,
        `[6:north =111111]`,
        `[7:west =000010]`,
        `[8:south =111111]`.
3. Use Loads of `[0:type=2(=traction)]` to apply a uniform transverse traction on the east edge. That is,
        `[1:side# =2]`,
        `[6:Z_trac =10]`,
        all others being zero. This gives a resultant force of $P_o = 1$.
4. Mesh using $<20>$ modules through the length and $<4>$ transverse modules and a small amount of damping. Set
        `[0:Global_ID =32]`,
        `[1:X_mods =20]`,
        `[2:Y_mods =4]`,
        `[a:damping =.001]`.
        `[Press s]` to render the mesh.
        For clearer viewing, `[press o]` to set the orientation as $\phi_x = -60$, $\phi_y = 0, \phi_z = +25$.

COLLECTING THE DATA
1. Analyze the problem as Implicit nonlinear. Set
        `[n:P(t)name =force.51b]`,
        `[0:type =1]`,
        `[1:scale{P} =100]`,
        `[2:scale{Grav} =0.0]`,

```
[5:time inc =1.0],
[6:#steps =10],
[7:snaps@ =1],
[8:eigns@ = -1],
[9:node =103].
[a:algorithm =1] (full Newton–Raphson).
[Press s] to perform the time integration.
```
2. View a movie of the displaced Shapes with
```
[1:vars =111],
[2:xMult =1],
[3:snapshot =1],
[5:rate =0],
[6:size =1.0],
[7:max# =44],
[8:divs =10],
[9:slow =50].
```
[Press s] to process the snapshots and [press s] again to show the movie.

Observe that the out-of-plane deflection is considerable – the simply supported BCs did not restrain the rotation. Nonetheless, the plate finds a new equilibrium in the upright position; if a small brief disturbance is now applied to the plate, it will return to this position, indicating that the equilibrium is also stable.

3. View the stress Contours with
```
[0:type =1],
[1:vars =21],
[2:xMult =1],
[3:snapshot =11],
[5:rate =0].
```
This gives the stresses at the final load plotted on the original (undeformed) shape.

Observe, by zooming, that the (σ_{xx}, [Var:Smxx]) membrane stress is mostly uniform (keep in mind that there are still some residual dynamic effects) and tensile.

Observe, by contrast, that the (σ_{xx}, [Var:Sfxx]) bending (flexural) stress is mostly zero.

4. Return to the BCs section and change
```
[7:west =000000],
```
all others remaining the same. This constrains the rotation so that the plate is cantilevered.

Remesh and redo the analysis.

View a movie of the displaced Shapes and observe that the out-of-plane deflection is still considerable, with the rotation constraint being operative only near the clamp.

5. View the Traces of the displacement history. Set
```
[0:style type =1],
```

```
[1:vars =1],
[2:xMult =1],
[5:rate =0],
[6:node# =103],
[7:elem# =1],
[8:elem type =4].
```
[Press s] to view the responses. Individual panels can be zoomed by [pressing z #], where the panels are numbered 1 through 6, top row first. Observe the nonlinear response with respect to time/load. In particular, observe that the out-of-plane displacement (w, [Var:W]) history is concave down, indicating a stiffening effect.

Store the out-of-plane displacement (w, [Var:W]) history by zooming on it and renaming the file <<qed.dyn>> something appropriate. Do the same for the horizontal displacement (u, [Var:U]) history.

6. View the stress Contours with
```
[0:type =1],
[1:vars =21],
[2:xMult =1],
[3:snapshot =11],
[5:rate =0].
```
Observe, by zooming, that the (σ_{xx}, [Var:Smxx]) membrane stress has a complicated distribution, even having some compressive regions near the fixed end. Observe, by contrast, that the (σ_{xx}, [Var:Sfxx]) bending (flexural) stress has the typical distribution for a cantilevered beam (zero at the tip, maximum at the fixed end).

ANALYZING THE DATA
1. Plot the deflection data against load. Describe their major characteristics.
2. A simple model for elastic behavior is developed in Section 7.4, Model IV, and leads to

$$P_w = \frac{3EI}{L^3}\left[1 + \frac{81}{60}\frac{w_1^2}{L^2}\right]w_1, \ u_L = \left[0 - \frac{3}{5}\frac{w_1^2}{L^2} - \frac{1}{4}\frac{w_1^4}{L^4}\right]L, \ K_T = \frac{3EI}{L^3}\left[1 + \frac{81}{20}\frac{w_1^2}{L^2}\right],$$

where w_1 is the tip deflection and P_w is the resultant load. Superpose, as continuous lines, these on the preceding data plots.
How well do they match the data?
3. Conjecture about the origin of the compressive (σ_{xx}, [Var:Smxx]) membrane stress.

Part II. Plate Constrained on Two Edges
The plate of Part I is now constrained to be simply supported at both ends with one end initially on rollers. The objective of the data collection and analysis is to observe the stiffening effects of the axial (membrane) loads.

CREATING THE MODEL

1. Create a [10 × 1 × 0.1] aluminum plate using the [Plate:quadrilateral] geometry. That is,

   [0:geometry =1],
   [1:X_2 =10],
   [2:Y_2 =0],
   [3:X_3 =10],
   [4:Y_3 =1],
   [5:X_4 =0],
   [6:Y_4 =1],
   [a:mat# =1].
   [b:thick =0.1].

2. Use BCs of [0:type=1(=ENWS)] to set the ends to be simply supported. That is,

   [4:sequence =snew],
   [5:east =100010],
   [6:north =111111],
   [7:west =000010],
   [8:south =111111].

   Note that the east side is allowed to slide axially.

3. Use Loads of [0:type=2(=traction)] to apply a transverse uniform pressure. Set

   [0:type =2],
   [1:side =5],
   [z:Z_press =0.1].

   Ensure that all other loads are zero. This gives a resultant load of 1. Note that this is not a true pressure because under large deflections the local resultant of a true pressure remains normal to the surface.

4. Mesh using <20> modules through the length and <4> transverse modules and a small amount of damping. Set

   [0:Global_ID =32],
   [1:X_mods =20],
   [2:Y_mods =4],
   [a:damping =.001].
   [Press s] to render the mesh.

COLLECTING THE DATA

1. Analyze the problem as Implicit nonlinear with

   [n:P(t)name =force.51b],
   [0:type =1],
   [1:scale{P} =500],
   [2:scale{Grav} =0.0],
   [5:time inc =1.0],
   [6:#steps =10],

```
[7:snaps@ =1],
[8:eigns@ =1],
[9:node =53].
[a:algorithm =1] (full Newton–Raphson).
```
Although the previous problem and the present one are different, for comparison, this choice of load scale makes the present resultant five times that of the
former. Also note that a vibration eigenanalysis will be performed at each time
step – this will give information about the stiffness behavior of the plate.
`[Press s]` to perform the analysis.

2. View a movie of the displaced Shapes. Set the parameters as
```
[1:vars =111],
[2:xMult =1.0],
[3:snapshot =1],
[5:rate =0],
[6:size =1.5],
[7:max# =44],
[8:divs =10],
[9:slow =50].
```
`[Press s]` to process the frame, and `[press s]` again to run the movie.
Confirm that the deflection is in agreement with expectation.

3. View the Traces of the deflection history. Set
```
[0:style type =1],
[1:vars =1],
[2:xMult =1],
[5:rate =0],
[6:node# =53],
[7:elem# =1],
[8:elem type =4].
```
The chosen node corresponds to the plate center.

Observe that there is a sizable axial (u, `[Var:U]`) displacement. Observe that
the transverse (w, `[Var:W]`) displacement history is concave down, indicating a
stiffening effect.

Store the out-of-plane (w, `[Var:W]`) displacement history by zooming on it and
renaming the file `<<qed.dyn>>` something appropriate. Also copy the eigenvalues file `<<stadyn.mon>>` with an appropriate name.

4. View the stress Contours with
```
[0:type =1],
[1:vars =21],
[2:xMult =1],
[3:snapshot =11],
[5:rate =0].
```
This gives the stresses at the final load plotted on the original (undeformed)
shape.

Observe, by zooming, that the (σ_{xx}, [Var:Smxx]) membrane stress has a complicated distribution, even having some compressive regions near the center.
Observe, by contrast, that the (σ_{xx}, [Var:Sfxx]) bending (flexural) stress has the typical distribution for a simply supported beam (zero at the ends, maximum at the center).

5. Return to the BCs section and change
   [5:east =000010],
   all others remaining the same. This constrains the in-plane displacement of the plate but retains the simply supported end conditions.
   Remesh and redo the analysis.
   View the Traces of the deflection history and observe that the displacements are considerably less than for the other BC case.
   Store the out-of-plane displacement and copy the eigenvalue file.
   View the stress Contours and observe that the (σ_{xx}, [Var:Smxx]) membrane stress is now predominantly tensile and nearly uniformly distributed.

ANALYZING THE DATA

1. Plot both deflection data sets against pressure. Describe their major characteristics.
2. A simple model based on the Ritz function $w(s) = w_1 \sin(\pi s/L)$ and elastica developments of Section 7.4, Models I and IV, gives

$$p_o W = \frac{EI}{L^3}\frac{\pi^4}{4}\left[1 + \left(\frac{\pi w_1}{L}\right)^2\right]\frac{\pi w_1}{L},$$

where w_1 is the central deflection, p_o is the uniform pressure, and W is the width.
Superpose, as a continuous line, this relationship on the data plot.
Which boundary conditions does the simple model best represent?
Conjecture what might account for this.

3. Plot both eigenvalue data sets against load ([col 3 vs col 2] in file <<stadyn. mon>>). Describe their major characteristics.
4. Superpose, as a continuous line, the simple model

$$\omega^2 = \frac{K_T}{M}, \qquad K_T = \frac{EI}{L^3}\frac{\pi^4}{2}\left[1 + 3\left(\frac{\pi w_1}{L}\right)^2\right], \qquad M = \rho h W L/2.$$

How well do they match?

Part III. Plate Constrained on all Edges

The explorations of Parts I and II saw membrane stresses generated as a result of the deformation and how generally these stresses were associated with a stiffening effect. Here the applied loads directly generate compressive membrane stresses, and the objective of the data collection and analysis is to investigate the loss of stiffness and its consequences.

Begin by launching QED from the command line in the working directory.

CREATING THE MODEL

1. Create a $[10 \times 5 \times 0.1]$ aluminum plate using the [Plate:quadrilateral] geometry. That is,

   [0:geometry =1],
   [1:X_2 =10],
   [2:Y_2 =0],
   [3:X_3 =10],
   [4:Y_3 =5],
   [5:X_4 =0],
   [6:Y_4 =5],
   [a:mat# =1],
   [b:thick =0.1].

2. Use BCs of [0:type=1(=ENWS)] to set all edges to be simply supported but allow in-plane motion. That is,

   [4:sequence =snew],
   [5:east =110010],
   [6:north =110100],
   [7:west =010010],
   [8:south =100100].

3. Use Loads of [0:type=2(=traction)] to apply a uniform compressive traction on the east edge. That is,

   [0:type =2],
   [1:side =2],
   [4:X_trac =-1],
   [8:Y_mom =1.0e-4],

   with all others being zero. The M_{yy} moment will act as a load imperfection to nudge the deflection in the transverse direction.

4. Mesh with <20> modules through the length and <10> transverse modules. That is,

   [0:Global_ID =32],
   [1:X_mods =20],
   [2:Y_mods =10],
   [a:damping =.001].
   [Press s] to render the mesh.

COLLECTING THE DATA

1. Analyze the plate for its linear Static response. Set

   [0:type =1],
   [1:scale{P} =20000],
   [2:scale{Grav} =0].
   [Press s] to perform the analysis.

2. View the displacement Contours with

   [0:type =1],
   [1:vars =1],

[2:xMult =1],
[5:rate =0].
Confirm by zooming that the dominant deformation is the in-plane (w, [Var:W]) shortening.
View the stress contours with
[1:vars =21].
Zoom on the first panel to confirm that the membrane (σ_{xx}, [Var:Smxx]) stress is a uniform compression. Also observe that there is a small bending (flexural) (σ_{xx}, [Var:Sfxx]) stress in the vicinity of the loaded edge.

3. Analyze the problem as Implicit nonlinear. Set
   [n:P(t)name =force.51b],
   [0:type =1],
   [1:scale{P} =20000],
   [2:scale{Grav} =0.0],
   [5:time inc =0.5],
   [6:#steps =20],
   [7:snaps@ =1],
   [8:eigns@ =1],
   [9:node =182],
   [a:algorithm =1].
   The selected node corresponds to the mid-3/4 point. [Press s] to perform the time integration.

4. View a movie of the displaced Shapes with
   [1:vars =111],
   [2:xMult =30] (this exaggerates the deflection),
   [3:snapshot =1],
   [5:rate =0],
   [6:size =1.5],
   [7:max# =44],
   [8:divs =10],
   [9:slow =50].
   [Press s] to process the frames and [press s] again to run the movie.
   Observe that the out-of-plane deflection, at its maximum, has a two-half-wave pattern and that this pattern occurs quite abruptly.

5. View the time Traces with
   [0:style type =1],
   [1:vars =1],
   [2:xMult =1],
   [5:rate =0],
   [6:node# =182],
   [7:elem# =1],
   [8:elem type =4].
   Observe the abrupt increase in the out-of-plane displacement (w, [Var:W]).
   Store the out-of-plane (w, [Var:W]) displacement history by zooming on it and renaming the file <<qed.dyn>> to something appropriate.

6. Launch DiSPtool and View the eigenvalues in <<stadyn.mon>>; the stored columns have the sequence $[t, P, \omega_1^2, \omega_2^2, \ldots,]$.
   Observe that one of the modes tends to zero with ω_1^2 and ω_2^2 crossing. This can be observed by [pressing 3] to zoom on the third panel followed by [pressing i] to insert another file, typing the file name <<stadyn.mon>> and the column information [1 3 1].

7. View the stress Contours with
   [0:type =1],
   [1:vars =21],
   [2:xMult =1],
   [3:snapshot =21],
   [5:rate =0].
   This gives the stresses at the final load plotted on the original (undeformed) shape.
   Observe, by zooming, that the (σ_{xx}, [Var:Smxx]) membrane stress has a complicated distribution but is mostly compressive. Record the maximum and minimum values from the legend and their approximate location.
   Observe that the (σ_{xx}, [Var:Sfxx]) bending (flexural) stress has a distribution pattern that follows the deflected shape.

ANALYZING THE DATA

1. A stress of 1.0 was originally applied to the plate, and at maximum time this is scaled to 20000; therefore, the time scale of 0 to 10 corresponds to a stress scale of 0 to 20000.
   Plot the deflection data against applied stress.
   Superpose the first two eigenvalues (suitably scaled) from <<stadyn.mon>>.
   What do the plots have in common?

2. The minimum buckling load for a simply supported rectangle of width W and thickness h is [30]

$$\sigma_{xxc} = -\frac{4D}{h}\left(\frac{\pi}{W}\right)^2, \qquad D = \frac{Eh^3}{12(1-v^2)}$$

   and is independent of the length L when $L > W$.
   Place this value on the preceding plot.

3. At maximum load, the applied membrane stress is 20000 compressive. Subtract this value from the recorded values.
   Conjecture on why the core of the plate has a tensile difference.

Partial Results

Partial results for the deflections of Part I are shown in Figure 5.6. The linear theory is shown as the thin straight lines (with $u_L = 0$). The deflected shape is shown in Figure 5.5 as is the shape for Part II.
   The partial results for Part III are shown in Figure 5.7.

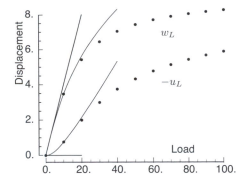

Figure 5.6. Partial results for the large deflection of a cantilevered plate.

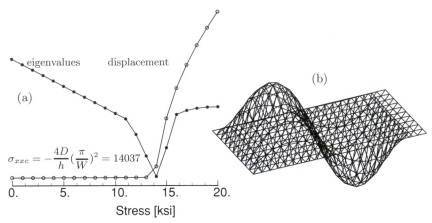

Figure 5.7. Partial results for a plate constrained on all edges: (a) displacement and vibration eigenvalues showing the significance of the critical stress, (b) Exaggerated ($\times$30) deformed shape.

Background Readings

Background material on nonlinear mechanics of solids is developed in References [21, 22, 26, 38, 82]; the nonlinear analysis of structures is covered in References [7, 30]. An accessible introduction to nonlinear mechanics is given in Reference [127] with an emphasis on dynamics.

The analytical treatment of the cases considered here is given in References [27, 30, 92]. The formulation of the Hex20 element for use in nonlinear problems is given in References [5, 26, 32].

5.2 Elastic-Plastic Response and Residual Stresses

This section explores the elastic-plastic response of components. Of particular interest is the effect of confined plastic flow that results in residual stresses on unloading. The bar in Figure 5.8 has confined plasticity around the hole.

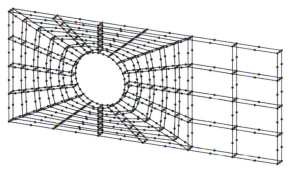

$a = 38\,\text{mm}\ (1.5\,\text{in})$
$L = 381\,\text{mm}\ (15.\,\text{in})$
$b = 152\,\text{mm}\ (6.0\,\text{in})$
$h = 12.7\,\text{mm}\ (0.5\,\text{in})$
aluminum

Figure 5.8. Bar with a hole meshed using Hex20 solid elements.

In preparation for the analyses, create a load history file with the [time load gravity] columns

```
0     0      0
2     0.6    0
10    1.0    0
12    1.1    0
14    0.0    0
15    0.0    0
19    1.6    0
20    1.6    0
```

and call it <<force.52>>.

Part I. Elastic-Plastic Behavior of a Bar

This exploration is a basic study in the constitutive behavior of a bar undergoing elastic-plastic deformation. A uniaxial specimen is loaded in tension, and the objective of the data collection and analysis is to observe the responses during a load–unload–reload cycle.

Begin by launching QED from the command line in the working directory.

CREATING THE MODEL

1. Create an aluminum bar of dimensions $[10 \times 0.5 \times 1]$ using the [Solid:block] geometry. That is,
   [0:geometry =1],
   [1:X_length =10],
   [2:Y_length =1.0],
   [3:Z_length =0.5],
   [a:mat# =1],
   [b:constit s/e =3],
   [c:Yield =30e3],
   [d:E_T =0.5e6].
   Position the offsets at zero.

2. Use BCs of [0:type=1(=ENWS)] to set the west boundary as fixed and the others as free. That is,
   [4:sequence =snew],
   [5:east =111111],
   [6:north =111111],
   [7:west =000000],
   [8:south =111111].
3. Use Loads of [0:type=2(=pressure)] to apply a uniform load on the east face. That is,
   [0:type =2],
   [1:face =1],
   [4:pressure =1.0];
   ensure that all other loads are zero.
4. Mesh using <10> elements through the length, <1> element through the height, and <1> element through the thickness direction. That is,
   [0:Global_ID =111],
   [1:X_dirn =10],
   [2:Y_dirn =1],
   [3:Z_dirn =1],
   [d:red_int =3];
   set other [specials] to zero.
   [Press s] to render the mesh. For clearer viewing, [press o] to set the orientation as $\phi_x = 20$, $\phi_y = -25$, $\phi_z = 0$. (If the whole mesh does not display, ensure that all tags are set to 1; that is, [press t # 1]).

COLLECTING THE DATA

1. Analyze the bar for its nonlinear Implicit elastic-plastic response. Set
   [n:P(t)name =force.52],
   [0:type =5],
   [1:scale{P} =30000],
   [2:scale{Grav} =0.00],
   [5:time inc =1.0],
   [6:# steps =20],
   [7:snaps@ =1],
   [8:eigns@ =-1],
   [9:node =61],
   [a:algorithm =1].
   [Press s] to perform the time integration.
   Note that a history of the currently yielded Elements/IPs is stored in <<simplex.ep>>.
2. View the time Traces of responses. To display the nodal average Lagrangian strain midway along the bar, set
   [0:style type =1],
   [1:vars =14],

```
[2:xMult =1.0],
[5:rate =0],
[6:node/IP =61],
[7:elem# =1].
```
[Press s] to display the strains. Observe that the axial strain mostly increases.
Display the Kirchhoff stress with
```
[1:vars =24].
```
Observe that the dominant stress is the axial stress and it mimics the applied load.
Display the Cauchy stress with
```
[1:vars =34].
```
Observe that it is similar to the Kirchhoff stress but slightly larger.
Display the plastic strain with
```
[1:vars =54].
```
Observe that the three normal components only increase and that at each time step their sum is zero.
3. Store the complete set of histories to <<stadyn.dyn>> by setting
```
[1:vars =59]
```
and [pressing s].
The format for the columns is $[t, E_{xx}, ..., \sigma_{xx}^K, ..., \sigma_{xx}^C, ..., E_{xx}^P, ...]$. Rename the file.

ANALYZING THE DATA

1. Plot the axial stresses ([cols 8 & 14]), the total strain ([col 2]), and the plastic strain ([col 20]), against time.
   What is similar and what is different between the time profiles?
   Why is the Cauchy stress slightly larger than the the Kirchhoff stress?
   Why does the plastic strain only increase?
   Why is $E_{xx}^P + E_{yy}^P + E_{zz}^P = 0$?
2. Plot the Kirchhoff stress against Lagrangian strain.
   What can be said about the unloading?
   What is the effect of work hardening on the reload part of the cycle?
3. Conjecture what would be different if the analysis were rerun with a smaller step size.

Part II. Confined Plasticity

In the previous exploration, the plastic flow was allowed to occur once the stress exceeded the yield stress. In a generally loaded body with various stress gradients, although the stress at a particular point might exceed the yield stress, the surrounding elastic material will constrain the amount of plastic flow that occurs. This gives rise (on unloading) to the very important phenomenon of *residual stresses*.

The objective of the data collection and analysis of this exploration is to determine a simple conceptual model for the formation of residual stresses. Begin by launching QED from the command line in the working directory.

CREATING THE MODEL

1. Create a solid aluminum block with a hole using the [Solid:hole] geometry.
   That is,
   [0:geometry =6],
   [1:X1_length =10.0],
   [2:X2_length =5.0],
   [3:Y_length =6.0],
   [4:Z_length =0.5],
   [5:radius =1.5],
   [6:X_pos =5.0],
   [7:Y_pos =3.0],
   [a:mat# =1],
   [b:constit s/e =3],
   [c:Yield =30e3],
   [d:E_T =0.5e6].

2. Use BCs of [0:type=1(=ENWS)] to set the west boundary as fixed and all others
   as free. That is,
   [4:sequence =snew],
   [5:east =111111],
   [6:north =111111],
   [7:west =000000],
   [8:south =111111].

3. Use Loads of [0:type=2(=pressure)] to apply a uniform pressure on the east
   side. That is,
   [0:type =2],
   [1:face =1],
   [3:pressure =1.0];
   ensure that all other loads are zero.

4. Mesh using <16> elements around the hole, <5> elements radially from the
   hole, <2> elements for the extender, and <1> element through the thickness
   direction. That is,
   [0:Global_ID =111],
   [1:hoop =16],
   [2:radial =5],
   [3:Z_dim =1],
   [4:X_2 dim =2],
   [d:red_int =3];
   set other [specials] to zero.
   [Press s] to render the mesh. For clearer viewing, [press o] to set the orientation as $\phi_x = 20$, $\phi_y = -25$, $\phi_z = 0$.

COLLECTING THE DATA

1. Analyze the bar for its linear Static response. Set
   [0:type =1],
   [1:scale{P} =1],

[2:scale{Grav} =0].
[Press s] to perform the analysis.

2. View the time Traces of responses. This is a linear problem so there are only two snapshots – the undeformed configuration and the loaded configuration; all histories will therefore appear as straight lines. To observe the Cauchy stress at the edge of the hole, set
[0:style type =1].
[1:vars =34],
[2:xMult =1.0],
[5:rate =0],
[6:node/IP =318],
[7:elem =1].
[Press s] to display the stresses.
Observe that the significant stress is $\sigma_{xx} = 4.3$; this means that the maximum stress at the edge of the hole is 4.3 times that of the nominally applied stress.

3. Analyze the bar for its nonlinear Implicit elastic response. Set
[n:P(t)name =force.52],
[0:type =1],
[1:scale{P} =10000],
[2:scale{Grav} =0.00],
[5:time inc =1.0],
[6:#steps =14],
[7:snaps@ =1],
[8:eigns@ =-1],
[9:node =318].
[a:algorithm =1].
[Press s] to perform the time integration.

4. View the time Traces of responses. To observe the Cauchy stress at the edge of the hole, set
[0:style type =1].
[1:vars =34],
[2:xMult =1.0],
[5:rate =0],
[6:node/IP =318],
[7:elem =1].
[Press s] to display the stresses.
Observe that the stress goes to zero on unloading.
Record the maximum value of the σ_{xx} ([Var:Sxx]) stress.
Repeat the measurement for all nodes on the bottom cross section with
[6:node/IP =294, 293, 295, 298, 302, 305, 308, 311, 313, 316, 318].
These are referred to as the elastic stresses.

5. Analyze the bar for its nonlinear Implicit elastic-plastic response. Set
[n:P(t)name =force.52],
[0:type =5],
[1:scale{P} =10000],

```
[2:scale{Grav} =0.0],
[5:time inc =1.0],
[6:#steps =14],
[7:snaps@ =1],
[8:eigns@ =-1],
[9:node =318].
[a:algorithm =1].
```
[Press s] to perform the time integration.

Note that a history of the currently yielded Elements/IPs is stored in <<simplex.ep>>.

6. View the history Traces. View the Cauchy stress traces for Node 318 with
```
[0:style type =1],
[1:vars =34],
[2:xMult =1],
[5:rate =0],
[6:node/IP =318],
[7:elem = ].
```
[Press s] to show the panel of responses,

Observe that the σ_{xx} ([Var:Sxx]) stress levels off even though the load continues to increase. Also observe that there is a compressive stress after unloading; this is the *residual stress*.

Record the maximum and residual stress values as given in <<simplex.dyn>>. Repeat the measurement for all nodes on the bottom cross section with
```
[6:node/IP =294, 293, 295, 298, 302, 305, 308, 311, 313, 316, 318].
```

7. View the nodal average Lagrangian strain traces for Node 318 with
```
[0:style type =1],
[1:vars =14],
[2:xMult =1],
[5:rate =0],
[6:node/IP =318],
[7:elem = ].
```
Observe that there is a positive E_{xx} residual strain. Record its value from the file <<simplex.dyn>>.

View the plastic strain histories with
```
[1:vars =54].
```
Observe that there is a positive E_{xx}^P residual strain. Record its value from the file <<simplex.dyn>>.

ANALYZING THE DATA
1. Plot the elastic and the elastic-plastic stresses against position.
2. Conjecture why the elastic stress concentration factor (ratio of maximum stress to nominal stress) is different for the low-load and large-load cases.

What is the effect of plasticity on the maximum stress?

What is the effect of plasticity on the distribution of stress on the cross section?

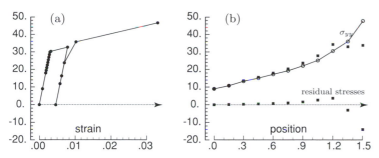

Figure 5.9. Partial results for the elastic-plastic behavior of components: (a) uniaxial stress–strain behavior of the bar, (b) stress distributions in the bar with a hole.

3. Plot the residual stresses against position.
   Conjecture why the distribution has both positive and negative values.
   The recorded residual elastic and plastic strains at the edge of the hole are both positive; conjecture why the residual stress is negative.
4. A simple model for residual stress assumes that the residual stress is the difference between the actual elastic-plastic stresses (at maximum load) and the elastic stresses if there were no plastic effects (also at maximum load).
   Compute the difference between these two sets of stresses and superpose on the preceding plot.
   How well is the simple model validated?
   Conjecture about the mechanism underlying this simple model.
5. Conjecture about what would happen to the stress distributions if the component were given another load cycle.
6. Conjecture how purposely induced residual stresses around a hole may be made to improve the fatigue life of components with holes.

Partial Results

The partial results for the uniaxial stress–strain behavior of Part I are shown in Figure 5.9(a). Observe that the load increments are smaller as yielding is approached. The partial results for the residual stresses of Part II are shown in Figure 5.9(b). The significant feature of the residual stresses is that they are both positive and negative.

Background Readings

Various aspects of plasticity theory are covered in References [57, 60]. The computer algorithms for solving plasticity problems are covered extensively in References [5, 21, 22]. Experimental methods for determining residual stresses are discussed in Reference [49].

5.3 Rubber Elasticity

Rubber and rubberlike materials exhibit large stretching (and hence strains) even under moderate loads, as shown in Figure 5.10. However, in contrast to the

$$\begin{aligned}
W &= 254\,\text{mm}\ (10\,\text{in})\\
b &= 102\,\text{mm}\ (4.0\,\text{in})\\
h &= 3.8\,\text{mm}\ (.15\,\text{in})\\
\alpha_{10} &= 60,\ \alpha_{01} = 0\\
\text{rubber}&
\end{aligned}$$

Figure 5.10. Rubber sheet under large deformations. Shape before and after tension load.

materials of Section 5.2, they remain elastic and recover all the deformations on unloading. This section explores the testing of rubberlike materials under stretching, compression, and shearing conditions.

The constitutive relation representing rubber behavior is nonlinear. The actual constitutive relation implemented in QED is a form of Mooney–Rivlin material that is nearly incompressible. The Kirchhoff stress is related to the deformation by

$$\sigma_{ij}^{K} = \beta_0 \delta_{ij} + \beta_1 C_{ij} + \beta_3 [C_{ij}]^{-1} + K[J_3 - 1]J_3[C_{ij}]^{-1}, \tag{5.1}$$

where $[C_{ij}]$ is the deformation gradient (see Section 7.1) and the coefficients are given by

$$\beta_0 = 2\alpha_{10}\frac{1}{I_3^{1/3}} + 2\alpha_{01}\frac{I_1}{I_3^{2/3}}, \qquad \beta_1 = -2\alpha_{01}\frac{1}{I_3^{2/3}}, \qquad \beta_3 = -\frac{2}{3}\alpha_{10}\frac{I_1}{I_3^{1/3}} - \frac{4}{3}\alpha_{01}\frac{I_2}{I_3^{2/3}}.$$

The coefficients α_{10} and α_{01} are the material parameters and the invariants are given by

$$I_1 = \lambda_1^2 + \lambda_2^2 + \lambda_3^2, \qquad I_2 = \lambda_1^2\lambda_2^2 + \lambda_2^2\lambda_3^2 + \lambda_3^2\lambda_1^2, \qquad I_3 = \lambda_1^2\lambda_2^2\lambda_3^2,$$

where λ_i are the principal stretches (ratio of new length to original length).

In preparation, create the three-column [time force gravity] load file with the contents

```
0       0       0
1       0.1     0
10      1.0     0
100     10.     0
```

and call it <<force.53>>. While QED is not running, modify the material file <<qed.mat>> to append the lines

```
4               ::max # matls
:
4               ::rubber
60  0     1.0e-4
```

Make sure that the maximum number of materials is specified correctly. The three numbers correspond to α_{10}, α_{01}, ρ, respectively. These are the properties of a neo-Hookean material.

Part I. Stretching Behavior

We begin by establishing the basic constitutive behavior of rubber over a strain range of about 100%. The objective of the data collection and analysis is to correlate the results for stress and strain with simple calculations and linear elastic behavior.

Launch QED from the command line in the working directory.

CREATING THE MODEL

1. Create a $[1.0 \times 11.0 \times 0.15]$ rubber strip using the [Solid:block] geometry. That is,
   [0:geometry =1],
   [1:X_length =1.0],
   [2:Y_length =11.0],
   [3:Z_length =0.15].
   Position the offsets at zero.
   Choose material # 4 with rubberlike constitutive behavior
   [a:mat# =4],
   [b:constit s/e =2].
2. Use BCs of [0:type=3(=ENWS_1)] to set the south boundary as fixed and the others free and the top layer as relatively stiff, allowed to move only vertically. That is,
   [0:type =3],
   [4:sequence =ewsn],
   [5:east =111111],
   [6:north =010000],
   [7:west =111111],
   [8:south =000000].
   The top layer of elements will automatically be given stiffness properties that are 1000× that of the base material.
3. Use Loads of [0:type=4(=resultant)] to apply a top face normal distributed force of resultant 1.0. That is,
   [0:type =4],
   [1:face =2],
   [4:X_force =0],
   [5:Y_force =1],
   [6:Z_force =0].
4. Mesh using <2> elements across the strip, <11> elements through the height, and <1> through the thickness direction. That is, set
   [0:Global_ID =111],
   [1:X_dirn =2],
   [2:Y_dirn =11],
   [3:Z_dirn =1].

Choose full integration by setting all [specials] to zero except
[d:red_intn=3].
[Press s] to render the mesh.
For clearer viewing, [press o] to set the orientation as $\phi_x = 25$, $\phi_y = 25$, $\phi_z =$
0. (If the whole mesh does not display, ensure that all tags are set to 1; that is,
[press t # 1].)

COLLECTING THE DATA

1. Analyze the problem using the nonlinear Implicit integration scheme. Set
   [n:P(t)name =force.53],
   [0:type =6],
   [1:scale{P} =40],
   [2:scale{G} =0],
   [5:time inc =1.0],
   [6:#steps =10],
   [7:snaps@ =1],
   [8:eigns@ =1],
   [9:node =197],
   [a:algorithm =1].
   The snapshots and monitors are stored at every time increment.
   [Press s] to perform the time integration.
   Note that the complete history data for the monitored node are in the file
   <<stadyn.dyn>>, and the complete history data for the (vibration) eigenvalues
   are in the file <<stadyn.mon>>; these generally will be overwritten during the
   postprocessing; therefore these will need to be copied to save the results.
2. View the deformed Shapes. Set the movie parameters as
   [1:vars =111],
   [2:xMult =1],
   [3:snapshot =1],
   [5:rate =0],
   [6:size =1],
   [7:max# =44],
   [8:divs =16],
   [9:slow =50].
   [Press s] to process the frames, and [press s] again to run the movie.
   Observe that the deformation is considerable.
3. View the displacement Traces for Node 197 (position $x = 0.5$, $y = 10$, $z = 0$)
   with
   [0:style type =1],
   [1:vars =1],
   [2:xMult =1],
   [5:rate =0],
   [6:node/IP =197],
   [7:elem# =1].

[Press s] to show the panel of responses; only the first three panels are relevant for solids.

Observe that the maximum y displacement corresponds to the strip being stretched to about two-and-a-half times its original length.

Store the complete set of histories for the middle of the strip to <<stadyn.dyn>> with

[1:vars =59],

[6:node/IP =101]

and [pressing s]. (A complete list of viewable variables can be seen by [pressing h] for the help menu.) The format for the columns is $[t, E_{xx}, ..., \sigma_{xx}^K, ..., \sigma_{xx}^C, ...,]$ Rename the file <<uniax.40>>

4. Return to the Analysis section and change the load scaling to

[1:scale{P} =1],

[2:scale{G} =0].

This will generate strains essentially in the linear range.

Rerun the analysis and go to View Traces. Store the complete set of histories for the middle of the strip to <<stadyn.dyn>> with

[1:vars =59],

[3:node/IP =101].

Rename the file <<uniax.01>>

5. Return to the Geometry section and Create a $[10 \times 4.5 \times 0.15]$ rubber sheet using the [Solid:block] geometry. That is,

[0:geometry =1],

[1:X_length =10],

[2:Y_length =4.5],

[3:Z_length =0.15].

The boundary conditions and loads are unchanged.

6. Mesh using $<10>$ elements through the length, $<9>$ through the height, and $<1>$ through the thickness direction. That is, set

[0:Global_ID =111],

[1:X_dirn =10],

[2:Y_dirn =9],

[3:Z_dirn =1].

Choose full integration by setting all [specials] to zero except [d:red_intn=3].

[Press s] to render the mesh.

For clearer viewing, [press o] to set the orientation as $\phi_x = 25$, $\phi_y = 25$, $\phi_z = 0$.

7. Analyze the problem using the nonlinear Implicit integration scheme. Set

[n:P(t)name =force.53],

[0:type =6],

[1:scale{P} =150],

[2:scale{G} =0],

[5:time inc =1.0],

```
[6:#steps =10],
[7:snaps@ =1],
[8:eigns@ =1],
[9:node =362],
[a:algorithm =1].
```
The snapshots and monitors are stored at every time increment.
[Press s] to perform the time integration.

8. View the deformed Shapes. Set the movie parameters as
```
[1:vars =111],
[2:xMult =1],
[3:snapshot =1],
[5:rate =0],
[6:size =1],
[7:max# =44],
[8:divs =16],
[9:slow =50].
```
[Press s] to process the frames and [press s] again to run the movie.
Confirm that the deformation is in agreement with expectation.

9. View the displacement Traces for Node 362 (position $x = 5$, $y = 2$, $z = 0$) with
```
[0:style type =1],
[1:vars =1],
[2:xMult =1],
[5:rate =0],
[6:node/IP =362],
[7:elem# =1].
```
[Press s] to show the panel of responses.
Record the maximum displacement.
The numbers for the traces can be found in the file <<stadyn.dyn>>; this will need to be copied if the results are to be saved.

10. Select the nodal average Lagrangian strain with
```
[1:vars =14].
```
Observe that the dominant strain is E_{yy}, that there is a very small E_{xx} but a correspondingly large thickness strain E_{zz}. Record the maximum strains.
Select the nodal average Kirchhoff stress with
```
[1:vars =24].
```
Observe that the dominant stress is concave down.
Select the nodal average Cauchy stress with
```
[1:vars= 34].
```
Observe that the dominant stress is concave up. Record the maximum stresses.

11. Launch DiSPtool and go to [View]. [Press p] for the parameters list and specify
```
[n:file =stadyn.mon],
[1:plot_1 =2 3],
[2:plot_2 =2 4],
```

```
[3:plot_3 =2 5],
[4:plot_4 =2 6].
```
[Press q R] for complete rereading of the file. This plots the eigenvalues against load.

Observe that the eigenvalues have very small initial values but increase rapidly with load. Also observe that they peak and then decrease with increasing load.

ANALYZING THE DATA

1. Plot the axial stress against axial strain from the file `<<uniax.40>>`, that is, [col 3 vs col 9].

   Superpose the incompressible behavior [32] of

   $$\sigma_{yy}^K = 2C_{00}\left[1 - \frac{1}{\lambda^3}\right], \qquad \lambda = \sqrt{1 + 2E_{yy}}, \qquad C_{00} = \alpha_{10} + \alpha_{01}.$$

   How well do the plots compare?
   Is it strange that the stress levels off even though the load continues to increase?

2. Plot the axial stress against axial strain and transverse strain against axial strain from the file `<<uniax.01>>`, that is, [col 3 vs col 9] and [col 3 vs col 2], respectively.

   For very small strains the Mooney–Rivlin constitutive relation for incompressible materials reduces to the linear Hooke's law with

   $$E = 3\mu, \qquad \mu = 2[\alpha_{10} + \alpha_{01}] = 2C_{00}, \qquad \nu = 0.5.$$

   How well do the plots compare with this linear behavior?

3. For the sheet specimen, let $dS_o = 2$ and $dS = dS_o + v_{max}$; then compare

   $$\frac{dS^2 - dS_o^2}{2dS_o^2} \quad \text{with} \quad E_{yy}.$$

   Use $W = \sqrt{1 + 2E_{xx}}\,W_o$, $h = \sqrt{1 + 2E_{zz}}\,h_o$ to estimate the deformed width and thickness, respectively. Compare the stress estimates

   $$\frac{P_1 \text{ scale}}{W \times h} \quad \text{with} \quad \sigma_{yy}.$$

   Comment on the comparisons.

Part II. Behavior Under Compression

A common application of rubber under compression is as supports, particularly to help suppress transmission of vibrations. In these applications the rubber is usually bulky. By contrast, consider a rubber sheet with a hole under nominally applied stretching conditions; the presence of the hole will cause some parts of the sheet to be in compression and invariably leads to wrinkling (local buckling) of the sheet.

The objective of the data collection and analysis is to determine the effect of the compressive stress on the stability behavior of rubber sheets. Begin by launching QED from the command line in the working directory.

CREATING THE MODEL

1. Create a $[10 \times 4.0 \times 0.15]$ rubber sheet using the [Solid:block] geometry. That is,

   [0:geometry =1],
   [1:X_length =10],
   [2:Y_length =4.0] (this is dimension b),
   [3:Z_length =0.15],

   with no offsets and

   [a:mat# =4],
   [b:constit s/e =2].

2. Use BCs of [0:type=1(=ENWS)] to set the south boundary as fixed, the north restrained to move vertically and others free. That is,

   [0:type =1],
   [4:sequence =ewsn],
   [5:east =111111],
   [6:north =010000],
   [7:west =111111],
   [8:south =000000].

   In comparison with Part I, there is no stiff top layer because the applied load levels will be relatively small.

3. Use Loads of [0:type=4(=resultant)] to apply a top face normal distributed force of resultant -1.0. That is,

   [0:type =4],
   [1:face =2],
   [4:X_force =0],
   [5:Y_force =-1],
   [6:Z_force =0].

4. Mesh using

   [0:Global_ID =111],
   [1:X_dirn=10],
   [2:Y_dirn=8],
   [3:Z_dirn=1],

   with other parameters unchanged.

   [Press s] to render the mesh.

COLLECTING THE DATA

1. Analyze the problem using the nonlinear Implicit integration scheme. Set

   [n:P(t)name =force.53],
   [0:type =6],
   [1:scale{P} =2],
   [2:scale{G} =0],
   [5:time inc =1.0],
   [6:#steps =10],
   [7:snaps@ =1],
   [8:eigns@ =1],

[9:node =197],
[a:algorithm =1].
[Press s] to perform the time integration.

2. View the displacement Traces for Node 326 (position $x = 5$, $y = 2$, $z = 0$) with
[0:style type =1],
[1:vars =1],
[2:xMult =1],
[5:rate =0],
[6:node/IP =326],
[7:elem# =1].
[Press s] to show the panel of responses. Observe that the displacements are small compared with those of Part I.
Select the nodal average Lagrangian strain with
[1:vars =14].
Select the nodal average Kirchhoff stress with
[1:vars =24].
Select the nodal average Cauchy stress with
[1:vars =34].
Observe that the only significant strain and stress are E_{yy} and σ_{yy}, respectively, with little difference between the Cauchy and Kirchhoff stresses. Also, observe that all the histories are linear.

3. Launch DiSPtool and go to [View]. [Press p] for the parameters list and specify
[n:file =stadyn.mon],
[1:plot_1 =2 3],
[2:plot_2 =2 4],
[3:plot_3 =2 5],
[4:plot_4 =2 6].
[Press q R] for complete rereading of the file. This plots the eigenvalues against load.
Observe that the eigenvalues go through zero; also observe that the plot of ω_1^2 versus load is linear.
Save the file <<stadyn.mon>>.

4. The file <<stadyn.tm4>> contains the vibration mode shapes recorded at the same rate as the eigenvalues in <<stadyn.mon>>. They are arranged in sequence as each set of 10 (lowest) against time. It has the same storage format as <<stadyn.snp>> so the postprocessing of QED can be used to view it.
Copy <<stadyn.tm4>> to the file <<stadyn.snp>>.
To get an animated version of all the mode shapes (morphed one into the other) View the deformed Shapes and set the movie parameters as
[1:vars =111],
[2:xMult =1],
[3:snapshot =1],
[5:rate =0],
[6:size =1.5],

[7:max# =200],
[8:divs =16],
[9:slow =50].
[Press s] to process the frames, and [press s] again to run the movie.
Confirm that most of the modes correspond to out-of-plane deflections.

5. Return to the Geometry section and change the dimensions to
[1:X_length =10],
[2:Y_length =2.83],
[3:Z_length =0.15].
Remesh and repeat the analysis, having changed the load scale to
[1:scale{P} =4].
Save the file <<stadyn.mon>>.

ANALYZING THE DATA

1. Plot the lowest eigenvalue ω_1^2 against load ([col 3 vs col 2]) for each geometry.
   Describe their major characteristics.
   Conjecture on the significance of a zero eigenvalue.
2. Do the data support the contention that the instability load varies as $1/b^2$?
3. Estimate the initial slopes $\partial \omega^2 / \partial P$.
   Do they vary as $1/b^2$?
   Where might knowledge of the initial slope be useful?

Part III. Behavior Under Shear

For a similar reason as given in Part II for testing sheets in compression, we are also interested in testing under shear conditions. The objectives are the same here as in Part II.

Begin by launching **QED** from the command line in the working directory.

CREATING THE MODEL

1. Create a $[10 \times 4.0 \times 0.15]$ rubber sheet using the [Solid:block] geometry.
   That is,
   [0:geometry =1],
   [1:X_length =10],
   [2:Y_length =4.0] (this is dimension b),
   [3:Z_length =0.15],
   with no offsets and
   [a:mat# =4],
   [b:constit s/e =2].
2. Use BCs of [0:type=1(=ENWS)] to set the south boundary as fixed, the north restrained to move horizontally and others free. That is,
   [0:type =1],
   [4:sequence =ewsn],
   [5:east =111111],

[6:north =100000],
[7:west =111111],
[8:south =000000].

In comparison with Part I, there is no stiff top layer because the applied load levels will be relatively small.

3. Use Loads of [0:type=4(=resultant)] to apply a top face normal distributed force of resultant −1.0. That is,
[0:type =4],
[1:face =2],
[4:X_force =1],
[5:Y_force =0],
[6:Z_force =0].

4. Mesh using
[0:Global_ID =111],
[1:X_dirn=10],
[2:Y_dirn=8],
[3:Z_dirn=1],

with other parameters unchanged.
[Press s] to render the mesh.

COLLECTING THE DATA

1. Analyze the problem using the nonlinear Implicit integration scheme. Set
[n:P(t)name =force.53],
[0:type =6],
[1:scale{P} =10],
[2:scale{G} =0],
[5:time inc =1.0],
[6:#steps =10],
[7:snaps@ =1],
[8:eigns@ =1],
[9:node =197],
[a:algorithm =1].
[Press s] to perform the time integration.

2. View the displacement Traces for Node 326.
View the nodal average Lagrangian strain, Kirchhoff stress, and Cauchy stress by selecting
[1:vars =14, 24, 34],
respectively. Observe that the only significant strain and stress are E_{xy} and σ_{xy}, respectively, with little difference between the Cauchy and the Kirchhoff stresses. Also observe the concave behavior of all the normal components.

3. Launch DiSPtool and go to [View]. [Press p] for the parameters list and specify
[n:file =stadyn.mon],
[1:plot_1 =2 3],

```
[2:plot_2 =2 4],
[3:plot_3 =2 5],
[4:plot_4 =2 6].
```
[Press q R] for complete rereading of the file. This plots the eigenvalues against load.

Observe that the eigenvalues go through zero; furthermore, observe that the plot of ω_1^2 versus load is not linear as in Part II.

Save the file <<stadyn.mon>>.

4. Copy the file <<stadyn.tm4>> to the file <<stadyn.snp>> and View the movie of deformed Shapes to get an animated version of all the mode shapes. Confirm that most of the modes correspond to out-of-plane deflections.

5. Return to the Geometry section and change the dimensions to
```
[1:X_length =10],
[2:Y_length =2.83],
[3:Z_length =0.15].
```
Remesh and repeat the analysis, having changed the load scale to
```
[1:scale{P} =16].
```
Save the file <<stadyn.mon>>.

ANALYZING THE DATA

1. Plot the lowest eigenvalue ω_1^2 against load ([col 3 vs col 2]) for each geometry.

   Describe their major characteristics.

   Conjecture about the significance of the nonlinear plot.

   Do the data support the contention that the instability load varies as $1/b^2$?

2. The initial slopes $\partial \omega^2 / \partial P$ appear to be insensitive to load. Reasoning that pure shear is equivalent to the principal stress condition $\sigma_2 = -\sigma_1$, conjecture about how this would explain the result.

   Conjecture about how the hypothesis would be tested.

Partial Results

Partial results for the stability behavior in Parts II and III are shown in Figure 5.11. The stretched shape of Part I is shown in Figure 5.10.

Background Readings

The constitutive relations for rubberlike materials are discussed in References [5, 26, 27, 38]. Explicit expressions for the tangent modulus are given in Reference [22]. The computational implementation in QED is based on References [5, 26].

5.4 Nonlinear Vibrations

In this section, we explore some aspects of nonlinear vibrations. A common form of nonlinearity in structural dynamics comes from stiffness change that is due to large

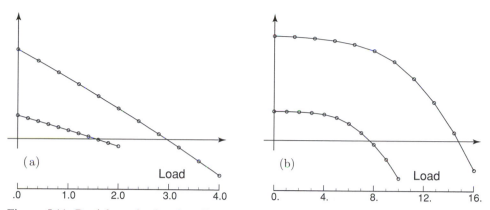

Figure 5.11. Partial results for the vibration eigenvalues during the large deformation of rubber: (a) compression, (b) shear.

deflections. The governing equation representing such systems is

$$M\ddot{u} + C\dot{u} + K[1 + \alpha u^{\beta}]u = P(t).$$

When β is an even power, the stiffness behavior in tension and compression is the same (symmetric), whereas odd powers of β make it nonsymmetric, as shown in Figure 5.12(a). Vibrations with limit cycles are explored using the Van der Pol equation

$$M\ddot{u} - \epsilon[1 - \alpha u^2]\dot{u} + Ku = P(t).$$

The "damping" in this equation is alternatively positive and negative.

Both equations are implemented as part of QED's gallery of nonlinear equations, as discussed in Section 7.7.

Part I. Free Vibration With Stiffening

A linear system vibrated at a given frequency will induce a response with the identical frequency. This is not true of nonlinear systems; indeed, the frequency depends very much on the amplitude of the excitation and response.

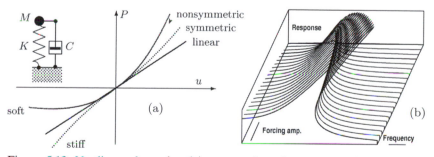

Figure 5.12. Nonlinear dynamics: (a) symmetric and nonsymmetric return force behavior, (b) forced-frequency-response amplitude showing a fold instability.

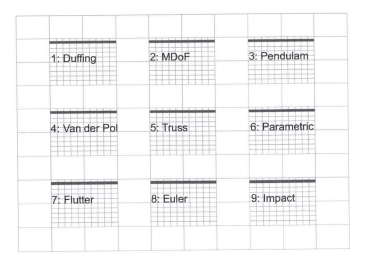

Figure 5.13. Opening screen for QED's gallery of ODEs.

The objective of the data collection and analysis is to determine the amplitude dependence of the frequency response. Begin by launching **QED** from the command line in the working directory.

COLLECTING THE DATA

1. Analyze the problem using [Gallery_ODE:1] to select the Duffing equation from the opening screen of Figure 5.13. Go to the [Geometry] and set
   [k:stiff =1.0],
   [c:damp =0.001],
   [m:mass =1.0],
   [1:a =3.0],
   [2:b =2.0].
   Ensure that autoscaling is on by [toggling x].

2. Go to [ICs] (initial conditions) and set
   [6:u0 =0.01],
   [7:Du0 =0.0],
   [8:Dt =.05],
   [9:#step =1000].

3. Go to [Loads] and set all the loads to zero by
   [1:P0 =0.0],
   [3:P0 =0.0],
   [5:P0 =0.0],
   [8:P0 =0.0].
   The other numbers can be set to 1.0.

4. Go to [Scales] to set the fixed scales and trajectory (state space against time) parameters. Set
[U:u_max =0.5],
[u:u_min =-0.5],
[V:Du_max =0.5],
[v:Du_min =-0.5],
[T:t_max =50],
[t:t_min =0].
Other parameters can be set as unity. The orientation of the trajectory can be changed by [pressing o] and setting it as $\phi_x = 20$, $\phi_y = 30$, $\phi_z = 0$.

5. Return to [ICs] and [press y] to look at the phase plane gallery. [Press 1] ("ell") to superpose the linear response; observe that both are nearly identical. Zoomed views are obtained by [pressing z#].

6. Estimate and record the frequency of vibration – get this information from the number of periods in the window, e.g., $f \approx 1/[50/8] = 0.16$ Hz. This is also expected from the linear model $f = \omega_o/(2\pi) = \sqrt{K/M}/(2\pi) = 0.159$.
Repeat the frequency measurement for different initial displacements ranging over 0.01 to 3.00. That is, in succession, set
[6:u0 =.01, .05, .10, .50, 1.0, 2.0, 3.0, 4.0].
These are DataSet I.
At the [6:u0 =.50] level, [toggle x] to set autoscaling off, [press y], then [press 1] ("ell") to superpose the linear response; observe the period contraction.

7. Before exiting the [ICs] section, set
[6:u0 =3.0],
[7:Du0 =0.0],
[8:Dt =.05],
[9:#step =2000].
Ensure that autoscaling is on by [toggling x].

8. Go to [Geometry] and increase the damping to
[c:damp =0.100].
[Press y] to look at the phase plane gallery. [Press w] to put all the response data into the file <<qed.dyn>>.

9. The following exercise shows how to use DiSPtool to create an amplitude spectrum.
Launch DiSPtool (from the command line) and [press v] to view the columns of data in <<qed.dyn>>. [Press p] to access the input parameters and set
[n:file =qed.dyn],
[1:plot_1 =1 2],
[2:plot_2 =1 3],
[3:plot_3 =1 4],
[4:plot_4 =1 5],
[6:line#1#2 =1 2000].

[Press q] to quit and confirm that the same information is displayed in both QED and DiSPtool.

[Press s] to access the STORE columns menu. Choose to store a two-column file of time [col 1] and amplitude [col 2] by setting

[n:filename =disp.1],
[1:1st col =1],
[2:2nd col =2],
[3:3rd col =0],
[4:#pts =2000],
[5:#thin =1].

[Press s] to store the columns to disk.

Go to the [Time Domain:One channel] section and [press p] to input the stored file. That is, set

[n:filename =disp.1],
[1:dt =0.05],
[3:#fft =2048],
[4:MA filt# =1],
[5:MA filt begin =1],
[6:plot#1#2 =1 2048],
[w:window =0].

[Press q] and then [press f] to see the frequency transform. An expanded view is obtained by [pressing z 1] and then [pressing E] multiple times. Observe that there are no significant amplitudes below a frequency of about 0.15 Hz.

[Press q s] to store the amplitude spectrum. Set

[n:filename =amp.1],
[1:1st col =11],
[2:2nd col =14],
[3:3rd col =0],
[4:#pts =1000],
[5:#thin =1].

[Press s] to store the columns to disk.

10. Switch to QED and go to the [Geometry] section. Change the nonlinear strength to

[1:a =0.0];

this makes the system linear. [Press y w] to put all the response data into the file <<qed.dyn>>. Repeat the DiSPtool procedure and save the amplitude spectrum as [n:filename =amp.2].

ANALYZING THE DATA

1. For DataSet I, plot amplitude against frequency.
2. Superpose on the plot (as a continuous line) the simple model relationship

$$ u_o = \sqrt{\frac{4}{3\alpha} \left[\frac{\omega^2}{\omega_o^2} - 1 \right]^{1/2}} , \qquad \omega_o^2 = \frac{K}{M} = 1 , \qquad \omega = 2\pi f, $$

where u_o is the initial displacement.

How well do the data agree with this relationship?

3. Make plots of the nonlinear and linear amplitude spectrums, <<amp.1>> and <<amp.2>>, respectively.

Comment on the differences.

Comment on the connection between these plots and the plot of DataSet I.

Part II. Free Vibration With Softening

An arch, for example, on the downward swing of a vibration loses stiffness or softens. If the loss is large enough, an instability occurs (called a snap-through instability). The objective of the data collection and analysis is to determine the amplitude that causes the instability and relate it to the structural parameters.

If QED is not already running, begin by launching it from the command line in the working directory.

COLLECTING THE DATA

1. Analyze the problem using [Gallery_ODE:1] to select the Duffing equation. Go to [Geometry] and set

[k:stiff =1.0],
[c:damp =0.0001],
[m:mass =1.0],
[1:a −0.4],
[2:b =1.0].

2. Go to [Loads] and set all the loads to zero by

[1:P0 =0.0],
[3:P0 =0.0],
[5:P0 =0.0],
[8:P0 =0.0].

The other numbers can be set to 1.0.

3. Go to [Scales] to set the fixed scales and trajectory (state space against time) parameters. Set

[U:u_max =2.0],
[u:u_min =-3.0],
[V:Du_max =2.0],
[v:Du_min =-2.0],
[T:t_max =50],
[t:t_min =0].

Other parameters can be set as unity. The orientation of the trajectory can be changed by [pressing o].

4. Go to [ICs] (initial conditions) and set

[6:u0 −0.8],
[7:Du0 =0.0],
[8:Dt =.05],
[9:#step =1000].

Set autoscaling off by [toggling x]. [Press y] to look at the phase plane gallery. [Press l] ("ell") to superpose the linear response; observe the period elongation of the nonlinear system.

5. [Press q x] to toggle autoscaling back on. [Press y] and record the frequency of vibration – get this information from the number of periods in the window, e.g., $f \approx 1/[50/7.5] = 0.15\,\text{Hz}$. Record the peak amplitude excursions u^+, u^-, where u^+ is the initial amplitude and u^- is the maximum (vertical) scale on the plot.

Repeat the frequency measurement for different initial displacements ranging over 0.8 to 1.2. That is, in succession, set

[6:u0 =0.8, 0.9, 1.0, 1.1, 1.2].

This is DataSet II.

6. Verify that there is no vibration for initial amplitudes greater than [6:u0 =1.26]. Set

[6:u0 =1.25].

Set autoscaling off by [toggling x]. [Press y] then [press l] ("ell") to superpose the linear response; observe the significant negative excursion.

7. Return to the [Geometry] section and decrease the nonlinear strength to

[1:a =0.02].

Return to [ICs] section and incrementally change the initial amplitude [6:u0 =??] until the response is no longer oscillatory. Record the peak amplitude excursions u^+, u^- at the last oscillatory response.

Repeat the analysis, having increased the nonlinear strength in [Geometry] to

[1:a =0.08].

This is DataSet III.

ANALYZING THE DATA

1. For DataSet II, form the amplitude number $A = (u^+ - u^-)/2$ (note that u^- is a negative number) and plot frequency^4 against A.

Form the offset amplitude $A_o = (u^+ + u^-)/2$ and plot against A^2.

Comment on the trend of the plots.

2. Superpose on the first plot (as a continuous line) the simple model relationship

$$\left(\frac{\omega}{\omega_o}\right)^4 = 1 - 2\alpha^2 A^2, \qquad \omega_o^2 = \frac{K}{M} = 1, \qquad \omega = 2\pi f.$$

Superpose on the second plot (as a continuous line) the simple model relationship

$$A_o = -\tfrac{1}{2}\alpha A^2.$$

How well do the data agree with these relationships?

3. For DataSet III, form the maximum amplitude number $A = (u^+ - u^-)/2$ and plot against the nonlinearity strength as $1/\alpha$.

Superpose on the plot the simple model relationship

$$A = \frac{1}{\sqrt{2}\,\alpha}.$$

This relationship is based on the preceding simple model with $\omega \to 0$.
How well do the data agree with this relationships?

Part III. Vibration With Limit Cycles

In some dynamic problems, we refer to equilibrium points. For example, a damped pendulum will eventually come to rest in the hang-down position. At this equilibrium point, both the velocity and acceleration are zero. In some systems, especially those connected to some external energy source, the equilibrium "point" is actually an orbit referred to as a *limit cycle*. In this equilibrium configuration, both the velocity and acceleration are nonzero, but if the system is disturbed, it comes back (eventually) to this configuration.

The objective of the data collection and analysis is to determine the connection between the strength of the nonlinearity and the amplitude of the limit cycle. If QED is not already running, begin by launching it from the command line in the working directory.

COLLECTING THE DATA

1. Analyze the problem using [Gallery_ODE:4] to select the Van der Pol equation. Go to [Geometry] and set
   [k:stiff =1.0],
   [c:damp =-0.2],
   [m:mass =1.0],
   [1:a =-0.4].
   Ensure that autoscaling is on by [toggling x].
2. Go to [Loads] and set all the loads to zero by
   [1:P =0.0],
   [3:P =0.0],
   [5:P =0.0],
   [8:P =0.0].
   The other numbers can be set to 1.0.
3. Go to [ICs] (initial conditions) and set
   [6:u0 =10.],
   [7:Du0 =0.0],
   [8:Dt =0.05],
   [9:#step =2000].
   [Press y] to look at the phase plane gallery. Observe that the system achieves a *limit cycle* of constant amplitude and that the initial large displacement spirals into it. Change the initial displacement to
   [6:u0 =0.4].

[Press y] and observe that the system again achieves a limit cycle of constant amplitude, but this time the initial small displacement spirals out to it.
Record the maximum amplitude of the limit cycle and its frequency (estimated from the number of periods in the window).

4. Go to the [Geometry] section and decrease the nonlinear strength to
[1:a =-0.1].
[Press y] and observe that the maximum amplitude of the limit cycle has increased. Record this amplitude and the frequency.
Increase the nonlinear strength to
[1:a =-1.6].
[Press y] and observe that the maximum amplitude of the limit cycle has decreased. Record this amplitude and the frequency.

ANALYZING THE DATA

1. Plot the amplitude and frequency data against nonlinearity as $1/\sqrt{\alpha}$.
   Comment on the trend of the plots.
2. Superpose on the first plot (as a continuous line) the simple model relationship

$$A = \frac{2}{\omega_o\sqrt{\alpha}}, \qquad f = \omega_o/(2\pi), \qquad \omega_o^2 = \frac{K}{M} = 1.$$

How well do the data agree with these relationships?

Part IV. Forced Vibration

A linear analysis of forced vibration predicts (for the no-damping case) an infinite displacement at resonance. This obviously could not occur in reality because of the nonlinearities associated with the large deflections. For example, the clamped end of a cantilever beam would prevent it from displacing too far.

The objective of the data collection and analysis is to correlate the forced-frequency-response amplitude with the forcing frequency. It would be useful to have Figure 5.12(b) in mind as the data are interpreted. If QED is not already running, begin by launching it from the command line in the working directory.

COLLECTING THE DATA

1. Analyze the problem using [Gallery_ODE:1] to select the Duffing equation. Go to the [Geometry] section and set the Duffing equation parameters as
[k:stiff =1.0],
[c:damp =0.02],
[m:mass =1.0],
[1:a =3.0],
[2:b =2.0].
Ensure that autoscaling is on by [toggling x].

2. Go to [ICs] and set all initial conditions to zero and the time stepping as
[6:u0 =.0],
[7:Du0 =.0],

[8:Dt =0.05],
[9:#step =4000].

3. Go to [Scales] to set the fixed scales and trajectory (state space against time) parameters. Set
[U:u_max =1.0],
[u:u_min =-1.0],
[V:Du_max =1.0],
[v:Du_min =-1.0],
[T:t_max =200],
[t:t_min =0].
Other parameters can be set as unity. The orientation of the trajectory can be changed by [pressing o].

4. Go to [Loads] and set all loads to zero except for sinusoidal loading:
[3:P =0.001],
[4:omega =1.0].
[Press y] to look at the phase plane gallery and [press l] ("ell") to superpose the linear response. Observe that both are nearly identical and exhibit a growing amplitude. Call this load level P_o.

5. Apply loads $P = n^3 \times P_o$, that is,
[3:P =.001, .008, .027, .064, .125, .216, .329, .512].
Record the steady-state (long-term) amplitude of the response. If necessary, DiSPtool can be used to get a finer view of the long-term behavior but estimating off the scales is adequate.

6. Keeping the load scale at $512 \times P_o$, increase the frequency from $\omega = 1.0$ to $\omega = 4.0$ in steps of 0.2 and record the steady-state (long-term) amplitude of the response.

7. Keeping the load scale at $512 \times P_o$, decrease the frequency from $\omega = 1.0$ to $\omega = 0.1$ in steps of 0.1 and record the steady-state (long-term) amplitude of the response.

ANALYZING THE DATA

1. For the forced vibration at fixed frequency, plot the amplitude data against load scale.

2. Superpose on the plot the relationship

$$A = \left[\frac{P/M}{\frac{3}{4}\alpha\omega_o^2} \right]^{1/3}.$$

How well do the data agree with this relationship?

3. For the forced vibration at fixed amplitude, plot the response amplitude data against frequency.

4. Superpose on the plot, as a continuous line, the relationship

$$A = 0, \quad \omega < \omega_o; \qquad A = \left[\frac{\omega^2 - \omega_o^2}{\frac{3}{4}\alpha\omega_o^2} \right]^{1/2}, \quad \omega \geq \omega_o.$$

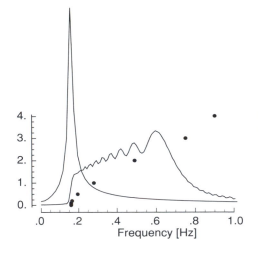

Figure 5.14. Partial results for the nonlinear vibration of a stiff system. The amplitude spectrums are scaled.

Comment on what is observed.
Conjecture on why there is a jump in the data around $\omega \approx 1.7\,\text{r/s}$.

Partial Results

The partial results for the frequency responses in Part I are shown in Figure 5.14. The forced-frequency amplitude responses of Part IV are shown in Figure 5.15.

Background Readings

Reference [58] is an excellent source of solutions to many nonlinear dynamic problems. The Runge–Kutta time-integration algorithm, which is used by QED to solve the nonlinear equations, is covered in Reference [101], which also provides the source code.

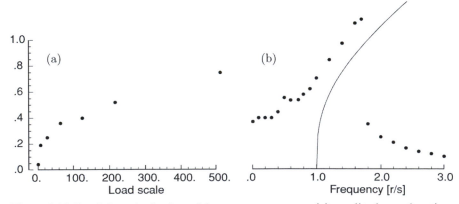

Figure 5.15. Partial results for forced-frequency responses: (a) amplitude as a function of load level, (b) amplitudes during the frequency scans.

Figure 5.16. Cantilevered beam under gravity.

5.5 Nonlinear Vibrations Under Gravity

Gravity loading on a structure causes stresses. As explored in Section 3.3, these stresses in turn can affect the vibration. Additionally, and especially for flexible structures, the gravity loads can cause a significant change of shape, further affecting the vibration behaviors. Both of these effects are fundamentally nonlinear. This section explores some of the effects of gravity on the free vibrations of a structure. The model used for the investigation is a long (flexible) cantilevered beam made of plastic and is shown in Figure 5.16.

The free vibration is initiated by a *pluck* load that slowly causes an initial displacement, followed by a rapid release. The gravity load is also applied over time. In preparation, create a load-history file, call it <<force.55>>, with the three-column [time force gravity] histories:

```
0.0      0.0      0.0
1.0      1.0      1.0
1.1      0.0      1.0
10.0     0.0      1.0
100.     0.0      1.0
```

These load histories will automatically be interpolated by StaDyn/NonStaD as needed. Also, they will be scaled appropriately as part of the analyses.

While QED is not running, modify the materials file <<qed.mat>> to add another plastic with the properties

```
4            ::2nd plastic
345000   127777   1.1e-4
```

Make sure to change the total number of materials.

Part I. Nonlinear Analysis

The cantilevered beam is placed in stand-up and hang-down positions where the gravity has a compressive and tensile stressing effect, respectively. For fixed gravity, the magnitude of the stress is affected by the length of the beam. The objective of the data collection and analysis is to determine the sensitivity of the vibration frequency to the length of the beam.

Begin by launching QED from the command line in the working directory.

CREATING THE MODEL

1. Create a cantilevered beam with the dimensions $[40 \times 1.0 \times 0.125]$ using the
   [Frame:1D_beam] geometry. That is,
   [0:geometry =1],
   [1:X_length =40],
   [4:# X_span =1],
   [a:mat#1 =4],
   [b:thick1 =0.125],
   [c:depth1 =1.0000].
2. Use BCs of [0:type=1(=AmB)] to fix the left end. That is,
   [0:type =1],
   [5:end_A =000000],
   [6:others =111111],
   [7:end_B =111111].
3. Use Loads of [0:type=1(=near xyz)] to apply a vertical point load $P_y = 1.0$ at
   the tip. That is,
   [0:type =1],
   [1:X_pos =40.],
   [2:Y_pos =0.0],
   [4:X_load =0.0]],
   [5:Y_load =1.0].
   Set all other loads and positions to zero.
4. Mesh using $<40>$ elements and treat the beam as 2D. That is,
   [0:Global_ID =22],
   [1:X_elems =40],
   [a:damping =0.5e-4],
   [b?:gravity_X =386].
   Gravity acts along the beam x direction and is inputted by [pressing b x] and
   then the value. [Press s] to render the mesh.
   If necessary, orient the mesh by [pressing o].

COLLECTING THE DATA

1. Analyze the problem for its nonlinear response using the Implicit integration
   scheme. That is, set
   [n:P(t)name =force.55],
   [0:type =1],
   [1:scale{P} =0.02],
   [2:scale{Grav} =1.0],
   [5:time inc =0.025],
   [6:#steps =600],
   [7:snaps@ =1],
   [8:eigns@ =-1],
   [9:node =41],
   [a:algorithm =1].
   [Press s] to perform the time integration.

2. View Traces of the displacement responses at the beam tip. That is,
[0:style type =1].
[1:vars =1],
[2:xMult =1],
[5:rate =0],
[6:node# =41],
[7:elem# =1],
[8:elem type =3].
[Press s] to show the results; individual panels can be zoomed by [pressing z #], where the panels are numbered 1 through 6, top row first. (This can also be done while in zoom mode.) Observe the oscillatory behavior.
Store the transverse response [Var:V]. The simplest way to store the response is to zoom on the appropriate panel, panel #2 in this case; this will automatically store the data in a file called <<qed.dyn>> which can then be renamed.

3. Launch DiSPtool and go to [Time:One]. [Press p] for the parameters menu and specify
[n:file =qed.dyn],
[1:dt =0.025],
[3:#fft pts =2048],
[4:MA filt# =1],
[5:MA filt begin =1],
[6:plot#1#2 =1 2048],
[w:window =0].
[Press q r] to see the processed file. [Press f] to get the frequency-domain transform. Zoom on the [Linear Amplitude] plot and record the frequency of the spectral peak (use the cursor keys [</>] for this).

4. Return to the Analysis section in QED and change the gravity scale to
[2:scale{Grav} =-1.0].
Rerun the analysis and zoom the trace of the transverse response. Switch to DiSPtool and [press R] to refresh the file. Record the frequency of the spectral peak.

5. Return to the Geometry section and change the length of the beam to
[1:X_length =28.3].
In the Loads section, change the location of the load to
[1:X_pos =28.3].
Remesh and redo the analysis, zoom the trace of the transverse response, and record the frequency of the spectral peak. Do this for both positive and negative gravity.

6. Repeat the preceding step so that data are collected for the lengths
[1:X_length =40, 28.3, 20, 14.1, 10].

ANALYZING THE DATA

1. Make a plot of all the frequencies against $1/L^2$.
Connect the data points for positive gravity and for negative gravity, separately.

2. Identify the length at which gravity has little effect.
   Identify the length at which negative gravity causes the frequency to go to zero.
   What is the meaning of a zero or negative frequency?

Part II. Linear Analysis

The objective of the data collection and analysis of this exploration is to develop a simple model for the effect of gravity on the free vibrations of the cantilevered beam. In this context, simple means linear and low order.

The same model as in Part I is used, so if QED is not already running, begin by launching it from the command line in the working directory.

COLLECTING THE DATA

1. In the Geometry section, change the length of the beam to
   [1:X_length =28.3],
   and in the Loads section, change the location of the load to
   [1:X_pos =28.3].
   Remesh.

2. Analyze the problem for its linear Transient response. Set
   [n:P(t)name =force.55],
   [0:type =1],
   [1:scale{P} =0.02],
   [2:scale{G} =1.0],
   [5:time inc =0.025],
   [6:#steps =600],
   [7:snaps@ =1],
   [9:node =41].
   [Press s] to perform the time integration.

3. View the Trace of the displacements of the tip.
   Zoom the transverse response, switch to DiSPtool, and record the frequency of the spectral peak.

4. Change the prestress scale to
   [2:scale{Grav} =-1.0],
   and rerun the analysis. Compare the response and frequency with those of the previous case. Note that the two are virtually identical and that the gravity load has no effect on the vibrations.

5. Change the Transient analysis to a prestress analysis. That is, set
   [n:P(t)name =force.55],
   [p:--> =point],
   [0:type =2],
   [2:scale[K_G] =1.0],
   [5:time inc =0.025],
   [6:#steps =600],
   [7:snaps@ =1],
   [9:node = 41].

[Press p] to set the point excitation to
[1:X_pos =28.3],
[2:Y_pos =0.0],
[4:X_load =0.00],
[5:Y_load =0.02].
Set other loads to zero. [Press q s] to perform the time integration.
6. View the Trace of the displacements of the tip.
   Zoom the transverse response, switch to DiSPtool, and record the frequency of the spectral peak.
7. Change the gravity scale to
   [2:scale[K_G] =-1.0].
   Rerun the analysis and record the frequency of the spectral peak.
   Observe that these two cases show the effect of gravity.
8. Remesh the beam using just a single element. Repeat the analysis for both gravity scales, and observe that these two cases also show the effect of gravity.

ANALYZING THE DATA
1. For the single-length case $L = 28.3$, compare the frequencies from the nonlinear analysis, the prestress linear analysis, and the single-element prestress linear analysis. What conclusions can be drawn?
2. A Ritz SDoF analysis with $v(x) = v_o[x^2/L^2]$ leads to the energies

$$\mathcal{U}_E = \frac{1}{2} \int EI\Big[\frac{\partial^2 v}{\partial x^2}\Big]^2 dx = \frac{1}{2}v_o^2 4EI/L^3,$$

$$\mathcal{U}_G = \frac{1}{2} \int \bar{F}_o\Big[\frac{\partial v}{\partial x}\Big]^2 dx = \frac{1}{2}v_o^2 \rho Ag/3,$$

$$\mathcal{T} = \frac{1}{2} \int \rho A[\dot{v}]^2 dx = \frac{1}{2}\dot{v}_o^2 \rho AL/5.$$

These in turn lead to the governing equation for free vibrations as

$$[\rho AL/5]\ddot{v}_o + [4EI/L^3 \pm \rho Ag/3]v_o = 0.$$

This gives an estimate of the frequency behavior as

$$f = \frac{\omega}{2\pi} = \frac{1}{2\pi}\sqrt{\frac{20}{L^4}\frac{EI}{\rho A} \pm \frac{5}{3}\frac{g}{L}}.$$

How well does this simple model capture the essential behaviors of a structure vibrating under gravity?

Part III. Vibration of Natural Conformations

For the stand-up and hang-down positions of Parts I and II, gravity acted along the length of the beam, and so the beam remained essentially straight. Here we explore the case in which gravity acts transverse to the beam, thus causing a large deflection to a new equilibrium position. The vibrations are about this deflected shape, which is referred to as a *natural conformation*.

The objective of the data collection and analysis is to determine the effect the changed shape has on the vibration behavior. Begin by launching QED from the command line in the working directory.

CREATING THE MODEL
1. Create a cantilevered beam with the dimensions $[40 \times 1.0 \times 0.125]$ using the [Frame:1D_beam] geometry. That is,
   [0:geometry =1],
   [1:X_length =40],
   [4:# X_span =1],
   [a:mat#1 =4],
   [b:thick1 =0.125],
   [c:depth1 =1.0000].
2. Use BCs of [0:type=1(=AmB)] to fix the left end. That is,
   [0:type =1],
   [5:end_A =000000],
   [6:others =111111],
   [7:end_B =111111].
3. Use Loads of [0:type=1(=near xyz)] to ensure that no loads are applied. That is, set
   [0:type =1],
   and set all loads to zero.
4. Mesh using <40> elements and treat the beam as 2D. That is,
   [0:Global_ID =22],
   [1:X_elems =40],
   [a:damping =0.5e-4],
   [by:gravity_Y =-386].
   Gravity acts transverse to the beam. [Press s] to render the mesh.

COLLECTING THE DATA
1. Analyze the problem for its nonlinear response using the Implicit integration scheme. That is, set
   [n:P(t)name =force.55],
   [0:type =1],
   [1:scale{P} =0.0],
   [2:scale{Grav} =1.0],
   [5:time inc =0.025],
   [6:#steps =250],
   [7:snaps@ =1],
   [8:eigns@ =5] (records the vibration eigenvalues every fifth time step),
   [9:node =41],
   [a:algorithm =1].
   [Press s] to perform the time integration.

2. View the deformed Shapes. Set the movie parameters as
   [1:vars =111],
   [2:xMult =1],
   [3:snapshot =1],
   [5:rate =0],
   [6:size =1.5],
   [7:max# =444],
   [8:divs =4],
   [9:slow =10].
   [Press s] to process the snapshots, [Press V] to shift the model up, and
   [press s] again to run the movie.
   Confirm that the deformation is in agreement with expectation.
3. View Traces of the displacement responses at the beam tip. That is,
   [0:style type =1],
   [1:vars =1],
   [2:xMult =1],
   [5:rate =0],
   [6:node# =41],
   [7:elem# =1],
   [8:elem type =3].
   Observe the damped oscillatory behavior of the tip, indicating that the beam
   finds a new equilibrium position.
4. Launch DiSPtool to [View] the vibration eigenvalues in <<stadyn.mon>>.
   [Press v p] for the parameters menu and specify
   [n:file =stadyn.mon],
   [1:plot_1 =1 2],
   [2:plot_2 =1 3],
   [3:plot_3 =1 4],
   [4:plot_4 =1 5],
   [6:line#1 #2 =1 52].
   [Press q] to see the histories; the stored columns have the sequence
   $[t, P, \omega_1^2, \omega_2^2, \ldots,]$. Zoom on the second panel and observe that the eigenvalue
   increased; this indicates that the beam increased its stiffness during the deformation.
   Estimate the limiting value of ω_1^2 (use the cursor keys [</>] for this).
5. Switch back to QED and View the Distributions of stress with
   [1:vars =61],
   [2:xMult =1],
   [3:snapshot =999] (select last snapshot).
   [Press s] to display the panels. The stresses are for the deformed configuration
   but are displayed on the original beam configuration.
   Observe by zooming that the axial stress ([Var:Smxx]) is zero at both ends of
   the beam and tensile in between. Observe that the shear stress ([Var:Smxy])

and moment stress ([Var:Sfzz]) are maximum at the fixed end and zero at the free end.

Record the maximum value of axial, shear, and moment stresses.

6. A new mesh will now be created of the beam in its deformed configuration. View the gallery of deformed shapes with

[1:vars =51],
[2:xMult =1],
[3:snapshot =999] (this will choose the last snapshot),
[7:rate =0].

[Press s] to display the panels, and then [press m1] to create a mesh with the deformed shape. The file is called <<qed2.sdf>> and is a complete SDF.

Launch PlotMesh and confirm that the mesh is as expected.

7. Copy <<qed2.sdf>> to <<qed.sdf>>; the Analysis and View capabilities of QED are now available.

8. Analyze the problem for its linear Static response. Set

[0:type =1],
[1:scale{P} =0],
[2:scale{Grav} =1].

[Press s] to perform the analysis.

View the Distributions of stress with

[1:vars =61],
[2:xMult =1].

Observe that the stresses are plotted on the shape and they have the same distribution characteristics as seen for the nonlinear case.

Record the maximum value of axial, shear, and moment stresses.

9. Analyze the problem for its linear Vibration response. Set

[0:type =1],
[2:scale[K_G] =0] (no gravity effect),
[8:#modes =16],
[a:algorithm =1].

[Press s] to perform the eigenanalysis. Look in the file <<stadyn.out>> to see the computed frequencies. Record the first frequency information.

Turn gravity effects on with

[2:scale[K_G] =1],

and redo the eigenanalysis. Record the first frequency information from <<stadyn.out>>.

ANALYZING THE DATA

1. How well do the stresses compare between the linear and nonlinear models?
2. How well do the frequencies compare between the linear and nonlinear models?
3. Consider a beam of fixed length bent plastically to a nonstraight state and the stresses then annealed. Conjecture about the effect the geometry has on the frequency behavior.

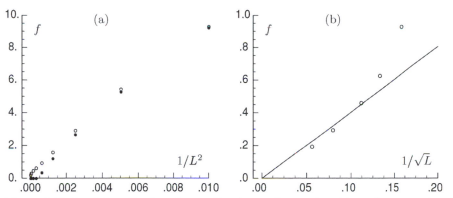

Figure 5.17. Partial results for a beam vibrating under gravity: (a) nonlinear results, (b) behavior in the limit of large length.

Partial Results

The partial results for Part I are shown in Figure 5.17. The behavior in the limit of large length with tension-causing gravity is shown in Figure 5.17(b). The simple model predicts that this limiting behavior should vary as $1/\sqrt{L}$.

Background Readings

The effect of gravity is seldom covered in books on dynamics because its interesting effects are observed only as part of a nonlinear analysis. Thus the pertinent references are those that deal with the geometric stiffness matrix; References [18, 30, 102, 129, 134] establish the geometric stiffness matrix for structural members and plates, and References [5, 26, 32] establish it for the Hex20 element. Related experimental results are given in Reference [25].

5.6 Impact

A significant source of dynamic loading is due to the short-duration contact between two solids. The modeling of these problems is inherently nonlinear and therefore computationally expensive. The explorations in this section show how the forces can be estimated and the stress histories determined without resorting to a full nonlinear FE analysis.

5.6.1 Contact Force Histories During Impact

The force history generated between two bodies in contact depends, *inter alia*, on the relative velocities of the bodies and the contacting geometries. The effect of the geometries is captured or summarized in the form of a *contact law*. We consider two main contact laws: Hertzian contact and flush contact. The explored geometries are shown in Figure 5.18.

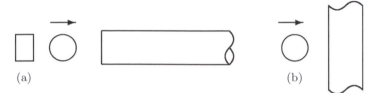

Figure 5.18. Impact geometries: (a) sphere or flat body on a rod, (b) sphere on a beam or plate.

Hertzian contact is for rounded bodies for which the contact region is highly localized. The force between two contacting spheres, the striker and the target, is given by

$$P = K\alpha^{\frac{3}{2}} = K(u_s - u_t)^{\frac{3}{2}},$$

$$K = \frac{4}{3}\sqrt{\frac{R_s R_t}{R_s + R_t}}\left(\frac{k_s k_t}{k_s + k_t}\right), \qquad k_s = \frac{E_s}{1 - v_s^2}, \qquad k_t = \frac{E_t}{1 - v_t^2},$$

where the subscripts s and t stand for striker and target, respectively; R is the radius, E is Young's modulus, and v is Poisson's ratio. The relationship is nonlinear, and therefore the problem is nonlinear. Flush contact indicates that the contact region is extended and the velocity of the striker is transmitted instantaneously to the target.

The explorations here are different than in other sections in that no FE model is created. Instead, we use QED's built-in gallery of nonlinear equations (as discussed in Section 7.7) to perform the analyses.

Part I. Longitudinal Impact of a Sphere on a Rod

We begin with a sphere impacting a rod along its axis. The objective of the data collection and analysis is to correlate the impact history parameters with the velocity.

Launch QED from the command line in the working directory.

COLLECTING THE DATA
1. Analyze the problem using [Gallery_ODE:9] to select impact models.
    Go to [Scales] and set the fixed scales as
    [U:u_max =1.0],
    [u:u_min =0.0],
    [V:Du_max =200],
    [v:Du_min =0.0],
    [T:t_max =300e-6],
    [t:t_min =0.0];
    let all others be zero.
2. Go to [ICs] and set the initial conditions as
    [8:Dt =1.0E-6],
    [9:#step =300].
    This gives a time-integration window of $300\,\mu$s.

3. Go to [Geometry] and set the parameters of the impacting bodies as a steel sphere impacting an aluminum rod. That is, set
   [v:vel =100],
   [0:type =2],
   [1:E =10e6],
   [2:rho =2.5e-4],
   [3:dim1(r|h) =0.5],
   [4:dim2(-|b) =],
   [6:E =30E6],
   [7:rho =7.5e-4],
   [8:rad =0.5],
   [9:con_rad =0.5] (contact radius).
   Note that the contact radius of the sphere ([9:con_rad]) can be different than the sphere size radius.
   Ensure that the autoscaling toggle is turned on, that is, [x:auto scale=on].
4. [Press g] to observe the gallery of responses. The results are displayed as u_s, $\dot{u}_s, u_t, \dot{u}_t, P$, ?. Individual panels can be zoomed by [pressing z #], where the panels are numbered 1 through 6, top row first.
   Record the maximum force ([var 5]), the time of its occurrence, and the total contact time.
   If the zoom does not give sufficiently precise numbers, then [press w] to put the histories into the file <<qed.dyn>> and launch DiSPtool to [View] them. The histories are stored as $t, u_s, \dot{u}_s, u_t, \dot{u}_t, P$.
5. Repeat the test and data recording for velocities of
   [v:vel =1, 3, 10, 30, 100, 300, 1000, 3000].

ANALYZING THE DATA
1. Plot, using log–log scales, the maximum force against velocity, the rise time against velocity, and the contact time against velocity.
2. Superpose, as continuous lines, the simple model formulas (see Section 7.6, Model I)

$$P_{\max} = K^{2/5} \left[\tfrac{5}{4} M_s V_s^2 \right]^{3/5}, \qquad T = \pi \sqrt{\frac{M_s}{K}} \left[\frac{K}{M_s V_s^2} \right]^{1/10}, \qquad M_s = \rho_s \tfrac{4}{3} \pi R_s^3.$$

How well do they match?

Part II. Flush Impact of a Rod
Following on from Part I, the contact radius is now changed (without changing the mass), and the objective of the data collection and analysis is to observe the limiting effect of the radius as the contact becomes flush.

If QED is not already running, begin by launching it from the command line in the working directory.

COLLECTING THE DATA

1. Analyze the problem using [Gallery_ODE:9] to select impact models.
   Go to [Geometry] to set the velocity and contact radius as
   [v:vel =100],
   [9:con_rad =0.50].
2. [Press g] to observe the gallery of responses. Store the results by [pressing w] to put the histories into the file <<qed.dyn>> and then rename this file.
3. Change the contact radius to
   [9:con_rad =0.5e1, 0.5e2, 0.5e3, 0.5e4, 0.5e5, ...],
   each time recording the full data.
   Note that it will be necessary to appropriately reduce *dt* off the [ICs] menu, going as small as
   [8:Dt=0.5E-8]
   for the largest radius; change the number of time steps [9:#step=??] accordingly.
   If the number of steps is made very large, e.g., greater than 10000, then close QED, change the maximum number of lines in <<qed.cfg>>, and relaunch QED.

ANALYZING THE DATA

1. Plot the load history for each of the different contact radii. Observe how the initial rise time gets shorter.
2. Superpose, as a continuous line, the simple model formula (see Section 7.6, Model I)

$$ P(t) = \frac{V_s\,E A}{c_o} e^{-\beta t}, \qquad \beta = \frac{E A}{c_o M_s}, \qquad M_s = \rho_s \tfrac{4}{3} \pi R_s^3, $$

where $c_o = \sqrt{E/\rho}$ and $A = \pi r_t^2$, both for the rod.
How well do they match?

Part III. Plate Impact

A somewhat surprising fact is that the differential equation governing the impact of a plate by a sphere has the same form as that of the impact of a rod by a sphere. The objective of the data collection and analysis is to determine the correlating parameter between a plate and a rod.

Begin by launching QED from the command line in the working directory.

COLLECTING THE DATA

1. Analyze the problem using [Gallery_ODE:9] to select impact models.
   Go to [Geometry] and set the parameters for the impact of a steel ball on an aluminum rod. That is,
   [v:vel =100],
   [0:type =2],
   [1:E =10e6],
   [2:rho =2.5e-4],

  [3:dim1(r|h) =0.5],
  [4:dim2(-|b) =],
  [6:E =30E6],
  [7:rho =7.5e-4],
  [8:rad =0.5],
  [9:con_rad =0.5] (contact radius).
2. Go to [ICs] and set
  [8:Dt =0.5E-6],
  [9:#step =400],
  to have a time integration window of $200 \, \mu$s.
3. [Press g] to observe the gallery of responses. Record the maximum force and store the results by [pressing w] to put the histories into the file <<qed.dyn>>; rename this file.
4. Return to the Geometry section and rerun the analysis with rod radii of
  [3:dim1(r|h) =0.075, 0.100, 0.125, 0.150, 0.175].
  In each case, store the complete histories.
5. Change the target structure to a plate. Set
  [0:type =4],
  [3:dim1(r|h) =0.125].
  Here [3:dim1] is the plate thickness.
6. [Press g] to observe the gallery of responses. Record the maximum force and complete histories.
7. Rerun the analysis with plate thicknesses of
  [3:dim1(r|h) =0.075, 0.100, 0.125, 0.150, 0.175].
  In each case, store the complete histories.

ANALYZING THE DATA

1. Plot the maximum load against radius (R) or thickness (h) as appropriate.
2. Show that a shift of the data corresponding to

$$h = \left[\pi^2/6\right]^{1/4} R = 1.13 \, R$$

  aligns both sets of data.
3. Plot all the force histories, and argue (through interpolation) that the preceding shift would give histories that overlap.
4. Conjecture about the physical origin of $h = 1.13 \, R$.

Part IV. Multiple Impact of a Beam

A sphere impacting a rod experiences a hard contact because the rod is extended in the direction of the impact. As a consequence, the sphere eventually rebounds. When a slim beam is impacted, the contact region (being of small effective mass) is given a larger velocity than the sphere, and separation occurs. The change of momentum of the sphere is small, and it therefore continues in the same direction with a slightly reduced but constant speed. As more of the beam is put in motion, it slows down, and eventually the sphere catches up with it and another impact occurs. This process could repeat itself multiple times.

The objective of the data collection and analysis is to identify the variable(s) contributing to the multiple impact. Begin by launching QED from the command line in the working directory.

COLLECTING THE DATA

1. Analyze the problem using [Gallery_ODE:9] to select impact models.
   Go to [Geometry] to set the parameters for a steel ball impacting an aluminum beam. Set
   [v:vel =100],
   [0:type =3],
   [1:E =10e6],
   [2:rho =2.5e-4],
   [3:dim1(r|h) =0.5],
   [4:dim2(-|b) =0.5],
   [6:E =30E6],
   [7:rho =7.5e-4],
   [8:rad =0.5],
   [9:con_rad =0.5] (contact radius).
   For the beam, [3:dim1] is the height and [3:dim2] is the depth.
2. Go to [ICs] and set
   [8:Dt =1.0E-6],
   [9:#step =500],
   to have a time integration window of 500 μs.
3. [Press g] to observe the gallery of responses. Observe the double impact in Panel 5.
   [Press w] to record the complete histories to the file <<qed.dyn>>.
4. Return to the [Geometry] section and change the dimensions of the beam to
   [3:dim1(r|h) =1.00],
   [4:dim2(-|b) =0.25],
   and record the complete histories.
5. Return to the [Geometry] section and change the dimensions of the beam to
   [3:dim1(r|h) =0.25],
   [4:dim2(-|b) =1.00],
   and record the complete histories.

ANALYZING THE DATA

1. Plot, on the same graph but shifted vertically, the force history for each beam geometry.
   What can be said about the double impacts?
   Is it necessary to have separation to have the multiple-impact effect?
2. Conjecture why a sphere impacting a plate does not have multiple impacts.

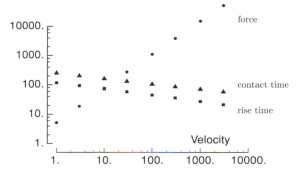

Figure 5.19. Partial results for the dependence of maximum force, rise time, and contact time on the impacting velocity for a ball striking a rod. Times are in microseconds.

Partial Results

The partial results for Part I are shown in Figure 5.19. It can be concluded that the contact time is relatively insensitive to the velocity.

5.6.2 Disk Impacting a Half-Plane

Even if both impacting bodies remain elastic and their dynamic behavior is linear, the problem is nonetheless nonlinear because the contact law is nonlinear. What we explore here is the use of QED's gallery of nonlinear differential equations to determine the force history, and then we use a linear FE analysis to determine the stress histories and distributions in the bodies. The specific problem of the impact of a disk/annulus on a half-plane is considered (see Figure 5.20).

The Hertz theory for the contact of spheres as used earlier is based on the assumption that the stresses close to the contact point are very large compared with those elsewhere. This is not true for the contact of disks; a good discussion of the static contact of disks is given in Reference [56]. Consequently the static theory for the contact of disks and cylinders is not applicable in this impact case. A simple model that relates the force between two contacting disks is

$$P = K\alpha^{\frac{3}{2}} = K(u_s - u_t)^{\frac{3}{2}},$$

$$K = \tfrac{4}{3}E_e h\sqrt{2R_e}[1 - \tfrac{4}{5}(u_s - u_t)/L], \qquad \frac{1}{E_e} \equiv \frac{1}{E_s} + \frac{1}{E_t}, \qquad \frac{1}{R_e} \equiv \frac{1}{R_s} + \frac{1}{R_t}.$$

The adjustable parameter $L = \gamma R_e$ is typically set with $\gamma \approx 0.8$.

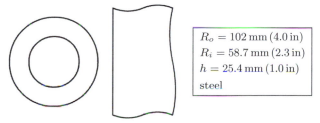

$$\begin{array}{|l|}
\hline
R_o = 102\,\text{mm}\,(4.0\,\text{in}) \\
R_i = 58.7\,\text{mm}\,(2.3\,\text{in}) \\
h = 25.4\,\text{mm}\,(1.0\,\text{in}) \\
\hline
\text{steel} \\
\hline
\end{array}$$

Figure 5.20. Annulus impacting a half-plane.

Part I. Determining the Force Histories

The objective of the data collection and analysis is to generate the force histories to be used in the disk–annulus impact problems and to determine the effect of total mass on the impact force history.

Begin by launching QED from the command line in the working directory.

COLLECTING THE DATA

1. Analyze the problem using [Gallery_ODE:9] to select impact models. Go to [Scales] and set the fixed scales as
   [U:u_max =1.0],
   [u:u_min =0.0],
   [V:Du_max =100],
   [v:Du_min =0.0],
   [T:t_max =1000e-6],
   [t:t_min =0.0];
   let all others be zero.

2. Go to [Geometry] to set the parameters for a steel disk impacting an aluminum half-plane. Set
   [v:vel =100],
   [0:type =5],
   [1:E =10e6],
   [2:rho =2.5e-4],
   [3:dim1(r|h) =400] (plane radius),
   [4:dim2(-|b) =1.0] (disk & plane thickness),
   [6:E =30E6],
   [7:rho =7.5e-4],
   [8:rad =4.0],
   [9:con_rad =4.0] (contact radius).

3. Go to [ICs] and set
   [8:Dt =2.0E-6],
   [9:#step =500],
   to have a time-integration window of $500\,\mu$s. [Press g] to observe the gallery of responses. Observe that the impact history is symmetric.

4. [Press w] to record the complete histories to the file <<qed.dyn>>. Launch DiSPtool and go to [View]. [Press p] for the parameters list and specify
   [n:file =qed.dyn],
   [1:plot_1 =1 2],
   [2:plot_2 =1 3],
   [3:plot_3 =1 4],
   [4:plot_4 =1 5].
   [Press q] and confirm that the histories are the same as in QED.
   [Press s] to access the [STORE columns] menu. Choose to store a two-column file of time [col 1] and force [col 6] by setting
   [n:filename =force.dk1],

```
[1:1st col =1],
[2:2nd col =6],
[3:3rd col =0],
[4:#pts =500],
[5:#thin =1].
```
[Press s] to store the columns to disk.

5. Switch to **QED** and return to the [Geometry] section. Change the density of the disk to
```
[7:rho =5.0e-4].
```
This will simulate a disk with a hole whose volume is 1/3 of the complete disk. [Press g] to see the results and then [press w] to record the complete histories.

Switch to **DiSPtool**, [press R] for complete rereading of the file. [Press s] to access the [STORE columns] menu and change the new force file name to
```
[n:filename =force.dk2].
```
[Press s] to store the columns to disk.

ANALYZING THE DATA

1. On a log–log graph, plot the maximum force against mass and the contact time against mass. Estimate the slope of the graphs.
2. Based on the simple model for the contact of disks, the maximum force and contact time are given by, respectively,

$$P_{\max} = K^{2/5} \left[\frac{5}{4} M_s V_o^2 \right]^{3/5}, \qquad T = 3.2 \sqrt{\frac{M_s}{K}} \left[\frac{K}{M_s V_o^2} \right]^{1/10}.$$

How well do the slopes agree with the simple model results of 3/5 and 2/5, respectively?

Part II. Determining the Stresses in a Solid-Disk Impactor

The first force history of Part I is now applied to a solid disk modeled as a circular plate. The objective of the data collection and analysis is to ascertain the validity of the simple model. Begin by launching **QED** from the command line in the working directory.

CREATING THE MODEL

1. Create a steel disk using the [Plate:ellipse] geometry. That is,
```
[0:geometry =3],
[1:X_rad =4],
[2:Y_rad =4],
[a:mat# =2],
[b:thick =1].
```
2. Use BCs of [0:type=1(=ENWS)] to set all sides as free. That is,
```
[0:type =1],
[4:sequence =snew],
```

[5:east =111111],
[6:north =111111],
[7:west =111111],
[8:south =111111].

3. Use Loads of [1:type=1(=near xyz)] to apply a point force on the east side. That is,

[0:type =1],
[1:X_pos =4],
[2:Y_pos =0],
[4:X_load =-1],
[5:Y_load =0];

all other loads and positions can be set to zero.

4. Mesh using <12> modules in the radial direction and <48> in the hoop direction. Set

[0:Global_ID =22],
[1:R_mods =12],
[2:H_mods =48],
[4:minR% =10] (sets radius of wedge elements),
[5:exp =1] (sets bias of radial elements),
[a:damping =.01].

[Press s] to render the mesh. If necessary, [press o] to set the orientation as $\phi_x = 0$, $\phi_y = 0$, $\phi_z = 0$.

COLLECTING THE DATA

1. Analyze the problem for its linear Transient response. Set

[n:P(t)name =force.dk1],
[0:type =1],
[1:scale{P} =1.0],
[2:scale{G} =0.0],
[5:time inc =5.0e-6],
[6:#steps =200],
[7:snaps@ =1],
[9:node =300].

[Press s] to perform the time integration.

2. View the Traces of the velocities for Node 300 (position $x = 0$, $y = 0$, $z = 0$; other node positions can be obtained using the PlotMesh utility) with

[0:style type =1],
[1:vars =1],
[2:xMult =1],
[5:rate =1],
[6:node# =300],
[7:elem# =1],
[8:elem type =4].

[Press s] to show the panel of responses.

Observe that the horizontal (rebound) velocity is essentially twice that of the initial velocity in Part I; this is because the present analysis has the disk initially at rest.

View the accelerations with
[5:rate =2].

Observe that the initial portion of the trace mimics the applied load history.

3. View the Contours of velocities with
[0:type =1],
[1:vars =1],
[2:xMult =1],
[3:snapshot =11],
[5:rate =1].

The snapshot is at time $t = 50\,\text{s}$. Observe, by zooming, that the velocity is mostly concentrated at the load application point.

In succession, zoom the contours at
[3:snapshot =21, 32, 41, 51, 61, 71, 81, ...].

Observe how the velocity becomes essentially uniform in the disk. What special observations can be made about [3:snapshot =57]?

View the Contours of stress with
[0:type =1],
[1:vars =21],
[2:xMult =1],
[3:snapshot =11],
[5:rate =0].

Zoom on the sequence
[3:snapshot =11, 21, 32, 41, 51, 61, 71, 81, ...].

Observe that the stresses, unlike the velocities, do not appear to achieve a uniform state.

ANALYZING THE DATA

1. Conjecture why the stresses, unlike the velocities, do not appear to achieve a uniform state.
2. To what extent does the disk behave as a concentrated mass?

Part III. Determining the Stresses in an Annulus Impactor

The second force history of Part I is now applied to an annulus modeled as a circular plate with a hole. The objective of the data collection and analysis is similar to that of Part II. Begin by launching QED from the command line in the working directory.

CREATING THE MODEL

1. Create a steel annulus using the [Hole:annulus] geometry. That is,
[0:geometry =3],
[1:inner_rad =2.31],

[2:outer_rad =4.00],
[a:mat# =2],
[b:thick =1].

2. Use BCs of [0:type=1(=bottom)] to set all edges as free. That is,
[0:type =1],
[4:bottom =111111].

3. Use Loads of [1:type=1(=near xyz)] to apply a point force on the east side. That is,
[0:type =1],
[1:X_pos =4],
[2:Y_pos =0],
[4:X_load =-1],
[5:Y_load =0];
all other loads and positions can be set to zero.

4. Mesh using <8> modules in the radial direction and <48> in the hoop direction. Set
[0:Global_ID =22],
[1:hoop divs =48],
[2:radial divs =8],
[a:damping =.01].
[Press s] to render the mesh.

COLLECTING THE DATA

1. Analyze the problem for its linear Transient response. Set
[n:P(t)name =force.dk2],
[0:type =1],
[1:scale{P} =1.0],
[2:scale{G} =0.0],
[5:time inc =5.0e-6],
[6:#steps =200],
[7:snaps@ =1],
[9:node =300].
[Press s] to perform the time integration.

2. View the Traces of the velocities for Node 338 (position $x = 0$, $y = 2.31$, $z = 0$; other node positions can be obtained using the PlotMesh utility) with
[0:style =1],
[1:vars =1],
[2:xMult =1],
[5:rate =1],
[6:node# =338],
[7:elem# =1],
[8:elem type =4].
[Press s] to show the panel of responses. Observe that the horizontal velocity of the line of symmetry is essentially twice that of the initial velocity in Part I.

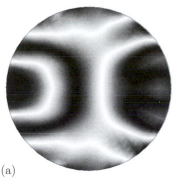
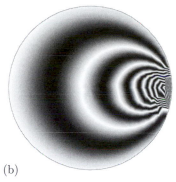

(a)   (b)

Figure 5.21. Partial results for the impact of a disk on a half-plane; time is when the force reaches its maximum: (a) contours of $\dot{u}$, (b) contours of von Mises stress σ_{vm}.

    View the velocity histories at other nodes such as
    [3:node# =208],
    (position $x = 2.31$, $y = 0.0$, $z = 0$). Observe that the velocity does not achieve a constant value but instead oscillates about a mean approximately that of [3:node# =338].

3. View the Contours of velocities with
    [0:type =1],
    [1:vars =1],
    [2:xMult =1],
    [3:snapshot =11],
    [5:rate =1].
    The snapshot is at time $t = 50$ s.
    Observe, by zooming, that the velocity is mostly concentrated at the load application point.
    In succession, zoom the contours at
    [3:snapshot =21, 32, 41, 51, 61, 71, 81, ...].
    Observe how the velocity never becomes uniform but the vertical line of symmetry does achieve a constant velocity.
    View the Contours of stress with
    [0:type =1],
    [1:vars =21],
    [2:xMult =1],
    [3:snapshot =11],
    [5:rate =0].
    Zoom on the sequence
    [3:snapshot =11, 21, 32, 41, 51, 61, 71, 81, ...].
    Observe that the stresses exhibit stress concentration effects around the edge of the hole.

ANALYZING THE DATA

1. Conjecture on the significance that the vertical line of symmetry achieves a constant velocity.

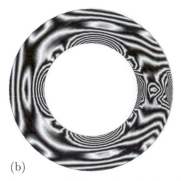

(a)   (b)

Figure 5.22. Partial results for the impact of an annulus on a half-plane; time is when the force reaches its maximum: (a) contours of $\dot{u}$, (b) contours of von Mises stress σ_{vm}.

2. To what extent does the annulus behave as a concentrated mass? What is the effect of the hole?
3. Conjecture about the use of a simple model for which the load history is assumed triangular with

$$P_{\max} = \frac{2MV_o}{T_{\max}}, \qquad T_{\max} = \alpha 4R/c_o,$$

where α is an adjustable parameter.

Partial Results

The partial results for Parts II and III are shown in Figures 5.21 and 5.22. The velocity and stress distributions in the two models are quite different, and yet their force histories are not very different.

Background Readings

Other contact laws are given in References [46, 56] as well as in Section 7.6. The impact of a beam is given in Reference [110], and an interesting variation dealing with fruit is given in Reference [109].

6 Stability of the Equilibrium

An important property or quality of the equilibrium of a structure is the *stability of the equilibrium*; that is, its sensitivity to small disturbances. If, after the small disturbance has ended, the structure returns to its original position, then the equilibrium state is said to be *stable*; on the other hand, if the small disturbance causes an excessive response, then the equilibrium state is *unstable*. An important consideration is to where the unstable structure goes – this is called the *postbuckling* behavior. The postbuckling behavior is typically highly nonlinear, undergoing large displacements and sometimes incurring plasticity effects. Figure 6.1(a) shows an example of a collapsed frame.

In all stability analyses, there is an important parameter associated with the *unfolding* of the instability. For example, the axial compressive load in the buckling of a column or the velocity in an aeroelastic flutter problem. Imperfections of load or geometry also play a significant role in unfolding the instability. Identifying this parameter and observing its effect is one of the keys to understanding the stability of a system.

The explorations in this chapter consider the stability of both the static and dynamic equilibrium. The first exploration uses imperfections (of loading and geometry) to illustrate the concept of sensitivity to the unfolding parameter. The second exploration introduces eigenanalysis as a tool to determine the buckling loads and mode shapes of a perfect structure; Figure 6.1(b) shows the first three buckled mode shapes of a ring with uniform pressure around the circumference. The third exploration elaborates on the connection between an eigenanalysis and a large-deflection analysis with imperfections. An eigenanalysis, however, does not give information about the postbuckle state; the fourth exploration considers that aspect and, in particular, shows the effect of plasticity. The fifth and sixth explorations look at the stability of motions.

The quasi-static linear loading of a system is controlled by

$$[[K_E] + [K_G(\bar{\sigma}_o)]]\{u\} = \{P\},$$

where $[K_E]$ is the assembled global elastic stiffness matrix and $[K_G]$ is the assembled global geometric stiffness matrix. The notation $[K_G(\bar{\sigma}_o)]$ emphasizes that the

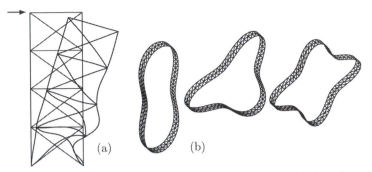

Figure 6.1. Stability of structures: (a) collapsed shape of a plane frame, (b) first three buckled mode shapes for a ring.

geometric stiffness is dependent on the member stresses. The buckling eigenvalue problem can be established as

$$[K_E + \lambda K_G]\{u\} = \{0\}.$$

The meaning of the eigenvalue λ is a scale to be applied to the actual load distribution to cause that mode of buckling. The solution $\{u\}$ is called the buckling mode shape. A buckling eigenanalysis is one of the basic stability analysis methods, but this predicts only the critical load and gives little information about the postbuckling behavior.

The presence of imperfections in combination with buckling loads, in general, causes the structure to undergo large deflections and sometimes large strains. Then the problem must be analyzed in a fully nonlinear (and sometimes dynamic) fashion based on the governing equation

$$\lceil M \rfloor \{\ddot{u}\} + [C]\{\dot{u}\} = \{P\} - \{F(u)\}.$$

This must be solved incrementally and iteratively over time, as explored in Chapter 5.

The buckling characteristics of structures can be represented as

$$\text{buckling load} \propto \sqrt{\frac{\text{elastic stiffness}}{\text{geometric stiffness}}} \qquad \text{or} \qquad \lambda = \alpha \sqrt{\frac{K_E}{K_G}}.$$

In constructing simple models for particular structures, it is often just a matter of identifying and approximating the appropriate stiffness terms.

6.1 Introduction to Elastic Stability

In many of the earlier explorations such as those of Sections 3.3 and 5.5, it was pointed out that stress can cause a change of stiffness. When the change is negative, then the stiffness can be driven to zero and the structure collapses. The explorations in this section investigate this aspect of the instability behavior of structures.

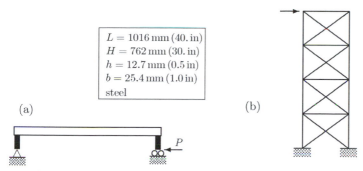

Figure 6.2. Stability of structures: (a) beam with eccentric loading, (b) braced plane frame with lateral loading.

A simple beam is used to illustrate the basic concepts, and then a braced frame is used to show the application to more complex structures; these are shown in Figure 6.2.

The loads will be applied over time although slow enough that no inertia effects are generated. In preparation, create a load-history file called <<force.61>> with the three-column [time amplitude amplitude] contents:

```
0.00    0.0    0.0
10.0    1.0    0.0
15.0    1.5    0.0
30.0    0.0    0.0
```

The second amplitude would correspond to a gravity load history. The load history will be automatically interpolated by StaDyn/NonStaD as needed.

Part I. Eccentric Loading of a Beam

As the loads on a structure change, so too does the equilibrium configuration. The sequence of such configurations is called the *equilibrium path*. The essence of an instability is that the current equilibrium path cannot be maintained and a new path is found. The new path may be near or remote; in the latter, a dynamic event occurs, and it may be stable or unstable; in the latter, collapse occurs.

The objective of the data collection and analysis of this exploration is to use an eccentrically loaded beam to elaborate on these ideas. Begin by launching QED from the command line in the working directory.

CREATING THE MODEL

1. Create a $[40 \times 0.5 \times 1.0]$ steel beam with an eccentric load using the [Frame:2D_frame] geometry. That is,

   [0:geometry =2],
   [1:X_length =40],
   [2:Y_height =1],
   [4:# X_bays =1],

```
[5:# Y_floors =1],
[7:brace =0].
```
Set the cross-sectional properties as
```
[a:mat#1 =2],
[b:thick1 =0.5],
[c:depth1 =1.0],
[d:mat#2 =2],
[e:thick2 =0.5],
[f:depth2 =1.0].
```

2. Use BCs of [0:type=1(=AmB)] to pin both ends. That is,
```
[0:type =1],
[5:corners_A =000001],
[6:others =111111],
[7:corners_B =100001].
```
Note that the right end is free to move horizontally.

3. Use Loads of [0:type=1(=near xyz)] to apply a horizontal force at the right boundary. That is,
```
[0:type =1],
[1:X_pos =40],
[2:Y_pos =0],
[4:X_load =-1];
```
all other loads are zero.

4. Mesh using <40> elements through the length and <1> vertical element. Set
```
[0:Global_ID =22],
[1:X_elems =40],
[2:Y_elems =1],
[a:damping =.001].
```
[Press s] to render the mesh.
If necessary, [press o] to set the orientation as $\phi_x = 0$, $\phi_y = 0$, $\phi_z = 0$.

COLLECTING THE DATA

1. Analyze the problem as Implicit nonlinear. Set
```
[n:P(t)name =force.61],
[0:type =1],
[1:scale{P} =2000],
[2:scale{Grav} =0.0],
[5:time inc =1.0],
[6:#steps =30].
[7:snaps@ =1],
[8:eigns@ =-1],
[9:node =22],
[a:algorithm =1]. (full Newton–Raphson)
```
[Press s] to perform the time integration.

2. View a movie of the displaced Shapes with
   [1:vars =111],
   [2:xMult =1],
   [3:snapshot =1],
   [5:rate =0],
   [6:size =1.5],
   [7:max# =44],
   [8:divs =10],
   [9:slow =50].
   [Press s] to process the snapshots, and [press s] again to show the movie. Observe that the out-of-plane deflection is considerable and that the original shape is recovered on unloading.
3. View the Traces of the displacement history. Set
   [0:style type =1],
   [1:vars =1],
   [2:xMult =1],
   [5:rate =0],
   [6:node# =22],
   [7:elem# =1],
   [8:elem type =3].
   [Press s] to show the traces. Individual panels can be zoomed by [pressing z #], where the panels are numbered 1 through 6, top row first. Observe that the transverse deflection [Var:V] varies nonlinearly with time, even though the load varies linearly with time.
4. Return to the Analysis section and change the time step to
   [5:time inc =0.5],
   [6:#steps =30],
   [9:node =22].
   This will record just the loading stage. [Press s] to perform the time integration.
   Store the <<stadyn.dyn>> file with a name such as <<dyn1.100>>.
5. Return to the Geometry section and change the eccentricity by setting
   [1:X_length =40],
   [2:Y_height =0.5].
   Remesh and rerun the analysis and store <<stadyn.dyn>> as <<dyn1.050>>.
6. Repeat this process for two more values of eccentricity,
   [1:X_length =40],
   [2:Y_height =0.1, 0.05]
   and store the results as <<dyn1.010>>, <<dyn1.005>>, respectively.

ANALYZING THE DATA

1. Plot the transverse deflection against load for the four cases of eccentricity. The format for the stored <<stadyn.dyn>> files is $[t, P_1, P_2, P_3, u, v, w, \ldots,]$.

Observe that the plots have three stages: Initially there is little deflection; this is followed by large deflection for a small increase in load, and, finally, there is a region of stiffening.

2. The three stages are accentuated as the eccentricity decreases.

Conjecture what the plot would look like in the limit of very small (but not zero) eccentricity.

Conjecture what the plot would look like if the eccentricity were zero.

Part II. Stiffening Effect of Axial Loads

Following on from the previous exploration, the eccentricity is made zero, but now the objective of the data collection and analysis is to determine the change of stiffness caused by the axial load. This is done by means of a static poke test and a vibration eigenanalysis.

Launch QED from the command line in the working directory.

CREATING THE MODEL

1. Create a $[40 \times 0.5 \times 1.0]$ steel beam using the [Frame:1D_beam] geometry. That is,

[0:geometry =1],
[1:X_length =40],
[4:# X_span =1].

Set the cross-sectional properties as

[a:mat# =2],
[b:thick1 =0.5],
[c:depth1 =1.0].

2. Use BCs of [0:type=1(=AmB)] to pin both ends. That is,

[0:type =1],
[5:end_A =000001],
[6:others =111111],
[7:end_B =100001].

Note that the right end is free to move horizontally.

3. Use Loads of [0:type=1(=near xyz)] to apply an axial force at the right boundary. That is,

[0:type =1],
[1:X_pos =40],
[4:X_load =1];

all other loads are zero. This is the prestress load P_o that will be scaled during the analysis.

4. Mesh using <40> elements through the length. Set

[0:Global_ID =22],
[1:X_elems =40],
[a:damping =.001].

[Press s] to render the mesh.

COLLECTING THE DATA

1. Analyze the beam for its linear Static response with prestress. Set
   `[0:type =2]`,
   `[2:scale[K_G] =0.0]`.
   `[Press p]` to set the transverse perturbation load δP as
   `[1:X_pos =20]`,
   `[5:Y_load =1]`;
   all other loads are zero. This is the *poke* load; measuring the response at the
   poked location will give a measure of the transverse stiffness.
   `[Press q s]` to perform the analysis.
2. Look in the file `<<stadyn.out>>` to see the complete displacement results.
   Record the transverse displacement at Node 21 (at the center of the beam).
3. In QED, set the prestress scale to
   `[2:scale[K_G] =1000]`.
   `[Press s]` to perform the analysis and record the transverse displacement at
   Node 21. Observe that the deflection is smaller than before.
   Repeat the analysis for the additional scales
   `[2:scale[K_G] =2000, -500, -1000, -1500, -2000]`.
   Observe that the deflection increases for the compressive loads. Actually, the
   last load case seems to give an apparently meaningless result; conjecture why.
   Repeat the last load case with a scale of
   `[2:scale[K_G] =-1900]`.
   These are the data for loading Case I.
4. Analyze the beam for its Vibration response using the subspace iteration algorithm. Set
   `[0:type =1]`
   `[2:scale[K_G] =0.0]`,
   `[8:#modes =16]`,
   `[a:algorithm =1]`.
   `[Press p]` to set the transverse perturbation load δP as
   `[1:X_pos =20]`,
   `[5:Y_load =0]`.
   `[Press q s]` to perform the vibration analysis.
   Before exiting the analysis section, look in the file `<<stadyn.out>>` to see the
   frequencies for the first 16 modes. Record the value for the lowest mode.
5. View the motion of displaced Shapes with
   `[1:vars =11]`,
   `[2:xMult =1]`,
   `[3:mode =1]`,
   `[5:rate =0]`,
   `[6:size =1.5]`,
   `[7:max# =4]`,
   `[8:divs =20]`,
   `[9:slow =50]`.

[Press s] to show the animation.

Observe that the motion is out-of-plane.

Change

[3:mode =#]

to animate the other modes.

6. Rerun the vibration analysis for different values of prestress scale of
[2:scale[K_G] =2000, -500, -1000, -1500, -2000].
Record the frequency of the lowest mode (make a note if the frequency changes from real to imaginary).
These are the data for loading Case II.

ANALYZING THE DATA

1. The deflection of a spring is related to the load by $P = Kv$, where K is the spring stiffness. Using the loading Case I data, make a plot of $K = \delta P/v$ against $P = P_o \times$ scale.
What is the trend of the data?
What can be said about the scale value of -2000?

2. In simple model terms, the vibration frequency of a structure can be represented as

$$\text{frequency} = \sqrt{\frac{\text{stiffness}}{\text{mass}}} \qquad \text{or} \qquad \text{stiffness} \propto \text{frequency}^2.$$

Using the loading Case II data, make a plot of frequency squared (plot the imaginary value on the negative axis) against $P = P_o \times$ scale.
What is the trend of the data?
What can be said about the scale value of -2000?

3. Conjecture about the connection between the results of this part and those of Part I.

Part III. Elastic Collapse of a Frame

We now explore the structure shown in Figure 6.2. The loading is such that some members are in tension whereas others are in compression. Furthermore, there is a nonuniform distribution of axial stress among the members. The objective of the data collection and analysis is to show how a static analysis, a vibration analysis under prestress, and a nonlinear analysis with load imperfections each give different insights into the collapse of this frame.

Begin by launching **QED** from the command line in the working directory.

CREATING THE MODEL

1. Create a four-story steel frame using the [Frame:2D_frame] geometry. That is,
[0:geometry =2],
[1:X_length =40],

```
[2:Y_height =30],
[4:# X_bays =1],
[5:# Y_floors =4],
[7:brace =2].
```
Set the cross-sectional properties as
```
[a:mat#1 =2],
[b:thick1 =0.5],
[c:depth1 =1.0],
[d:mat#2 =2],
[e:thick2 =0.5],
[f:depth2 =1.0].
```

2. Use BCs of [0:type=1(=AmB)] to fix the bottom corners. That is,
```
[0:type =1],
[5:corners_A =000000],
[6:others =111111],
[7:corners_B =000000].
```

3. Use Loads of [0:type=1(=near xyz)] to apply a horizontal force at the top left corner. That is,
```
[0:type =1],
[1:X_pos =0],
[2:Y_pos =120],
[4:X_load =1];
```
all other loads are zero.

4. Mesh using <40> elements for all members. Set
```
[0:Global_ID =22],
[1:X_elems =40],
[2:Y_elems =40],
[a:damping =.001].
```
[Press s] to render the mesh.
If necessary, [press o] to set the orientation as $\phi_x = 0$, $\phi_y = 0$, $\phi_z = 0$.

COLLECTING THE DATA

1. Analyze the frame for its Static response. Set
```
[0:type =1],
[1:scale{P} =3000],
[2:scale{Grav} =0.0].
```
[Press s] to perform the analysis.
View the Distributions of member stresses with
```
[1:vars =61],
[2:xMult =1].
```
Zoom on the axial stress [Var:Smxx] and observe that the west verticals and northeast braces are in tension whereas the east verticals and northwest braces are in compression.

2. Analyze the frame for its Vibration response using the subspace iteration algorithm. Set
   [0:type =1],
   [2:scale[K_G] =0.0],
   [8:# modes =16],
   [a:algorithm =1].
   [Press q s] to perform the vibration analysis.
   Before exiting the analysis section, look in the file <<stadyn.out>> to see the frequencies for the first 16 modes. Record the value for the lowest mode.
   View the motion of displaced Shapes with
   [1:vars =11],
   [2:xMult =1.5],
   [3:mode =1],
   [5:rate =0],
   [6:size =1.0],
   [7:max# =4],
   [8:divs =20],
   [9:slow =50].
   [Press s] to show the animation.
   Observe that all motion is transverse to the member, that is, the joints hardly move. Also observe that some members exhibit more motion than others; this is an indication of lower stiffness.
   Change
   [3:mode =#]
   to animate the other modes. A comparative gallery of modes can be observed with
   [1:vars =51],
   [2:xMult =1.5],
   [3:mode =1],
   [5:rate =0].

3. Rerun the Vibration analysis for different values of prestress scale of
   [2:scale[K_G] =1000, 2000, 3000, 4000, ...]
   until the frequency of one of the modes becomes imaginary. Adjust the maximum scale so that the frequency of the first mode just becomes imaginary.
   View the gallery of mode Shapes for each of the load scales and observe the changing pattern of the deflected shapes. This is an indication of a changing stiffness pattern. Record (by sketch or otherwise) the mode shape at the largest scale value; observe that its motion is primarily located around the second joint on the east side.

4. Analyze the problem as Implicit nonlinear. Set
   [n:P(t)name =force.61],
   [0:type =1],
   [1:scale{P} =3000],

```
[2:scale{Grav} =0.0],
[5:time inc =0.5],
[6:#steps =30].
[7:snaps@ =1],
[8:eigns@ =1],
[9:node =750],
[a:algorithm =1].
```
A vibration eigenanalysis will be (automatically) performed at each time step so as to monitor the stiffness of the frame. The monitored node is at the load point.
[Press s] to perform the time integration.

5. View a movie of the displaced Shapes with
```
[1:vars =111],
[2:xMult =1],
[3:snapshot =1],
[5:rate =0],
[6:size =1.0],
[7:max# =44],
[8:divs =10],
[9:slow =50].
```
[Press s] to process the snapshots, and [press s] again to show the movie. Observe that the collapsed shape resembles the vibration mode shape at maximum load scale.

ANALYZING THE DATA

1. Launch DiSPtool and [View] the vibration eigenvalues in <<stadyn.mon>>; the stored columns have the sequence $[t, P, \omega_1^2, \omega_2^2, \ldots,]$. Observe that the eigenvalues tend to zero.

   Estimate the load level at which the frequency gets close to zero; does this load correlate with any other results?

2. Zoom on the second panel and [press V] multiple times to vertically expand the plot. Observe that there is a small positive value of ω_1^2.

   Conjecture on the meaning of this small value.

   Conjecture what would happen if the value were negative.

3. Conjecture about the role(s) of the braces. In particular, conjecture what would happen if the bracing were changed to one and then none.

4. Conjecture what would happen if the load were removed.

Partial Results

The partial results for the deflections in Part I are shown in Figure 6.3(a), and the poke stiffness behavior of Part II is shown in Figure 6.3(b). The collapsed shape of Part III is shown in Figure 6.1.

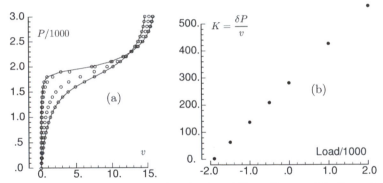

Figure 6.3. Partial results for the elastic stability of beams: (a) effect of eccentricity, (b) effect of axial loads.

Background Readings

General treatments of stability are given in References [96, 120, 121, 139]; more analytical treatments are given References [45, 77], and computational aspects are covered in References [37, 94, 95, 125]. A comprehensive review of experimental work is given in Reference [114]. Reference [104] discusses additional aspects of numerical solutions when the instability involves mode jumping.

6.2 Eigenanalysis of Buckling

An eigenanalysis of buckling can directly determine the critical loads and avoids the nonlinear analyses as used in the previous section. However, it has little information about the postbuckle state such as its stability. This section uses an eigenanalysis to determine the critical buckling loads for some thin-walled structures.

6.2.1 Buckling of Closed Sections

This QED assignment is a computer exploration of the static stability analysis of thin-walled structures using the eigenanalysis approach. A cantilevered box-beam structure is loaded transversely and the effect of the aspect ratio is investigated. The exploration is preceded by the stability analysis of a flat simply supported plate. These structures are shown in Figure 6.4.

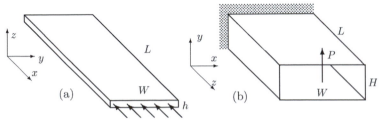

Figure 6.4. Buckling of structures: (a) flat plate, simply supported on all edges, with in-plane membrane loading, (b) cantilevered box beam with transverse load.

Part I. Simply Supported Plate with Uniform Load

We begin with a simply supported plate with uniform membrane load. The objective of the data collection and analysis is to use a sensitivity study to determine how the buckling loads depend on the aspect ratio (length-to-width ratio) of the plate.

Launch QED from the command line in the working directory.

CREATING THE MODEL

1. Create an aluminum plate of dimensions $[10 \times 5 \times 0.1]$ using the [Plate: quadrilateral] geometry. That is,

   [0:geometry =1],
   [1:X_2 =10.],
   [2:Y_2 =0.],
   [3:X_3 =10.],
   [4:Y_3 =5.],
   [5:X_4 =0.],
   [6:Y_4 =5.],
   [a:mat# =1],
   [7:thick =0.1].

2. Use BCs of [0:type=1(=ENWS)] to set simply supported BCs on all sides. That is,

   [0:type =1],
   [4:sequence =snew],
   [5:east =110010],
   [6:north =110100],
   [7:west =010010],
   [8:south =100100].

3. Use Loads of [0:type=2(=traction)] to apply normal tractions on the east side. That is,

   [0:type =2],
   [1:side# =2],
   [4:X_trac =-1];

   set all other loads to zero.

4. Mesh using <20> modules through the length and <20> transverse modules. That is,

   [0:Global_ID =32],
   [1:X_mods =20],
   [2:Y_mods =10];

   all [specials] can be set to zero.

   [Press s] to render the mesh. For clearer viewing, [press o] to set the orientation as $\phi_x = -55, \phi_y = 0, \phi_z = -20$.

COLLECTING THE DATA

1. Analyze the plate for its linear Static response. Set

   [0:type =1],

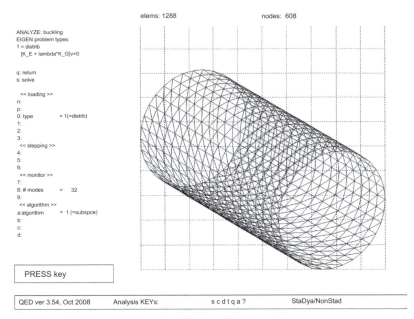

elems: 1288   nodes: 608

ANALYZE: buckling
EIGEN problem types:
1 = distrib
   [K_E + lambda*K_G]v=0

q: return
s: solve

  << loading >>
n:
p:
0: type   = 1(=distrb)
1:
2:
3:
  << stepping >>
4:
5:
6:
  << monitor >>
7:
8: # modes   =   32
9:
  << algorithm >>
a:algorithm   = 1 (=subspce)
b:
c:
d:

PRESS key

QED ver 3.54, Oct 2008   Analysis KEYs: s c d t q a ?   StaDya/NonStad

Figure 6.5. Opening screen for QED buckling analysis.

> [1:scale{P} =1.0],
> [2:scale{Grav} =0.0].
> [Press s] to perform the analysis.

2. View the Contours of displacement with

   [0:type =1],
   [1:vars =1],
   [2:xMult =1],
   [5:rate =0],

   and verify that the dominant action is axial compression.
   Look at the stress contours with

   [1:vars =21]

   and observe (by zooming the panels to see the contour legends) that the membrane stresses are $\sigma_{xx} = -1.0$, $\sigma_{yy} = 0.0$, $\sigma_{xy} = 0.0$, and are uniformly distributed.

3. Analyze the plate for its Buckling response using the subspace iteration algorithm; the opening screen should look like Figure 6.5. Set

   [1:type =1],
   [8:# modes =16],
   [a:algorithm =1].

   [Press s] to perform the eigenanalysis.

4. Before exiting the buckling analysis section, record the positive eigenvalues stored in the file <<stadyn.out>>; these are the aspect ratio = 2.0 load data. The meaning of the eigenvalue is a scale to be applied to the actual load distribution to cause that mode of buckling.

5. View the deformed Shapes. For comparative viewing, select the gallery of modes with
   [1:vars =51],
   [2:xMult =1.0],
   [3:mode =1],
   [5:rate =0].
   The gallery begins at [3:mode=?]. The deformed shapes can be exaggerated by changing [2:xMult=?] to something larger than 1.0.
   Record (by sketch or otherwise) the first 10 mode shapes; these are the aspect ratio = 2 shape data.
   For additional clarity, animate the mode shapes with
   [1:vars =11],
   [2:xMult =1],
   [3:mode =1],
   [5:rate =0],
   [6:size =1.5],
   [7:max# =4],
   [8:divs =32],
   [9:slow =32],
   and/or view the displacement contours.
6. Return to the Geometry section and change the dimensions of the plate to [5 × 5 × 0.1]. That is, set
   [1:X_2 =15.],
   [2:Y_2 =0.],
   [3:X_3 =15.],
   [4:Y_3 =5.],
   [5:X_4 =0.],
   [6:Y_4 =5.].
   Remesh and redo the buckling eigenanalysis.
   Record the positive eigenvalue and shape of the first 10 modes. These are the aspect ratio = 1 data.
7. Repeat the preceding analyses for the following sized plates:
   [2.5 × 5 × 0.1],
   [7.5 × 5 × 0.1],
   [12.5 × 5 × 0.1],
   [15 × 5 × 0.1].
   Each time, remesh and redo the buckling eigenanalysis, and record the positive eigenvalue and shape of the first 10 modes.
   These correspond to aspect ratios of 0.5, 1.5, 2.5, and 3.0, respectively.

ANALYZING THE DATA

1. Make a plot of all the eigenvalues against aspect ratio. Use an identifying symbol to distinguish the different mode-shape data.

2. Connect the data points for a particular mode shape across all aspect ratios. Observe that some modes cross.
3. How does the buckling load depend on the aspect ratio of the plate?
4. For aspect ratios greater than unity, the minimum buckling load is approximately independent of this ratio; determine how well the minimum buckling load is approximated by

$$\sigma_{xx} = -\frac{4D}{h}(\frac{\pi}{W})^2, \qquad D = \frac{Eh^3}{12(1-v^2)}.$$

5. Conjecture why the simply supported plate (in contrast to a beam–column structure) exhibits a limiting buckling load.

Part II. Simply Supported Plate with Nonuniform Load
Operating structures typically have nonuniform distributions of loads causing nonuniform stress distributions. Here we explore the effect of nonuniform stresses by applying an end shearing load to the plate of Part I. The objective of the data collection and analysis is similar to that of Part I.

If QED is not already running, begin by launching it from the command line in the working directory.

CREATING THE MODEL
1. Create an aluminum plate of dimensions $[10 \times 5 \times 0.1]$ using the [Plate: quadrilateral] geometry. That is,
   [0:geometry =1],
   [1:X_2 =10.],
   [2:Y_2 =0.],
   [3:X_3 =10.],
   [4:Y_3 =5.],
   [5:X_4 =0.],
   [6:Y_4 =5.],
   [a:mat# =1],
   [7:thick =0.1].
2. Use Loads of [0:type=2(=traction)] to apply a shear traction on the east side. That is,
   [0:type =2],
   [1:side# =2],
   [4:X_trac =0],
   [5:Y_trac =1];
   ensure that all other loads are set to zero.
3. Use BCs of [0:type=1(=ENWS)] to set simply supported boundary conditions on all sides but also allow the shear deformation. That is,
   [0:type =1],
   [4:sequence =snew],
   [5:east =110011],

```
[6:north =110101],
[7:west =000011],
[8:south =110101].
```

4. Mesh using <20> modules through the length and <10> transverse modules. That is,
```
[0:Global_ID =32],
[1:X_mods =20],
[2:Y_mods =10],
```
with all [specials] being zero.
[Press s] to render the mesh.

COLLECTING THE DATA

1. Analyze the plate for its linear Static response. Set
```
[0:type =1],
[1:scale{P} =1.0],
[2:scale{Grav} =0.0].
```
[Press s] to perform the analysis.

2. View the Contours of stress with
```
[0:type =1],
[1:vars =21],
[2:xMult =1],
[5:rate =0].
```
Observe (by zooming the panels to see the contour legends) that the membrane stresses are nonuniformly distributed; in particular, observe that the σ_{xx} ([Var:Smxx]) stress is maximum at the west side and varies from tension to compression south to north. Observe that the σ_{xy} ([Var:Smxy]) shear stress is nearly constant along the length and is parabolic across the width.

3. Analyze the plate for its Buckling response using the subspace iteration algorithm. Set
```
[0:type =1],
[8:# modes =16],
[a:algorithm =1].
```
[Press s] to perform the eigenanalysis.

4. View the deformed Shapes. For comparative viewing, select the gallery of modes with
```
[1:vars =51],
[2:xMult =2.0],
[3:mode =1],
[5:rate =0].
```
Record (by sketch, comment, or otherwise) the shape of the first six modes with positive eigenvalues; these are the aspect ratio = 2 data.

5. Repeat the preceding analyses for the following sized plates:
$[5 \times 5 \times 0.1]$,
$[15 \times 5 \times 0.1]$.

Each time, remesh and redo the buckling eigenanalysis, and record the positive eigenvalue and shape of the first six modes with positive eigenvalues; these correspond to aspect ratios of 1 and 3, respectively.

ANALYZING THE DATA

1. Make a plot of all the eigenvalues against aspect ratio. Use an identifying symbol to distinguish the different mode-shape data.
2. Connect the data points for a particular mode shape across all aspect ratios. Comment on the construction of this plot versus the corresponding plot in Part I.
   Are the data trends similar to those of Part I?
3. How does the buckling load depend on the aspect ratio of the plate?

Part III. Box-Beam Structures

The magnitude and position of a given load obviously affect the magnitude of the stresses and hence the buckling loads. But for cross sections with the same area and second moment of area, it is not expected that the stresses are affected by the particular distribution of material. Of interest here is whether or not the buckling loads are affected.

Two and three webbed box beams are loaded transversely, and the objective of the data collection and analysis is to determine the effect of the web spacing on the buckling behavior. Begin by launching **QED** from the command line in the working directory.

CREATING THE MODEL

1. Create a thin-walled aluminum box beam with the dimensions length $L = 10$, width $W = 8$, depth $H = 4$ using the [Closed:box beam] geometry. That is,
   [0:geometry =5],
   [1:length =10],
   [2:width =8],
   [3:depth =4].
   Set the flange and web thicknesses as $h = 0.1$ and the stringer areas as $A = 0.0001$. That is,
   [a:mat# =1],
   [b:flange =0.1],
   [c:webs =0.1],
   [d:area_st=0.0001] (stringers will not be created),
   [e:end plate =1].
2. Use BCs of [0:type=1(=L/R plane)] to set the BCs such that the left plane is fixed and the right plane is free. That is,
   [0:type =1],
   [8:L_bc =000000],
   [9:R_bc =111111].

3. Use Loads of [0:type=3(=end w/fr)] to apply a vertical load at the coordinates $x = 4$, $y = 2$, via a rigid frame attached to the end plate. That is,

   [0:type =3)],
   [1:X_pos =4.0],
   [2:Y_pos =2.0],
   [5:Y_load =1000].

   Set all other loads as zero.

4. Mesh using <20> modules through the length and <40> modules around the hoop direction. That is,

   [0:Global_ID =32],
   [1:length =20],
   [2:hoop =40],
   [3:V/H bias =1],
   [7:Z_rate =20];

   all [specials] can be set to zero.

   [Press s] to render the mesh.

   For clearer viewing, set the orientation ([press o]) as $\phi_x = +25$, $\phi_y = +30$, $\phi_z = 0$. If desired, toggle the top and end plates off by [pressing t 3 0] and [t 6 0], respectively. Other substructures can similarly be turned on or off.

COLLECTING THE DATA

1. Analyze the structure for its linear Static response. Set

   [0:type =1],
   [1:scale{P} =1.0],
   [2:scale{Grav} =0.0].

   [Press s] to perform the analysis.

2. View the Contours of stress with

   [0:type =1],
   [1:vars =21],
   [3:xMult =1],
   [5:rate =0].

   Observe that, except at corners, the membrane shear stress [Var:Smxy] is nearly uniformly distributed in each vertical panel (the webs). Observe also that the membrane axial stress [Var:Smxx] varies along the length and has a neutral axis midway up the vertical panel such that the top is in compression and the bottom in tension. For clearer viewing, surfaces can be toggled on/off by [pressing t # 1/0]; in particular, toggle the end plate off by [pressing t 6 0].

   At midlength of the beam, record the maximum [Var:Smxx] stress at the top and the [Var:Smxy] stress at center, both in the vertical panel. (If desired, use fixed scales, [0:type =2], to bracket the contours and thereby get a finer estimate. Alternatively, view the distributions with [1:vars =93].)

3. Analyze the structure for its Buckling response using the subspace iteration algorithm. Set

   [0:type =1],

[8:# modes =32],
[a:algorithm =1].
[Press s] to perform the eigenanalysis.
(If the analysis does not execute, check the <<stadyn.log>> file to see the symptom; in all probability, the memory allocation in <<stadyn.cfg>> will need to be increased.)
Before exiting the Analysis section, record the positive eigenvalues stored in the file <<stadyn.out>>. These are the load data for web spacing = 8.

4. View the deformed Shapes. For comparative viewing, select the gallery of modes with
[1:vars =51],
[2:xMult =100],
[3:mode =1],
[5:rate =0].
The gallery begins at [3:mode=?]. The deformed shapes can be exaggerated by changing [2:xMult=?] to something larger than 1.0.
Record (by sketch or otherwise) the shapes of the first six modes with positive eigenvalues. These are the shape data for web spacing = 8.
For additional clarity, animate the mode shapes (first exaggerate by 100), view the displacement contours, or do both.

5. Return to the Geometry section and change the model to
[0:geometry =6],
[c:webs =.0667],
and leave all other parameters unchanged. This gives the same cross-sectional area of material.
Remesh.

6. Repeat the Static analysis, and, at midlength of the beam, record the maximum [Var:Smxx] stress at the top and the [Var:Smxy] stress at center.

7. Repeat the Buckling analysis and record the positive eigenvalues stored in the file <<stadyn.out>> and (by sketch or otherwise) the shapes of the first six modes with positive eigenvalues.
These are the data for web spacing = 4.

ANALYZING THE DATA

1. Plot the two static stresses against distance between webs.
Does the spacing of the webs have much effect on the stresses?

2. Plot all the eigenvalues against distance between webs. If possible, connect points corresponding to the same mode shape.
Does the spacing of the webs have much effect on the buckling behavior?

3. Conjecture about the connection between the results of Part II and Part III.

Partial Results

The partial results for the buckling loads of Part I are shown in Figure 6.6(a). The data connected by dashed lines are for mode shapes with two waves in the y direction. The stresses in the box beams of Part III are shown in Figure 6.6(b).

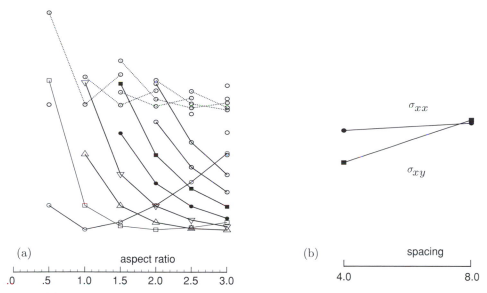

(a)

aspect ratio

(b)   spacing

Figure 6.6. Partial results for the eigenbuckling analysis: (a) effect of aspect ratio on the buckling behavior of a flat simply supported plate, (b) effect of web spacing on the stress behavior of the box beams.

6.2.2 Torsional Buckling of Open Sections

The emphasis of the explorations here is on the *torsional buckling* of structures; that is, buckling modes with significant twisting even though the applied load is axial. A cantilevered plate is loaded in compression but free to twist about its axis. Thin-walled structures of open cross section are quite susceptible to this type of buckling. In addition, variations of an I-beam (modeled as thin-walled folded-plate structures) are explored to determine the effect of the end constraints. These structures are shown in Figure 6.7.

Part I. Flat Plate
We begin with a cantilevered plate loaded in compression but free to twist about its axis. The objective of the data collection and analysis is to use a sensitivity study to

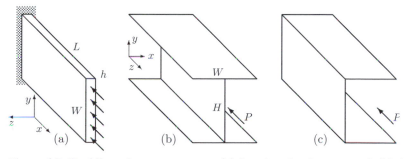

Figure 6.7. Buckling of open structures: (a) flat plate fixed at one end, (b) simply supported I-beam, (c) simply supported C-channel.

show how the buckling of the torsional mode depends on the structural parameters of length, width, and thickness.

Launch QED from the command line in the working directory.

CREATING THE MODEL

1. Create a $[20 \times 4 \times 0.4]$ aluminum plate using the [Plate:quadrilateral] geometry. That is,
   [0:geometry =1],
   [1:X_2 =20],
   [2:Y_2 =0],
   [3:X_3 =20],
   [4:Y_3 =4],
   [5:X_4 =0],
   [6:Y_4 =4],
   [7:thick =0.4],
   [9:mat# =1].

2. Use BCs of [0:type=1(=ENWS)] to set the west side as fixed, others free. That is,
   [0:type =1],
   [4:sequence =senw],
   [5:east =111111],
   [6:north =111111],
   [7:west =000000],
   [8:south =111111].

3. Use Loads of [0:type=2(=traction)] to apply a compressive traction on the east side. That is, set
   [0:type =2],
   [1:side# =2],
   [3:X_trac =-1];
   all other loads are zero.

4. Mesh using <40> modules through the length and <8> transverse modules. Set
   [0:Global_ID =32],
   [1:X_mods =40],
   [2:Y_mods =8],
   and all [specials] as zero.
   Render the mesh by [pressing s]. For clearer viewing, [press o] to set the orientation as $\phi_x = -55, \phi_y = 0, \phi_z = -20$.

COLLECTING THE DATA

1. Analyze the plate for its Static response. Set
   [0:type =1],
   [1:scale{P} =1.0],

[1:scale{Grav} =0.0].
Perform the analysis by [pressing s].

2. View the Contours of displacement with
[0:type =1],
[1:vars =1],
[2:xMult =1],
[5:rate =0],
and verify that the dominant displacement is axial compression.
Look at the stress contours with [1:vars=21] and observe (by zooming the panels to see the contour legends) that, except at corners, the membrane stresses are $\sigma_{xx} = -1.0$, $\sigma_{yy} = 0.0$, $\sigma_{xy} = 0.0$.

3. Analyze the plate for its Buckling response using the subspace iteration algorithm. Set
[8:# modes =24],
[a:algorithm =1].
[Press s] to perform the eigenanalysis.

4. View the deformed Shapes.
For comparative viewing, select the gallery of modes with
[1:vars =51],
[2:xMult =1],
[3:mode =1].
The gallery begins at [3:mode=#], and the deformed shapes can be exaggerated by changing [2:xMult=] to something larger than 1.0.
For additional clarity, animate the mode shapes by
[1:vars =11],
[2:xMult =1],
[3:mode =1],
[5:size =1],
[6:max# =4],
[7:rate =0],
[8:divs =16],
[9:slow =40].

5. Search for the first few torsional modes; these are characterized by zero displacement along the middle line.
Record the buckling eigenvalue located after [mode_#] in the gallery. The meaning of the eigenvalue is a scale to be applied to the actual load distribution to cause that mode of buckling.

6. Return to the Geometry section and change the dimensions of the plate to
[1:X_2 =40],
[3:X_3 =40],
[4:Y_3 =4],
[6:Y_4 =4],
[7:thick =0.4].

Remesh.

Redo the Buckling eigenanalysis and record the lowest torsional eigenvalue.

7. Change the dimensions of the plate to

[1:X_2 =20],
[3:X_3 =20],
[4:Y_3 =2],
[6:Y_4 =2],
[7:thick =0.4].

Remesh.

Redo the Buckling eigenanalysis and record the lowest torsional eigenvalue.

8. Change the dimensions of the plate to

[1:X_2 =20],
[3:X_3 =20],
[4:Y_3 =4],
[6:Y_4 =4],
[7:thick =0.2].

Remesh.

Redo the Buckling eigenanalysis and record the lowest torsional eigenvalue.

ANALYZING THE DATA

1. A simple model using $\phi(x) = \phi_L[x/L]$ as the Ritz function for twisting and $w(x, y) = \phi_L[xy/L]$ as the Ritz function for the prestress effect gives the energies

$$\mathcal{U}_E = \tfrac{1}{2} \int_L GJ\Big[\frac{\partial \phi}{\partial x}\Big]^2 dx = \tfrac{1}{2}\phi_L^2 G\tfrac{1}{3}Wh^3/L,$$

$$\mathcal{U}_G = \tfrac{1}{2} \int_A \sigma_{xx}h\Big[\frac{\partial w}{\partial x}\Big]^2 dx\,dy = \tfrac{1}{2}\phi_L^2 \sigma_{xx}\tfrac{1}{12}hW^3/L,$$

where $J = \tfrac{1}{3}Wh^3$ for a thin sheet in torsion. These in turn lead to the stiffness and eigenvalue estimates of

$$K_E = G\tfrac{1}{3}Wh^3/L, \qquad K_G = \sigma_{xx}\tfrac{1}{12}hW^3/L, \qquad \lambda = \frac{K_E}{K_G} = \frac{4G}{\sigma_{xx}}\frac{h^2}{W^2}.$$

This interesting result says that the torsional buckling is independent of the length.

2. Plot the buckling loads against length L and superpose the simple model as a continuous line.

   How well are the buckling data matched by the simple model?

3. Plot the buckling loads against width W and superpose the simple model as a continuous line.

   How well are the buckling data matched by the simple model?

4. Plot the buckling loads against thickness h and superpose the simple model as a continuous line.

   How well are the buckling data matched by the simple model?

5. Conjecture why the torsional buckling loads seem to occur in clusters.

Part II. I-Beam Folded-Plate Structure

Thin-walled structures supporting axial loads typically have end plates to permit the load transfer. In this exploration, an I-beam (modeled as a thin-walled folded-plate structure) is investigated to determine the effect of the end constraints on both the bending and torsional modes. The objective of the data collection and analysis is to use a sensitivity study to determine how the buckling loads depend on the length of the beam.

Begin by launching QED from the command line in the working directory.

CREATING THE MODEL

1. Create a thin-walled aluminum I-beam using the [Open:I_beam] geometry; give it the dimensions $L = 40$, width $W = 4$, depth $H = 4$. That is,
   [0:geometry =2],
   [1:length =40],
   [2:width =4],
   [3:depth =4].
   Set the flange and web thicknesses as $h = 0.4$, the stringer areas as negligible with $A = 0.0001$, and attach an end plate. That is,
   [a:mat# =1].
   [b:flange =0.4],
   [c:web =0.4],
   [d:area =0.0001],
   [e:end plate =1].
2. Use BCs of [0:type=4(=pivot)] so that the ends are on pivot supports; this will allow the imposition of the simple support conditions. That is, set
   [0:type =4],
   [6:X_pos =2.0],
   [7:Y_pos =2.0],
   [8:L_piv_bc =000110],
   [9:R_piv_bc =001110].
3. Use Loads of [0:type=3(=w/end_fr)] to apply an axial end load at the centroid $x = 2$, $y = 2$. That is,
   [0:type =3],
   [1:X_pos =2.0],
   [2:Y_pos =2.0],
   [6:Z_load =-1].
   All other loads are zero.
4. Mesh using <40> modules through the length and <8> modules for the flange and web. That is,
   [0:Global_ID =32],
   [1:length =40],
   [2:flange =8],
   [3:web =8],
   [6:Z_rate =40].

Set all [specials] as zero. Render the mesh by [pressing s].
For clearer viewing, set the orientation ([press o]) as $\phi_x = +25$, $\phi_y = +35$, $\phi_z = 10$. If desired, toggle the top flange off by [pressing t 3 0]; other substructures can similarly be turned on or off.

COLLECTING THE DATA

1. Analyze the structure for its Static response. Set
   [0:type =1],
   [1:scale{P} =1.0],
   [2:scale{Grav} =0.0].
   Perform the analysis by [pressing s].

2. View the Contours of stress by
   [0:type =1],
   [1:vars =21],
   [2:xMult =1],
   [5:rate =0].
   Observe that, except at the ends, the axial stress [Var:Smxx] is uniformly distributed in each panel and is of value $\sigma_{xx} = P/A = -0.208$.

3. Analyze the structure for its Buckling response using the subspace iteration algorithm. Set
   [8:# modes =24],
   [a:algorithm =1].
   [Press s] to perform the eigenanalysis.
   (If the analysis does not execute, check the <<stadyn.log>> file to see the symptom; in all probability, the memory allocation in <<stadyn.cfg>> will need to be increased.)

4. View the deformed Shapes.
   For comparative viewing, select the gallery of modes with
   [1:vars =51],
   [2:xMult =1],
   [3:mode =1].
   The gallery begins at [3:mode=#]. The deformed shapes can be exaggerated by changing [2:xMult=] to something other than 1.0.
   Record the eigenvalues for the first six modes, identifying them as first x bending, first y bending, first torsion, second x bending, and so on. Some of the modes are crimpling modes (in which just the flange deforms); ignore these and concentrate on the global modes.

5. Return to the Geometry section and change the length to
   [1:length =80].
   Remesh and redo the Buckling analysis.

6. Return to the Geometry section and change the length to
   [1:length =160].
   Remesh and redo the Buckling analysis.

ANALYZING THE DATA

1. Plot all the eigenvalues against length using a distinguishing symbol for each mode.
2. Superpose, as continuous curves, the simple model results for the bending buckling (Euler buckling) of beams with simple support conditions

$$P_c = EI_{yy}(\frac{n\pi}{L})^2, \qquad P_c = EI_{xx}(\frac{n\pi}{L})^2, \qquad I_{xx} \approx \frac{1}{2}WH^2h, \qquad I_{yy} \approx \frac{1}{6}W^3h,$$

where n is the mode number, e.g., first x bending, second x bending, and so on. How well do the simple models represent the data?

3. Superpose, as continuous curves, the simple model results for the torsional buckling modes

$$P_c = \frac{A}{I_S}[GJ + EC_w(\frac{n\pi}{L})^2], \qquad J = \frac{1}{3}[2W + H]h^3, \qquad I_S = I_{xx} + I_{yy},$$

where the torsional warping constant for an I-beam is $C_w = \frac{1}{24}H^2W^3h$.
How well does the simple model represent the data?
How significant of a contribution does the warping term $EC_w(\cdots)$ make to the buckling load estimate?

Part III. Coupled Buckling Modes in a C-channel

During buckling of a member with a general thin-walled open cross section, the cross section undergoes both displacement and rotation (about the shear center). This couples the bending and torsion actions, and the buckling mode is then a combination of both twisting and bending. The rectangle of Part I and I-beam of Part II both have the shear center coinciding with the centroid and therefore did not exhibit this form of buckling. The C-channel of Figure 6.7(c) has its shear center to the left of the vertical, far removed from the centroid (which is to the right), and therefore is very susceptible to combined bending–torsion buckling.

The objective of the data collection and analysis is to investigate combined bending–torsion buckling in a C-channel. Begin by launching QED from the command line in the working directory.

CREATING THE MODEL

1. Create a thin-walled aluminum C-channel section using the [Open:C_chan_rect] geometry; give it the dimensions $L = 40$, width $W = 4$, and depth $H = 4$. That is,
   [0:geometry =3],
   [1:length =40],
   [2:width =4],
   [3:depth =4],
   [a:mat#=1].
   Set the flange and web thicknesses as $h = 0.4$, the stringer areas as negligible with $A = 0.0001$, and attach an end plate. That is,
   [b:flange =0.4],

    `[c:web =0.4],`
    `[d:area =0.0001],`
    `[e:end plate =1].`

2. Use BCs of `[0:type=4(=pivot)]` so that the ends are on pivot supports; this will allow the imposition of different types of support conditions. Initially, the structure will be cantilevered. That is, set
    `[0:type =4],`
    `[6:X_pos =2.0],`
    `[7:Y_pos =2.0],`
    `[8:L_piv_bc =000000],`
    `[9:R_piv_bc =111111].`

3. Use Loads of `[0:type=3(=w/end_fr)]` to apply an axial end load at position $x = 2$, $y = 2$. That is,
    `[0:type =3],`
    `[1:X_pos =2.0],`
    `[2:Y_pos =2.0],`
    `[5:Y_load =0].`
    `[6:Z_load =-1].`
    All other loads are zero.

4. Mesh using $<40>$ modules through the length and $<8>$ modules in the web and flange. That is, set
    `[0:Global_ID =32],`
    `[1:length =40],`
    `[2:flange =8],`
    `[3:web =8],`
    `[7:Z_rate =40].`
    Set all `[specials]` as zero. Render the mesh by `[pressing s]`.
    For clearer viewing, set the orientation (`[press o]`) as $\phi_x = +25$, $\phi_y = +35$, $\phi_z = 10$.

COLLECTING THE DATA

1. Analyze the structure for its Static response. Set
    `[0:type =1],`
    `[1:scale{P} =1.0],`
    `[2:scale{Grav} =0.0].`
    Perform the analysis by `[pressing s]`.

2. View the Contours of stress with
    `[0:type =1],`
    `[1:vars =21],`
    `[2:xMult =1],`
    `[5:rate =0].`
    For clarity, toggle the bottom flange off by `[pressing t 1 0]`.
    Observe that the axial stress `[Var:Smxx]`, although uniformly distributed along the length, varies significantly in the hoop direction. This is because the load is not applied at the centroid.

View the Contours of displacement with [1:vars=1]. Observe that there is relative rotation about the global y axis (ϕ_y, [Var:Ry]) of the two end plates. Zoom on this panel and record the maximum rotation of the $x = L$ face.

3. This exercise determines the centroid and the shear center for the cross section. Return to the Loads section and change the load position to
[1:X_pos =1.0],
[2:Y_pos =2.0].
Remesh, reanalyze, and record the maximum (ϕ_y, [Var:Ry]) rotation.
Return to the Loads section and change the load to
[5:Y_load =1].
[6:Z_load =0].
Remesh, reanalyze, and record the maximum (ϕ_z, [Var:Rz]) rotation.
Return to the Loads section and change the load position to
[1:X_pos =-2.0],
[2:Y_pos =2.0].
Remesh, reanalyze, and record the maximum (ϕ_z, [Var:Rz]) rotation.
Plot the two ϕ_y rotations against position and interpolate to find the centroid at the position of zero rotation. Plot the two ϕ_z rotations against position and interpolate to find the shear center at the position of zero rotation. Record these values.
Return to the Loads section and set
[0:type =3],
[1:X_pos =] (position of centroid),
[2:Y_pos =2.0],
[5:Y_load =0].
[6:Z_load =-1].
Change the BC pivot conditions to
[6:X_pos =] (position of centroid),
[7:Y_pos =2.0],
[8:L_piv_bc =000110],
[9:R_piv_bc =001110].
These are simply supported BCs.
Remesh, reanalyze, and confirm that there are no significant end rotations and the axial stress is uniform around the hoop direction.

4. Analyze the structure for its Buckling response. Set
[8:# modes =24],
[a:algorithm =1].
[Press s] to perform the eigenanalysis.

5. View the deformed Shapes. For comparative viewing, select the gallery of modes with
[1:vars =51],
[2:xMult =1],
[3:mode =1].
The gallery begins at [3:mode=#].

Observe that most of the modes are local buckling of the flange and only the first mode is global.

Observe that the first mode has both a vertical deflection and a rotation. This is the bending–torsion buckling mode of interest. Record the buckling load.

View the displacements Contours of the first mode with

```
[0:type =1],
[1:vars =21],
[2:xMult =1],
[3:mode =1],
[5:rate =0].
```

By zooming, record the vertical deflection (v, [Var:V]) and rotation (ϕ_z, [Var:Rz]) of the side panel.

6. Return to the geometry section and change the length to
   ```
   [1:length =80].
   ```
   Remesh and redo the Buckling analysis. Identify the first bending–torsion buckling mode and record the data.

7. Return to the geometry section and change the length to
   ```
   [1:length =160].
   ```
   Remesh and redo the Buckling analysis. Identify the first bending–torsion buckling mode and record the data.

ANALYZING THE DATA

1. How close are the measured centroid and shear center locations to the expected values given by

$$x_c = \frac{W^2}{2W + H}, \qquad e = -\frac{3W^2}{6W + H},$$

respectively, measured from the left edge along the line of symmetry?

2. Plot the buckling loads against length.
   Plot, against length, the amplitude ratios of the shear center estimated as $\phi_z/(v + \phi_z e)$.

3. The uncoupled bending buckling loads (Euler buckling) are given by

$$P_{cx} = EI_{xx}(\frac{n\pi}{L})^2, \qquad P_{cy} = EI_{yy}(\frac{n\pi}{L})^2, \qquad I_{xx} \approx \tfrac{1}{2}WH^2h, \qquad I_{yy} \approx \tfrac{1}{3}W^3h,$$

where n is the mode number. The uncoupled torsion buckling loads are given by

$$P_{c\phi} = \frac{A}{I_S}[GJ + EC_w(\frac{n\pi}{L})^2], \qquad C_w = \frac{H^2W^3[3W + 2D]h}{12[6W + D]}.$$

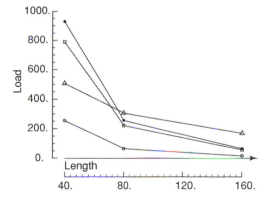

Figure 6.8. Partial results for the Buckling of an I-beam. Buckling loads against length.

The area moments are given by

$$A = [2W + H]h, \qquad J = \tfrac{1}{3}[2W + H]h^3, \qquad I_S = I_{xx} + I_{yy} + A[x_S^2 + y_S^2].$$

The moment of inertia I_S is about the shear center; thus $x_S = -|e| - |x_c|$, $y_S = 0$. Superpose, as continuous curves, these simple model results for P_{cx} and $P_{c\phi}$ on the preceding load plot.

Comment on the relationship between the data and the simple models.

4. References [87], [123] discuss this problem and show that the coupled modes are governed by the eigenvalue problem

$$\begin{bmatrix} P - P_{cx} & -x_s P \\ -x_s P A / I_s & P - P_{c\phi} \end{bmatrix} \begin{Bmatrix} v_o \\ \phi_o \end{Bmatrix} = 0.$$

The critical load and amplitude ratio are obtained by solving

$$[1 - x_s^2 A / I_s]P^2 - [P_{cx} + P_{c\phi}]P + P_{cx} P_{c\phi} = 0, \qquad \frac{\phi_o}{v_o} = \frac{P - P_{cx}}{x_s P}.$$

Superpose, as continuous curves, these simple model results on the preceding two plots.

How well does the simple model represent the data?

Partial Results

Partial results for the buckling loads for the I-beam of Part II are shown in Figure 6.8.

Background Readings

A comprehensive collection of eigenanalyses of various thin-walled structures is given in Reference [124]. The computational methods for accomplishing this are discussed in depth in References [5, 29, 101].

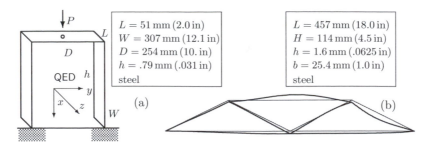

Figure 6.9. Models for postbuckling analyses: (a) folded-plate portal frame, (b) Warren truss (frame).

6.3 Stability and Load Imperfections

This section explores the postbuckling (large-deflection) nonlinear behavior of frame-type structures. A portal frame and a Warren truss (frame) [as shown in Figure 6.9(b)] are investigated for their sensitivity to load imperfections.

The load will be applied over time although slow enough that no inertia effects are generated. In preparation, create a load-history file; call it <<force.63>> with the three-column [time force gravity] contents:

```
0.      0.0   0
10.     1.0   0
20.     1.2   0
30.     1.4   0
```

This load history will automatically be interpolated by StaDyn/NonStaD as needed.

Part I. Nonlinear Deflection Analysis of the Portal Frame

The portal frame could be modeled using the [Frame:2D_frame] geometry; instead it will be modeled as a 3D folded-plate structure so as to better resemble the physical experiment [35]. The objective of the data collection and analysis is to recognize a connection between an eigenbuckling analysis and a large-deflection analysis with imperfections.

Begin by launching QED from the command line in the working directory.

CREATING THE MODEL

1. Create a steel portal frame (modeled as a folded-plate structure) using the [Open:C_chan_rect] geometry. That is, set
   [0:geometry =3],
   [1:length =2],
   [2:width =12.1],
   [3:depth =10],
   [a:mat# =2].

Set the flange, web, and stringer dimensions as
[b:flange =0.031],
[c:web =0.25],
[d:area =0.001],
[e:end plate =0].
The small stringer area means they will not be meshed.

2. Use BCs of [0:type=3(=X_max_fix)] to fix the complete right-hand side – all other edges and faces will be free. That is, set
[0:type =3)].

3. Use Loads of [0:type=1(=near xyz)] to place a point load at the center of the web. That is, set
[0:type =1],
[1:X_pos =0],
[2:Y_pos =5],
[3:Z_pos =1],
[4:X_load =1.0],
[5:Y_load =0.00];
all other loads are zero.

4. Mesh using
[0:global_ID =32],
[1:Length =4],
[2:Flange =16],
[3:Web =8],
[6:Z_rate =4],
with all [specials] set to zero.
[Press s] to render the mesh. For conventional viewing of the portal frame [as shown in Figure 6.9(a)], set the orientation ([press o]) as $\phi_x = +25$, $\phi_y = +30$, $\phi_z = -90$.

COLLECTING THE DATA

1. Analyze the structure for its Buckling response using the subspace iteration algorithm. Set
[8:#modes =12],
[a:algorithm =1].
[Press s] to perform the eigenanalysis.

2. View the deformed Shapes. For comparative viewing, select the gallery of modes with
[1:vars =51],
[2:xMult =1],
[3:mode =1],
[5:rate =0].
The gallery begins at [3:mode=?]. Observe the different buckling mode shapes and note that some of them are (almost) repeated values.

3. Return to the Loads section and add a small load eccentricity with
   `[4:X_load =1.0]`,
   `[5:Y_load =0.01]`.
   For future reference, refer to P_x as P_o and P_y as the load eccentricity.
   Remesh and redo the buckling analysis. Observe that the load eccentricity has caused only a small change in the eigenvalues (but mode shapes 2 and 3 are now confined to single legs).
   Record the lowest buckling load (the first number after mode_#); this will set the load level to be applied during the nonlinear deformation analysis.
4. Analyze the problem as Implicit Nonlinear. Set
   `[n:P(t)name =force.63]`,
   `[0:type =1]`,
   `[1:scale{P} =20]`,
   `[2:scale{G} =0]`,
   `[5:time inc =0.5]`,
   `[6:#steps =21]`,
   `[7:snaps@ =1]`,
   `[8:eigns@ =1]`,
   `[9:node =93]`,
   `[a:algorithm =1]`.
   `[Press s]` to perform the time integration.
   (Use `PlotMesh` to confirm that `[9:node=93]` is on the cross bar.)
5. Before exiting the analysis section, copy the file `<<stadyn.dyn>>` to `<<dynp.100>>`; this contains the response at the load point in the format $[t, P_1, P_2, P_3, u, v, w, i, \ldots,]$. Also copy the file `<<stadyn.mon>>` to `<<monp.100>>`; this contains the (vibration) eigenvalue information in the format $[t, P_1, \omega_1^2, \omega_2^2, \ldots,]$.
6. View a movie of the deformed Shapes with
   `[1:vars =111]`,
   `[2:xMult =1.0]`,
   `[3:snapshot =1]`,
   `[5:rate =0]`,
   `[6:size =1.5]`,
   `[7:max# =44 ]`,
   `[8:divs =8]`,
   `[9:slow =20]`.
   `[Press s]` to process the snapshot frames, `[press V]` to vertically position the model, and `[press s]` again to run the movie. Observe the rapid increase in deflections as the load (time) is increased.
7. View the Traces of displacement history with
   `[0:style type =1]`.
   `[1:vars =1]`,
   `[2:xMult =1]`,
   `[5:rate =0]`,
   `[6:node =93]`,

```
[7:elem# = ],
[8:elem type =4].
```
Observe the rapid increase in deflections as the load (time) is increased.
8. Repeat the analysis with different values of load eccentricity
```
[5:Y_load =0.100, .050,.010, .005, .001];
```
make sure to remesh each time and copy <<stadyn.dyn>> and <<stadyn.mon>> using appropriate data file names.

ANALYZING THE DATA
1. For each <<dynp.###>> data set, plot the v deflection against load. Comment on the trend of the plots.
2. For each <<dynp.###>> data set, make a plot of v/P_o versus v. This type of plot is known as a *Southwell's plot* [114] and has the interesting feature that the critical load can be obtained from the slope by $P_c = 1/$slope.
   Compute the different values of P_c.
3. How do the P_c values compare with those obtained from the buckling eigen-analysis?
4. Make a plot of the first two eigenvalues in <<monp.###>> against load. What can be said about the role of the imperfection?

Part II. Nonlinear Deflection Analysis of the Warren Truss

What makes the Warren truss interesting is that it exhibits what is called *asymmetric buckling*; that is, in contrast to the portal frame say, it is sensitive to the sign or direction of the imperfection; for one sign it is postbuckle stable, for the other it is postbuckle unstable. The objective of the data collection and analysis is the same as for Part I in that we wish to make a connection between an eigenbuckling analysis and a large-deflection analysis with imperfections.

CREATING THE MODEL
1. Create a steel Warren truss using the [Frame:2D_bridge] geometry. That is, set
```
[0:geometry =4],
[1:X_len =18],
[2:Y_hgt =4.5],
[4:# X_span =2].
[7:brace =0].
```
For the slants (props=1) and horizontals (props=2) set
```
[a:mat#1 =2],
[b:thick1 =0.0625],
[c:depth1 =1.0000],
[d:mat#2 =2],
[e:thick2 =0.0625],
[f:depth2 =1.0000].
```
2. Use BCs of [0:type=1(=AmB)] to pin the supports. That is, set
```
[0:type =1],
[5:end_A =000001],
```

[6:others =111111],
[7:end_B =100001].
Note that the second boundary is free to move horizontally.

3. Use Loads of [0:type=1(=near xyz)] to apply a vertical point load $P_y = -1.0$ at the top right corner. That is,
[0:type =1],
[1:X_pos =27.0],
[2:Y_pos =4.5],
[4:X_load =0.00]],
[5:Y_load =-1];
set all other loads and positions to zero. For future reference, refer to P_y as P_o.

4. Mesh using <10> elements for each member and constrain the behavior to be planar. That is,
[0:Global_ID =22],
[1:X_elems =10],
[2:D_elems =10];
set all [specials] to zero.
[Press s] to render the mesh. If necessary, orient the mesh to the plane:
[press o] and change the orientation to $\phi_x = 0$, $\phi_y = 0$, $\phi_z = 0$.

COLLECTING THE DATA

1. Analyze the problem for its linear Static response. Set
[0:type =1],
[1:scale{P} =1],
[2:scale{Grav} =0].
[Press s] to perform the analysis.

2. View the gallery of member stress Distributions with
[1:vars =61],
[2:xMult =0.5].
Zoom on the [Smxx] (σ_{xx}) panel to observe the distribution of axial stress in each member. In particular, observe which members have relatively large compressive stresses.

3. Analyze the problem for its Buckling response using the subspace iteration algorithm. Set
[0:type =1],
[8:# modes =10],
[a:algorithm =1].
[Press s] to perform the eigenanalysis.

4. View the gallery of deformed Shapes with
[1:vars =51],
[2:xMult =1.5],
[3:mode =1],
[5:rate =0].

Observe the different buckling mode shapes, focusing on those with positive buckling loads (first number after [mode_# ?] at the top of the panels). Observe that the lowest mode (second mode) has a significant rotation at the loaded joint.

Record the lowest positive buckling load and a note about its shape; the former will set the load level for the nonlinear deformation analysis, whereas the latter will set the direction of the load imperfection.

5. Return to the Loads section and apply a slightly imperfect vertical point load with

```
[0:type =1],
[1:X_pos =27.0],
[2:Y_pos =4.5],
[5:Y_load =-1],
[9:Z_mon =-0.1];
```

all other loads and positions being zero. The small moment is the load imperfection that will cause the joint to rotate clockwise.

Remesh.

6. Analyze the problem as nonlinear Implicit. Set

```
[n:P(t)name =force.63],
[0:type =1],
[1:scale{P} =50],
[2:scale{Grav} =0],
[5:time inc =0.5],
[6:#steps =21],
[7:snaps@ =1],
[8:eigns@ =1],
[9:node =54],
[a:algorithm =1].
```

[Press s] to perform the time integration.

7. Before exiting the analysis section, copy the files <<stadyn.dyn>> and <<stadyn.mon>> with names such as <<dynw.10m>> and <<monw.10m>>, respectively.

8. View a movie of the deformed Shapes with

```
[1:vars =111],
[2:xMult =5.0],
[3:snapshot =1],
[5:rate =0],
[6:size =1.5],
[7:max# =44],
[8:divs =8],
[9:slow =50].
```

[Press s] to process the snapshot frames, [press V] to vertically position the model, and [press s] again to run the movie. Observe the rapid increase in deflections as the load (time) is increased.

9. View the Traces of displacement history with
   `[0:style type =1],`
   `[1:vars =1],`
   `[2:xMult =1],`
   `[5:rate =0],`
   `[6:node# =54],`
   `[7:elem# = ],`
   `[8:elem type =3].`
   Observe the rapid increase in deflections as the load (time) is increased.
10. Return to the Loads section and change the imperfection to
    `[5:Y_load =-1],`
    `[9:Z_mon =-0.05].`
    Remesh and redo the analysis, and copy the files `<<stadyn.dyn>>` and `<<stadyn.mon>>` with names such as `<<dynw.##>>` and `<<monw.##>>`, respectively.
11. Repeat the analysis with smaller imperfections. That is, follow the sequence
    `[9:Z_mon =-0.1, -0.05, -0.02, -0.01,...].`
    Remesh after each change.
12. As a final case, return to the Loads section and change the imperfection to
    `[5:Y_load =-1],`
    `[9:Z_mon=+0.10].`
    This is the same as the original imperfection except that it is counterclockwise. Remesh and redo the analysis. Conjecture as to why the response in this case is so different from the other cases.

ANALYZING THE DATA

1. The format for the data in the `<<dynw.##>>` files is $[t, P_1, P_2, P_3, u, v, w, i, ...,]$.
   Use these data to make Southwell's plots of v/P versus v.
   Compute the slopes and then the different values of critical load as $P_c = 1/$ slope.
2. How do the P_c values compare with those obtained from the eigenanalysis?
3. Make a plot of the first two eigenvalues in `<<monw.##>>` against load.
   Conjecture about the role of the imperfection.
4. Conjecture as to why the response in the final case is so different from the other cases.

Partial Results

The partial results for the portal frame of Part I are shown in Figure 6.10. The South-well's plots have two straight lines with slopes corresponding to the buckling loads predicted by the eigenanalysis. The stable postbuckled shape for the Warren truss is shown in Figure 6.9.

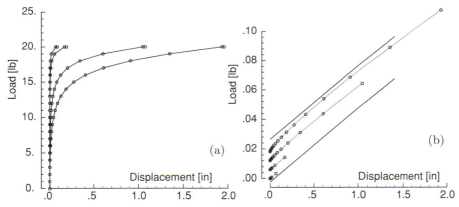

Figure 6.10. Partial results for the portal frame: (a) deflections for different amounts of load imperfections, (b) Southwell's plots (shifted for clarity).

Background Readings

The experimental version of the portal frame is given in Reference [35], which also contains a derivation of the Southwell's plot parameters. The experimental version of the Warren truss is given in Reference [107] and discussions of the experimental results are given in References [114, 120].

6.4 Elastic-Plastic Buckling

There are two aspects to the elastic-plastic buckling of columns. In the first one, the stress level is sufficiently high to cause yielding before buckling. In second one, after the elastic buckling occurs, the generated stresses are sufficient to then cause yielding in the postbuckle state. These are designated as short-column and long-column buckling, respectively, and both are explored here.

6.4.1 Buckling of Short Columns

This exploration studies the instability behavior of a short column when the compressive stresses are close to the yield stress. As explored in Sections 6.1 and 6.3, the instability is initiated through a load imperfection. The beam is shown in Figure 6.11.

In preparation, create a load-history file with the three-column [time force gravity] data

```
0       0      0
10      1.0    0
100     10     0
200     20     0
```

and call it <<force.64>>.

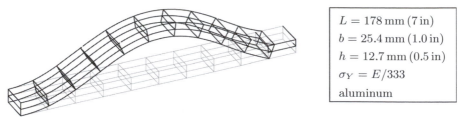

$$L = 178\,\text{mm}\ (7\,\text{in})$$
$$b = 25.4\,\text{mm}\ (1.0\,\text{in})$$
$$h = 12.7\,\text{mm}\ (0.5\,\text{in})$$
$$\sigma_Y = E/333$$
aluminum

Figure 6.11. Column under elastic-plastic buckling: Shape before and after deformation that resembles a beam in bending.

Part I. Preliminary Linear Behavior

We begin with a linear buckling analysis in which the objective of the data collection and analysis is to estimate a load level at which a buckling instability may occur. This will then be used to set the load levels for the nonlinear analyses of Parts II and III.

Launch QED from the command line in the working directory.

CREATING THE MODEL

1. Create a $[7 \times 0.5 \times 1.0]$ aluminum block using the [Solid:block] geometry. That is,
   [0:geometry =1],
   [1:X_length =7],
   [2:Y_length =0.5],
   [3:Z_length =1.0],
   [a:mat# =1],
   [b:constit s/e =1].

2. Use BCs of [0:type=1(=ENWS)] to set the west boundary as fixed and the east on rollers. That is,
   [0:type =1],
   [4:sequence =snew],
   [5:east =100000],
   [6:north =111111],
   [7:west =000000],
   [8:south =111111].
   This approximates beam BCs of fixed–pinned.

3. Use Loads of [0:type=2(=pressure)] to apply an end face pressure of -1.0. That is,
   [0:type =2],
   [1:face =1],
   [4:pressure =-1].
   Ensure that all other loads (including imperfections) are zero.

4. Mesh using <10> elements through the length, <2> through the height, and <1> through the thickness direction. That is, set
   [0:Global_ID =111],
   [1:X_dirn =10],

[2:Y_dirn =2],
[3:Z_dirn =1],
[d:red_intn =3].
The other [specials] can be set to zero.
[Press s] to render the mesh. For clearer viewing, [press o] to set the orientation as $\phi_x = 30$, $\phi_y = 25$, $\phi_z = 0$. (If the whole mesh does not display, ensure that all tags are set to 1; that is, [press t # 1].)

COLLECTING THE DATA
1. Analyze the bar for its linear Static response. Set
   [0:type =1],
   [1:scale{P} =1],
   [2:scale{Grav} =0].
   [Press s] to perform the analysis.
2. View the deformed Shapes with
   [1:vars =1],
   [2:xMult =1e6],
   [5:rate =0].
   [Press s] to see the deformed shape. Observe that the deformation is entirely axial.
3. Analyze the bar for its Buckling response using the subspace iteration algorithm.
   Set the number of modes and algorithm as
   [0:type =1],
   [8:# modes =12],
   [a:algorithm =1].
   [Press s] to perform the eigenanalysis.
   Before exiting the buckling analysis section, record the lowest positive eigenvalue stored in the file <<stadyn.out>>; the meaning of the eigenvalue is a scale to be applied to the actual load distribution to cause that mode of buckling.
4. View the gallery of deformed Shapes with
   [1:vars =51],
   [2:xMult =1],
   [3:mode =1],
   [5:rate =0].
   Additional modes can be observed by changing
   [3:mode=#].
   For additional clarity, animate the mode shapes with
   [1:vars =11],
   [2:xMult =1],
   [3:mode =1],
   [5:rate =0],
   [6:size =1.5],
   [7:max# =4],

```
[7:rate =0],
[8:divs =16],
[9:slow =50].
```

ANALYZING THE DATA

1. Use the recorded buckling eigenvalue to estimate the stress at which buckling will occur.
2. Will yielding occur for this stress level?

Part II. Elastic Buckling Behavior

Following on from Part I, we now explore the large-deflection nonlinear behavior of the bar. This large deflection will be initiated by applying a small eccentric load, and the objective of the data collection and analysis is to draw the connection between the nonlinear analysis and the linear buckling eigenanalysis.

If QED is not already running, begin by launching it from the command line in the working directory.

COLLECTING THE DATA

1. Analyze the problem using the nonlinear Implicit integration scheme. Set the parameters as
```
[n:P(t)name =force.64],
[0:type =1],
[1:scale{P} =90e3],
[2:scale{Grav} =0],
[5:time inc =0.5],
[6:#steps =20],
[7:snaps@ =1],
[8:eigns@ =1],
[9:node =102].
```
The snapshots and monitors are stored at every time increment.
[Press s] to perform the time integration.
Note that the complete history data for the monitored node is in the file <<stadyn.dyn>>, and the complete history data for the (vibration) eigenvalues are in the file <<stadyn.mon>>; these generally will be overwritten during the postprocessing; therefore these will need to be copied to save the results.

2. Launch DiSPtool and go to [View]. [Press p] for the parameters menu and specify
```
[n:file =stadyn.mon],
[1:plot_1 =2 3],
[2:plot_2 =2 4],
[3:plot_3 =2 5],
[4:plot_4 =2 6].
```
[Press q R] for complete rereading of the file. This plots the eigenvalues against load.
Observe that the eigenvalues decrease with the lowest going through zero. Estimate the load value where the eigenvalue goes through zero.

3. In QED, View the displacement Traces for Node 100 (position $x = 3.5$, $y = 0.5$, $z = 0$) with

    [0:style type =1],
    [1:vars =1],
    [2:xMult =1],
    [5:rate =0],
    [6:node/IP =100],
    [7:elem#].

    [Press s] to show the panel of responses; only the first three panels are relevant for solids.

    The numbers for the traces can be found in the file <<stadyn.dyn>>; this will need to be copied if the results are to be saved.

    Observe that the transverse displacements are negligible, being basically Poisson's ratio effect.

4. Return to the Loads section to change the loads to

    [4:pressure =-1],
    [8:Y_imp =0.002].

    This puts a small load imperfection in the transverse direction near the middle of the model.

    Remesh.

5. Repeat the nonlinear Implicit analysis.

    View the displacement Traces for Node 100 and observe that the transverse deflection is now significant.

    Switch to DiSPtool, [press R] to refresh the file, and observe that the vibration eigenvalue no longer goes through zero.

6. Repeat the analysis for different values of load imperfection:

    [8:Y_imp =0.2, 0.02, 0.002, 0.0002].

    Remesh after each change.

    Store the results from the files <<stadyn.dyn>> and <<stadyn.mon>> after each run.

ANALYZING THE DATA

1. The format for the file <<stadyn.dyn>> is

    $\{t; P_1, P_2, P_3; u, v, w; \dot{u}, \dot{v}, \dot{w}; i, \ldots,\}$.

    Plot the transverse deflection against load ([col 2 vs col 6]) for each eccentricity. Describe their major characteristics.

    Estimate the load limit point for zero imperfection, and call this the instability load.

2. The format for the file <<stadyn.mon>> is

    $\{t, P_1, \omega_1^2, \omega_2^2, \ldots,\}$.

    Plot the eigenvalue data against load, ([col 3 vs col 2]), for each eccentricity. Estimate the load value where the eigenvalue goes through zero and call this the neutral instability load.

3. How close are the instability load, the neutral instability load, and the buckling load of Part I?

Part III. Elastic-Plastic Behavior

Buckling is associated with loss of stiffness. After yielding occurs, a material loses stiffness depending on the amount of work hardening. In elastic-plastic buckling, both of these interact, which is explored here. The objective of the data collection and analysis is to determine the effect of the work-hardening slope on the stability behavior of elastic-plastic bars.

The only difference between the exploration here and that of Part II is the change of material properties allowing for the plasticity effects. If QED is not already running, begin by launching it from the command line in the working directory.

COLLECTING THE DATA

1. Return to the Geometry section to change the material properties.
   Choose the elastic-plastic constitutive relation by setting
   [a:mat# =1],
   [b:constit s/e =3],
   [c:Yield =30e3],
   [d:E_T =5e6].
   Go to the Loads section and reset the load eccentricity to zero with
   [8:Y_imp =0.0].
   Remesh.

2. Analyze the problem using the Implicit integration scheme with elastic-plastic behavior. Set
   [n:P(t)name =force.64],
   [0:type =5] (choose elastic-plastic analysis),
   [1:scale{P} =90e3],
   [2:scale{Grav} =0],
   [5:time inc =0.5],
   [6:#steps =20],
   [7:snaps@ =1],
   [8:eigns@ =1],
   [9:node =102].
   [a:algorithm =1].
   [Press s] to perform the time integration.
   Note that a list of the currently yielded IPs is stored in <<simplex.ep>>.

3. Launch DiSPtool and go to [View]. [Press p] for the parameters menu and specify
   [n:file =stadyn.mon],
   [1:plot_1 =2 3],
   [2:plot_2 =2 4],
   [3:plot_3 =2 5],
   [4:plot_4 =2 6].
   [Press q R] for complete rereading of the file. This plots the eigenvalues against load.

Observe that although the eigenvalues decrease with load there is a sudden jump in the vicinity of yielding. On further loading, the slope is the same as before the jump.

Save <<stadyn.mon>>.

4. Repeat the analyses for different values of work hardening.

   That is, return to the Geometry section to change the elastic-plastic constitutive parameters to

   [c:Yield =30e3],

   [d:E_T =8e6, 7e6, 6e6, 5e6, 4e6, 3e6, 2e6].

   Remesh after each change and save <<stadyn.mon>> after each analysis.

5. Return to the Geometry section to change the work hardening to

   [d:E_T =5e6].

   Go to the Loads section and set the load eccentricity to

   [8:Y_imp =0.02].

   Remesh and rerun the analysis with

   [1:scale{P} =70e3].

   This change of scale is made so as to better utilize the given number of load increments.

6. Switch to DiSPtool, go to [View] and [Press R] to refresh the <<stadyn.mon>> file.

   Observe that there is a jump near yielding but the bar never goes unstable.

7. Return to QED and View the deformed Shapes. Set the movie parameters as

   [1:vars =111],

   [2:xMult =1],

   [3:snapshot =1],

   [5:rate =0],

   [6:size =1.5],

   [7:max# =44],

   [8:divs =4],

   [9:slow =50].

   [Press s] to process the frames, and [press s] again to run the movie.

   Confirm that the deformation is in agreement with expectation.

ANALYZING THE DATA

1. Plot the eigenvalue data against load ([col 3 vs col 2]) for each value of work hardening.

2. Estimate the instability loads, and plot these values against work hardening. What is the threshold value of work hardening at which the bar is unstable immediately on yielding?

3. A simple model for the buckling of columns is to use the Euler buckling formulas but with the elastic-plastic tangent modulus

$$ P_c = \alpha \pi^2 \frac{E_T I}{L^2}, \qquad \sigma_c = \frac{P_c}{A} = \alpha \pi^2 \frac{E_T I}{AL^2}, \qquad A = bh, \quad I = \tfrac{1}{12}bh^3, $$

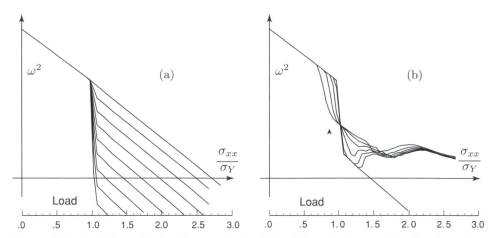

Figure 6.12. Partial results for the elastic-plastic buckling of a beam: (a) vibration eigenvalues, (b) the effect of load imperfections on the vibration eigenvalues.

where α depends on the particular boundary conditions – for the present case $\alpha = 2.05$ [29].

How well do the data support this simple model?

4. In what way does the load imperfection affect the instability?

How well does the simple model work in the presence of imperfections?

Partial Results

The partial results for the stiffness behavior in Part III are shown in Figure 6.12. The horizontal line in both plots corresponds to $\omega^2 = 0$, which corresponds to complete loss of stiffness. The deformed shape after yielding and buckling is shown in Figure 6.11.

6.4.2 Buckling of Long Columns

In this section, we explore the postbuckling stability behavior of a long column that yields at the postbuckle stage. The sequence of analyses is linear elastic, nonlinear elastic, and elastic-plastic. The beam is shown in Figure 6.13.

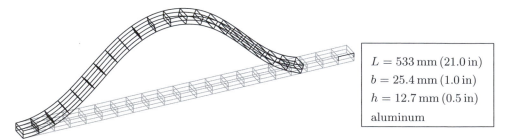

$$L = 533\,\text{mm} \;(21.0\,\text{in})$$
$$b = 25.4\,\text{mm} \;(1.0\,\text{in})$$
$$h = 12.7\,\text{mm} \;(0.5\,\text{in})$$

aluminum

Figure 6.13. Long column under axial compression causing plastic yielding. Shape before and after deformation that resembles a beam in bending.

In preparation for the analysis, create a load-history file with the three [time force gravity] columns

```
0     0       0
8     1.0     0
10    1.04    0
12    1.08    0
20    1.2     0
101   1.2     0
```

and call it <<force.64b>>.

Part I. Preliminary Linear Behavior

We begin with a linear prestress analysis in which the objective of the data collection and analysis is to assess the effect of compressive prestress on the stiffness behavior of structures. The stiffness information is obtained by performing poke tests.

Launch QED from the command line in the working directory.

CREATING THE MODEL

1. Create a solid aluminum block of dimensions $[21 \times 0.5 \times 1]$ using the [Solid:block] geometry. That is,
   [0:geometry =1],
   [1:X_length =21],
   [2:Y_length =0.5],
   [3:Z_length =1],
   [a:mat# =1],
   [b:constit s/e =1].
   Position the offsets at zero.

2. Use BCs of [0:type=1(=ENWS)] to set the west boundary as fixed and the east on rollers. That is,
   [4:sequence =snew],
   [5:east =100000],
   [6:north =111111],
   [7:west =000000],
   [8:south =111111].
   This approximates beam BCs of fixed–pinned.

3. Use Loads of type [1(=near xyz)] to apply a load of $P_x = -1$ at position $x = 20, \ y = 0.25$. That is,
   [0:type =1],
   [1:X_pos =20],
   [2:Y_pos =0.25],
   [3:Z_pos =0.5],
   [4:X_load =−1],
   [5:Y_load =0].

Ensure that all other loads are zero. This is the preexisting load that contributes to $[K_G]$ and will be scaled during the linear Static analysis.

4. Mesh using $<21>$ elements through the length, $<2>$ elements through the height, and $<1>$ element through the thickness direction. That is,

[0:Global_ID =111],
[1:X_dirn =21],
[2:Y_dirn =2],
[3:Z_dirn =1],
[d:red_int =3];

set other [specials] to zero.

[Press s] to render the mesh. For clearer viewing, [press o] to set the orientation as $\phi_x = 30$, $\phi_y = 25$, $\phi_z = 0$. (If the whole mesh does not display, ensure that all tags are set to 1; that is, [press t # 1].)

COLLECTING THE DATA

1. Analyze the bar for its linear Static response. Set

[0:type =2],
[2:scale[K_G] =1].

[Press p] to access the perturbation load menu and set

[1:X_pos =10],
[2:Y_pos =0.25],
[3:Z_pos =0.5],
[4:X_load =0],
[5:Y_load =100].

These are the perturbational loads to monitor the stiffness behavior of the bar. [Press q s] to perform the analysis.

Before exiting the static analysis section, record the largest transverse displacement stored at the bottom of the file <<stadyn.out>>.

2. Repeat the analysis with the prestress scales

[2:scale[K_G] =1, 1000, 2000, 3000, 4000, 5000, 6000].

Each time record the largest transverse displacement.

3. View the deformed Shapes with

[0:type =1],
[2:xMult =1.0],
[5:rate =0].

The deformed shapes can be exaggerated by changing [2:xMult =?] to something larger than 1.0.

Why did the bar (in the last case) deflect down when the load is acting up?

ANALYZING THE DATA

1. Plot the transverse deflection against load.
2. What is the effect of the compressive load?

Part II. Large-Deflection Elastic Behavior

Following on from Part I, we now explore the large-deflection nonlinear behavior of the bar. This large deflection is initiated by applying a small eccentric load, and the objective of the data collection and analysis is to determine the sensitivity of the transverse deflection to the eccentric load.

If QED is not already running, begin by launching it from the command line in the working directory.

COLLECTING THE DATA

1. Return to the Loads section to add a small load eccentricity in the transverse direction. That is, set
   ```
   [0:type =1],
   [1:X_pos =20],
   [2:Y_pos =0.25],
   [3:Z_pos =0.5],
   [4:X_load =-1],
   [5:Y_load =0.1].
   ```
   Remesh.
2. Analyze the problem using the Implicit integration scheme. Set
   ```
   [n:P(t)name =force.64b],
   [0:type =1],
   [1:scale{P} =5000],
   [2:scale{Grav} =0.00],
   [5:time inc =0.5],
   [6:#steps =30],
   [7:snaps@ =1],
   [8:eigns@ =-1],
   [9:node =197].
   [a:algorithm =1].
   ```
   [Press s] to perform the time integration.
   Note that the complete history data for the monitored node is in the file <<sta-dyn.dyn>>; this will be overwritten during the postprocessing phase. Therefore, when the analysis is complete, copy this file and give it a name such as <<dyn.1>>. The format for the columns is t, P_1, P_2, P_3, u, v, w, i, ...,.
3. View the deformed Shapes. Set the movie parameters as
   ```
   [1:vars =111],
   [2:xMult =1],
   [3:snapshot =1],
   [5:rate =0],
   [6:size =1.5],
   [7:max# =44],
   [8:divs =8],
   [9:slow =50].
   ```

[Press s] to process the snapshots and [press s] again to run the movie. Confirm that the deformation is in agreement with expectation.

4. View the history Traces. View the displacement traces for Node 197 (the node at position $x = 10$, $y = 0.25$, $z = 0.5$). That is, set

   [0:style type =1],
   [1:vars =1],
   [2:xMult =1],
   [5:rate =0],
   [6:node/IP =197],
   [7:elem# =1].

   [Press s] to show the panel of responses; only the first three panels are relevant for solids.

   Observe that the dominant displacement is the transverse [Var:V] displacement.

   View the nodal average strain history at Nodes 198, 197, 196. That is, change

   [1:vars =14],
   [6:node/IP =198], [=197], [=196],

   and [press s] to show the panel of responses. (A complete list of viewable variables can be seen by [pressing h] for the help menu.)

   Observe that the significant strain is E_{xx} ([Var:Exx]) and it exhibits the characteristics of a beam in bending (stretching on top, contraction on the bottom).

   View the nodal average Kirchhoff stress history at Nodes 198, 197, 196. That is, change

   [1:vars =24],

   and [press s] to show the panel of responses,

   Observe that the axial stress ([Var:Sxx]) is significant in comparison with the yield stress.

   View the nodal average Cauchy stress history at Nodes 198, 197, 196. That is, change

   [1:vars =34],

   and [press s] to show the panel of responses,

   Compare the relative magnitudes of the Kirchhoff and Cauchy stresses.

5. Repeat the analyses for different values of eccentric loading:

   [4:X_load =-1],
   [5:Y_load =0.1, 0.01, 0.001, 0.0001].

   Remesh after each change.

   Store the results from the file <<stadyn.dyn>> immediately after each analysis.

ANALYZING THE DATA

1. Plot the transverse deflection against load ([col 5 vs col 2] in the stored <<stadyn.dyn>> files) for each eccentricity.

   Describe their major characteristics.

2. Imagine the situation of almost-zero eccentricity; estimate the load value at which observable transverse deflections occur.
3. How well does this load compare with the simple model

$$P_c = 2\alpha\pi^2 E(\tfrac{1}{12}bh^3)/L^2,$$

with $\alpha = 2.05$, $L = 20$?
Why is $L = 20$ used in the simple model?

Part III. Large-Deflection Elastic-Plastic Behavior

As pointed out earlier, buckling is associated with loss of stiffness, and after yielding occurs, a material loses stiffness depending on the amount of work hardening. Thus it is possible that, although the elastic buckling is postbuckle stable, the additional decrease of stiffness that is due to yielding may be sufficient to make the structure unstable. This is the situation explored here. The objective of the data collection and analysis is to determine the effect of the work-hardening slope on the postbuckle stability of the bar.

The only difference between the exploration here and that of Part II is the change of material properties allowing for the plasticity effects. If QED is not already running, begin by launching it from the command line in the working directory.

COLLECTING THE DATA

1. Return to the Geometry section to change the material properties. Choose the elastic-plastic constitutive relation by setting
   [a:mat# =1],
   [b:constit s/e =3],
   [c:Yield =30e3],
   [d:E_T =8e6].
2. Change the Loads to
   [0:type =1],
   [1:X_pos =20],
   [2:Y_pos =0.25],
   [3:Z_pos =0.5],
   [4:X_load =-1],
   [5:Y_load =0.1].
   Remesh.
3. Analyze the problem using the nonlinear Implicit integration scheme for elastic-plastic behavior. That is, set
   [n:P(t)name =force.64b],
   [0:type =5],
   [1:scale{P} =5000],
   [2:scale{Grav} =0.00],
   [5:time inc =0.5],
   [6:#steps =30],

[7:snaps@ =1],
[8:eigns@ =-1],
[9:node =197],
[a:algorithm =1].
[Press s] to perform the time integration.

Note that a history of the currently yielded Elements/IPs is stored in <<simplex.ep>>. Store the results from the file <<stadyn.dyn>> before leaving the analysis section.

4. View the deformed Shapes. Set the movie parameters as
[1:vars =111],
[2:xMult =1],
[3:snapshot =1],
[5:rate =0],
[6:size =1.5],
[7:max# =44],
[8:divs =4],
[9:slow =50].
[Press s] to process the snapshots, and [press s] again to run the movie.
Confirm that the deformation is in agreement with expectation.

5. View the history Traces.
View the displacement traces for Node 197. That is, set
[0:style type =1],
[1:vars =1],
[2:xMult =1],
[5:rate =0],
[6:node/IP =197],
[7:elem# =1].
[Press s] to show the panel of responses; only the first three panels are relevant for solids.

The numbers for the traces can be found in the file <<stadyn.dyn>>; this will need to be copied if the results are to be saved.

View the nodal average Lagrangian strain, Kirchhoff stress, Cauchy stress, and plastic strain histories at Nodes 198, 197, 196. That is, change
[1:vars =14, 24, 34, 54],
[6:node/IP =198, 197, 196],
respectively.

Observe that the significant stress is σ_{xx} ([Var:Sxx]) and it exhibits the characteristics of a beam in bending.

6. Repeat the analyses for different values of work hardening. That is, return to the Geometry section to change the elastic-plastic constitutive parameters to
[c:Yield =30e3],
[d:E_T =8e6, 7e6, 6e6, 5e6].
Remesh after each change. Store the results from the file <<stadyn.dyn>> immediately after each analysis.

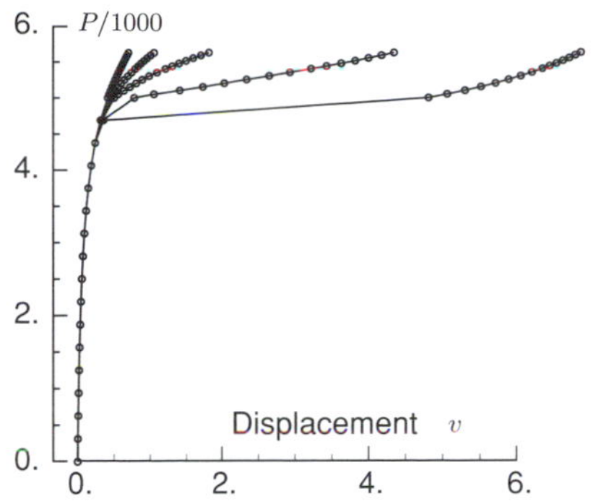

Figure 6.14. Partial results for the elastic-plastic compression of a long-column. load–deflection behavior.

ANALYZING THE DATA

1. Plot the transverse deflection against load ([col 5 vs col 2]) of the stored <<stadyn.dyn>> file) for each work-hardening value.
   Describe their major characteristics.
2. Conjecture about the behavior as the work hardening is made even smaller.

Partial Results

The partial results for the large deflection behavior of Part III are shown in Figure 6.14. That the final deflections for each case are different means that each have different stiffnesses. The deformed shape after buckling and yielding is shown in Figure 6.13.

Background Readings

Reference [57] gives a discussion of the theories proposed for the analysis of elastic-plastic buckling. A conceptual simple model is developed in Section 7.5 that helps explain why the postbuckling behavior is usually unstable for aluminum and steel.

6.5 Stability of Motion in the Large

An unstable structure invariably goes into motion and comes to rest only when it has found a new equilibrium position. When the deflections are large, then the event is referred to as *motion in the large*. This collection of explorations looks at different aspects of motion in the large. The first group considers the full modeling of an arch. This is computationally expensive, so the emphasis is on identifying significant features that should be incorporated in a simple model. The second group then does a more detailed exploration with the simple model.

6.5.1 Stability of an Arched Structure

Figure 6.15 shows an arch that is loaded transversely. The heavy lines are its shape during what is called a *snap-through* buckling; that is, the transverse load causes

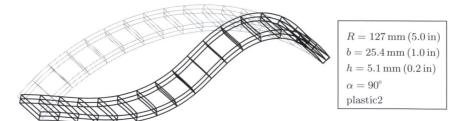

$$R = 127\,\text{mm}\ (5.0\,\text{in})$$
$$b = 25.4\,\text{mm}\ (1.0\,\text{in})$$
$$h = 5.1\,\text{mm}\ (0.2\,\text{in})$$
$$\alpha = 90°$$
plastic2

Figure 6.15. Snap-through instability of an arch modeled with Hex20 elements. Shape during snap-through.

the arch to snap to the other side, forming an inverted shape. This new equilibrium position–shape is remote from the initial configuration and therefore will serve as a good example of a motion in the large.

As a preliminary while QED is not running, edit the file <<qed.mat>> to add a fourth material with properties similar to a heavy plastic. That is, add the lines

```
4   ::plastic2
400000   150000   2.00e-4
```

Also change the first line in the file to reflect that there are now four materials.

In addition, run the following to create a smoothed ramp load history.

CREATING A SMOOTHED LOAD HISTORY
1. Launch DiSPtool from the command line.
2. Go to [Time domain, One channel], by [pressing to].
   [Press p] and ensure the window is off, [w:window = 0].
   [Press g] to get the [Signal Generation] parameters list and set
   [0:signal =5],
   [1:c1 =0.0],
   [2:c2 =4.0e-2],
   [3:c3 =0.0],
   [4:c4 =0.0],
   [5:dt =5.0e-5],
   [6:#fft pts =1024].
3. [Press q], then [press s] to save the file. Set
   [n:filename =force.65],
   [1:1st col =1],
   [2:2nd col =2],
   [3:3rd col =2],
   [4:#pts =1024],
   [5:#thin =1].
   [Press s] to actually store the file.

Part I. Preliminary Linear Vibration Behavior

One of the consistent themes running through the analyses of instability in this chapter is the connection between vibrational behavior and buckling. We begin the analysis of the arch with a vibrational analysis with the objective of the data collection and analysis to determine the effect of the loading on the vibration characteristics.

Launch QED from the command line in the working directory.

CREATING THE MODEL

1. Create a segment of a cylindrical plastic shell using the [Solid:arch] geometry. That is,
   ```
   [0:geometry =5],
   [1:length =1],
   [2:rad_out =5.0],
   [3:rad_in =4.8],
   [4:angle =90],
   [a:mat# =4],
   [b:constit s/e =1].
   ```
2. Use BCs of [0:type=4(=simple)] to set both edges as simply supported. That is,
   ```
   [0:type =4].
   ```
3. Use Loads of [0:type=1(=near xyz)] to apply a line load of $P_y = 1$ at position $x = 0.0$, $y = 4.9$ (a cylinder of three nodes will be captured). That is,
   ```
   [0:type =1],
   [1:X_pos =0],
   [2:Y_pos =4.9],
   [3:Z_pos =0.0],
   [4:X_load =0.02],
   [5:Y_load =-1.0],
   [6:X_load =0.0].
   ```
   Ensure that all other loads are zero. This is the preexisting load that contributes to $[K_G]$ and will be scaled during the linear vibration analysis. The small load, [4:X_load=0.02], contributes the load imperfection that will initiate the motion in the next part.
4. Mesh using <1> element through the length, <2> through the height, and <16> in the hoop direction. That is, set
   ```
   [0:Global_ID =111],
   [1:hoop =16],
   [2:thick =2],
   [3:length =1],
   [a:damp =2.0e-2],
   [d:red_intn =3].
   ```
   [Press s] to render the mesh.
   For clearer viewing, [press o] to set the orientation as $\phi_x = 30$, $\phi_y = 25$,

$\phi_z = 0$. (If the whole mesh does not display, ensure that all tags are set to 1; that is, [press t # 1].)

COLLECTING THE DATA

1. Analyze the bar for its Vibration response using the subspace iteration algorithm. That is, set

   [0:type =1],
   [2:scale[K_G] =0],
   [8:#modes =10],
   [a:algorithm =1].

   [Press s] to perform the eigenanalysis.

   Before exiting the Vibration analysis section, record the first four vibration frequencies stored in the file <<stadyn.out>>; record an imaginary frequency as negative.

2. View the deformed Shapes.

   For comparative viewing, select the gallery of modes with

   [1:vars =51],
   [2:xMult =0.01],
   [3:mode =1].

   The gallery begins at [3:mode=#], and the deformed shapes can be exaggerated through [2:xMult=?].

   For additional clarity, animate the mode shapes by

   [1:vars =11],
   [2:xMult =0.01],
   [3:mode =1],
   [5:rate =0],
   [6:size =1],
   [7:max# =4],
   [8:divs =16],
   [9:slow =50].

   [Press s] to show the animation.

   Record (by sketch or otherwise) a note about the shape of the first four modes.

3. Repeat the analysis (including recording the description of the deformed shape) changing the scale on the geometric matrix as

   [2:scale[K_G] =20, 40, 60, ...]

   until at least one of the frequencies becomes imaginary.

4. Analyze the bar for its Buckling response using the subspace iteration algorithm. That is, set

   [0:type =1],
   [8:#modes =10],
   [a:algorithm =1].

   [Press s] to perform the eigenanalysis.

Before exiting the Buckling analysis section, record the first four buckling loads stored in the file <<stadyn.out>>.

5. View the deformed Shapes.

For comparative viewing, select the gallery of modes with

[1:vars =51],
[2:xMult =1],
[3:mode =1],
[5:rate =0].

Record (by sketch or otherwise) a note about the shape of the first four modes.

ANALYZING THE DATA

1. Plot all the frequency data points on a frequency squared against load graph. Represent an imaginary frequency as a negative value.

Identify the data points according to their mode shape.

Connect the data points belonging to a particular mode shape. Observe that some modes cross.

2. What is the meaning of a zero frequency?

What is the meaning of a negative frequency?

3. How are these results related to the buckling analysis data?

Part II. Large-Deflection Elastic Behavior

Part I considered the initiation of the instability; this exploration considers the subsequent behavior. The objective of the data collection and analysis is to make the connection between the occurrence of an instability and the subsequent motion.

The same model as in Part I is used. If **QED** is not already running, begin by launching it from the command line in the working directory.

COLLECTING THE DATA

1. Analyze the problem using the nonlinear Implicit integration scheme.

Set

[n:P(t)name =force.65],
[0:type =1],
[1:scale{P} =65],
[2:scale{Grav} =0.0],
[5:time inc =0.5e-4],
[6:#steps =1000],
[7:snaps@ =2],
[8:eigns@ =10],
[9:node =160],
[a:algorithm =1].

[Press s] to perform the time integration. Note that this analysis will take a long time to run.

The complete history data for the monitored node are in the file <<stadyn.dyn>>; this will be overwritten during the postprocessing; therefore this will need to be copied if the results are to be saved.

The history data for the vibration eigenvalues and other monitors are recorded in the file <<stadyn.mon>> as

$$\{t;\ P;\ \omega_1^2,\ \omega_2^2,\ \ldots,\ \omega_{10}^2;\ \mathcal{U},\ \mathcal{T},\ \mathcal{W},\ \hat{x},\ \hat{u}\}.$$

Store the results from the files <<stadyn.dyn>>, <<stadyn.mon>>.

2. View the deformed Shapes. Set the movie parameters as
   [1:vars =111],
   [2:xMult =1],
   [3:mode =1],
   [5:rate =0],
   [6:size =1],
   [7:max# =4444],
   [8:divs =2],
   [9:slow =25].
   [Press s] to process the snapshots, and [press s] again to run the movie. Confirm that the deformation is in agreement with expectation.

3. View the history Traces. Select the displacement traces for Node 160 (original position $x = 0$, $y = 5.0$, $z = 0.5$) with
   [0:style type =1],
   [1:vars =1],
   [2:xMult =1],
   [5:rate =0],
   [6:node/IP =197],
   [7:elem# =1].
   [Press s] to show the panel of responses; only the first three panels are relevant for solids.
   The numbers for the traces can be found in the file <<stadyn.dyn>>; this will need to be copied if the results are to be saved.

4. Return to the Loads section and remove the eccentric loading with
   [4:X_load =-0.0],
   [5:Y_load =-1.0].
   Remesh and repeat the analysis.
   Store the results from the files <<stadyn.dyn>>, <<stadyn.mon>>.

5. In the Implicit analysis section increase the applied load scale to
   [1:scale{P} =75].
   Repeat the analysis and store the results from the files <<stadyn.dyn>>, <<stadyn.mon>>.

ANALYZING THE DATA

1. Plot the transverse deflection against load, ([col 6 vs col 2]) of <<stadyn.dyn>>, for each case.
   Describe their major characteristics.

2. Plot the first two eigenvalues against time, ([col 3 vs col 1] & [col 4 vs col 1]) of <<stadyn.mon>>, for each case.
   What is the difference between the cases?
3. Conjecture on the similarities and differences between analyses of Parts I and II.

Part III. Simple Modeling of an Arch

An important step in constructing a simple model is reducing the model to the minimum number of DoFs that nonetheless captures the phenomenon of interest. Here, we first replace the 3D solid arch with a 2D frame arch, then the 2D frame with four, then two, elements. The objective of the data collection and analysis is to lay the foundation for constructing a simple model.

Begin by launching **QED** from the command line in the working directory.

CREATING THE MODEL

1. Create a segment of a plastic arch using the [Frame:2D_arch] geometry. That is,
   [0:geometry =6],
   [1:radius =5],
   [2:angle =90],
   [a:mat# =4],
   [b:thick =.2],
   [c:depth =1.0].
2. Use BCs of [0:type=1(=AmB)] to set both edges as simply supported. That is,
   [0:type =1],
   [5:end_A =000001],
   [6:others =111111],
   [7:end_B =000001].
3. Use Loads of [0:type=1(=near xyz)] to apply a point load of $P_y = 1$ at position $x = 0.0, \ y = 2.0$. That is,
   [0:type =1],
   [1:X_pos =0],
   [2:Y_pos =2.0],
   [3:Z_pos =0.0],
   [4:X_load =0.0],
   [5:Y_load =-3.0],
   [6:Z_load =0.0],
   [9:Z_mom =0.02].
   Ensure that all other loads are zero. The small load, [9:Z_mom=0.02], contributes the load imperfection that will initiate the motion.
4. Mesh using <32> elements and treat the arch as 2D. That is, set
   [0:Global_ID =22],
   [1:X_elems =32],
   [a:damping =2.0e-2];
   other [specials] can be set to zero. [Press s] to render the mesh.

For clearer viewing, if necessary, [press o] to set the orientation as $\phi_x = 0$, $\phi_y = 0$, $\phi_z = 0$.

COLLECTING THE DATA

1. Analyze the bar for its Vibration response using the subspace iteration algorithm. That is, set
   [0:type =1],
   [2:scale[K_G] =0],
   [8:#modes =10],
   [a:algorithm =1].
   [Press s] to perform the eigenanalysis.
   Before exiting the Vibration analysis section, record the first four vibration frequencies stored in the file <<stadyn.out>>; record an imaginary frequency as negative.

2. View the deformed Shapes. For comparative viewing, select the gallery of modes with
   [1:vars =51],
   [2:xMult =0.01],
   [3:mode =1],
   [5:rate =1].
   For additional clarity, animate the mode shapes by
   [1:vars =11],
   [2:xMult =.01],
   [3:mode =1],
   [5:rate =0],
   [6:size =1],
   [7:max# =4],
   [8:divs =16],
   [9:slow =50].
   [Press s] to show the animation.
   Record (by sketch or otherwise) a note about the shape of the first four modes.

3. Analyze the bar for its Buckling response using the subspace iteration algorithm. That is, set
   [0:type =1],
   [8:#modes =10],
   [a:algorithm =1].
   [Press s] to perform the eigenanalysis.
   Before exiting the Buckling analysis section, record the first four buckling loads stored in the file <<stadyn.out>>.

4. View the deformed Shapes. For comparative viewing, select the gallery of modes with
   [1:vars =51],
   [2:xMult =1],
   [3:mode =1],

[5:rate =0].

Record (by sketch or otherwise) a note about the shape of the first four modes.

5. Analyze the problem using the nonlinear Implicit integration scheme. Set

[n:P(t)name =force.65],

[0:type =1],

[1:scale{P} =65],

[2:scale{Grav} =0.0],

[5:time inc =1.0e-4],

[6:#steps =600],

[7:snaps@ =10],

[8:eigns@ =10],

[9:node =17],

[a:algorithm =1].

[Press s] to perform the time integration.

The complete history data for the monitored node are in the file <<sta-dyn.dyn>>; this will be overwritten during the postprocessing; therefore this will need to be copied if the results are to be saved.

The history data for the vibration eigenvalues and other monitors are recorded in the file <<stadyn.mon>> as

$\{t;\ P;\ \omega_1^2,\ \omega_2^2,\ \ldots,\ \omega_{10}^2;\ \mathcal{U},\ \mathcal{T},\ \mathcal{W},\ \hat{x},\ \hat{u}\}.$

Store the results from the files <<stadyn.dyn>>, <<stadyn.mon>>.

6. View the deformed Shapes. Set the movie parameters as

[1:vars =111],

[2:xMult =1],

[3:mode =1],

[5:rate =0],

[6:size =1],

[7:max# =4444],

[8:divs =8],

[9:slow =25].

[Press s] to process the snapshots, and [press s] again to run the movie.

Confirm that the deformation is in agreement with expectation.

7. View the history Traces. Select the displacement traces for Node 17 (original position $x = 0$, $y \approx 2.0$, $z = 0.0$) with

[0:style type =1],

[1:vars =1],

[2:xMult =1],

[5:rate =0],

[6:node# =17],

[7:elem# =1],

[7:elem type =3].

[Press s] to show the panel of responses.

8. Return to the Mesh section and change the number of elements to

[1:X_elems =4].

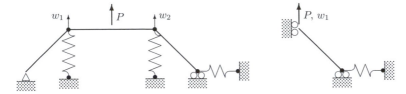

Figure 6.16. Two simple mechanical models of an arch using rigid links and linear springs.

Remesh and repeat the three analyses monitoring [9:node=3].
Store the results from the files <<stadyn.dyn>>, <<stadyn.mon>>.
9. Return to the Mesh section and change the number of elements to
[1:X_elems =2].
Remesh and repeat the three analyses monitoring [9:node=2].
Store the results from the files <<stadyn.dyn>>, <<stadyn.mon>>.

ANALYZING THE DATA
1. Plot the transverse deflection against load ([col 6 vs col 2]) of <<sta-
dyn.dyn>>, for each case.
Do they resemble those of Part II?
2. Plot the first two eigenvalues against time, ([col 3 vs col 1] & [col 4 vs
col 1]) of <<stadyn.mon>>, for each case.
Do they resemble those of Part II?
3. Conjecture on what happened as the number of elements was decreased.
4. Conjecture (in the context of Section 7.5) about the construction of a simple
model that captures the essence of the snap-through buckling of an arch. Specifically, compare and contrast the suitability of the proposed simple models shown
in Figure 6.16.
Conjecture on how the chosen model would be tested.

Partial Results

Partial results for the deflections and vibration eigenvalues are shown in Figure 6.17.
The deformed shape of the arch during the snap-through is shown in Figure 6.15.

6.5.2 Stability of a Truss Structure

The simple system explored here is that of the pinned truss shown in Figure 6.18(a).
The truss members have elasticity EA but no mass – all the mass is concentrated at
the apex. The equation of motion is

$$M\ddot{v}+C\dot{v} = P - F(v) = P - 2EA\sin\alpha_o\left[1+\frac{v}{L_o\sin\alpha_o}\right]\left[1-\frac{1}{\sqrt{1+2\frac{v}{L_o}\sin\alpha_o+(\frac{v}{L_o})^2}}\right].$$

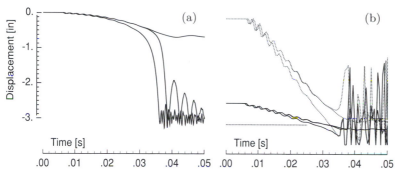

Figure 6.17. Partial results for the deflection of an arch: (a) deflection, (b) vibration eigenvalues.

This is implemented as one of QED's gallery of nonlinear differential equations. The static load deflection shown in Figure 6.18(b) indicates a highly nonlinear response.

The member force expression is too complicated for analytical manipulation; a polynomial approximation accurate over the complete load range is obtained by noting that the three zero-load points occur when v is 0, $-L_o \sin \alpha_o$, and $-2L_o \sin \alpha_o$, respectively, and leads to the simpler form

$$M\ddot{v} + C\dot{v} = P - EA \sin^3 \alpha_o \left[\bar{v}(1 + \bar{v})(2 + \bar{v}) \right], \qquad \bar{v} \equiv v/(L_o \sin \alpha_o).$$

The comparison between the exact from and this simpler form is also shown Figure 6.18(b).

To help construct a simple model, do the following as a preliminary analysis:

1. Compute the strain energy $\mathcal{U}_o = \frac{1}{2} 2EA\epsilon_o^2 L_o$ when the truss is forced to be flat, that is, use new length $L = L_o \cos \alpha_o$.
2. Compute the positive vertical deflection such that the truss has the strain energy $\mathcal{U}_o$, that is, use new length $L^2 = (v + L_o \sin \alpha_o)^2 + (L_o \cos \alpha_o)^2$.
3. Compute the velocity of the mass such that its kinetic energy is $\mathcal{T} = \frac{1}{2} M\dot{v}^2 = \mathcal{U}_o$.

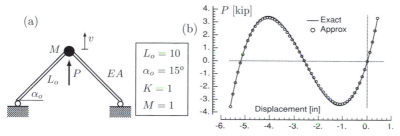

Figure 6.18. Truss under dynamic loading: (a) geometry and properties, (b) static force–deflection curve $P = F(v)$.

Part I. Quasi-Static Loading

We begin with a quasi-static loading of the truss, and the objective of the data collection and analysis is to identify the load at which the snap-through is initiated.

Launch QED from the command line in the working directory.

COLLECTING THE DATA

1. Analyze the problem using [Gallery_ODE:5] to select the truss equations.
   Go to [Geometry] and set the parameters as
   [k:stiff =1] ($K = EA/L_o$),
   [c:damp =.01],
   [m:mass =1],
   [1:a =15] (angle α_o),
   [2:b =10] (length L_o).
2. Go to [ICs] (initial conditions) and set
   [6:u0 =0.00],
   [7:Du0 =0.0],
   [8:Dt =0.10],
   [9:#step =4000].
3. Go to [Loads] to set a ramp load. That is, set
   [1:P =0.10],
   [2:dT =200];
   with all other loads set to zero.
   Ensure that autoscaling in on. The orientation of the trajectory (state space against time) can be changed by [pressing o]; values of [20, 30, 0] give a reasonable orientation. The scales for the trajectory are changed in the [Scales] section.
4. [Press y] to look at the phase plane gallery.
   Record the maximum deflection achieved.
   Repeat the measurement for the different positive (up) load levels
   [1:P =0.1, 0.2, 0.3, 0.4, 0.5].
   Repeat the measurement for the different negative (down) load levels
   [1:P =-0.02, -0.04, -0.06, -0.08 -0.10].
   If the oscillations have not subsided by the end of the time trace, then record the average of the excursions.
5. By varying the negative load
   [1:P =-?],
   find the value of load just before the snap-through occurs. Record the load and displacement.

ANALYZING THE DATA

1. Plot the load against deflection data.
2. Superpose the polynomial approximation.
3. Identify the load and deflection at which the snap-through instability occurs.
4. Conjecture why there is a large gap in the recorded displacement values.

Part II. Free-Vibration Behavior

In Part I the truss was quasi-statically "forced" to the instability point. But there are other ways to cause the instability; consider, for example, plucking the truss by applying a large upward force and then releasing. One can imagine that if the mass gains enough momentum it will snap through.

The objective of the data collection and analysis is to establish the initial displacement–velocity boundary of the instability region. If QED is not already running, begin by launching it from the command line in the working directory and Analyze the problem using [Gallery_ODE:5].

COLLECTING THE DATA

1. In the [Loads] section, set all loads to zero.
2. In the [ICs] section, set the initial conditions as those of an initial displacement. That is, set
   [6:u0 =2.0],
   [7:Du0 =0.0],
   [8:Dt =.10],
   [9:#step =4000].
3. [Press y] to look at the phase plane gallery. Observe that the response orbits two equilibrium points before coming to rest at one of them.
   By varying only the initial displacement
   [6:u0 =?],
   (with initial velocity being zero [7:Du0=0]), find the two values of initial deflection ($\pm v$) that make the system borderline stable (i.e., find only the near equilibrium directly). Record these values.
   By varying only the initial velocity
   [7:Du0 =?]
   (with initial displacement being zero [6:u0=0]), find the two values of initial velocity ($\pm \dot{v}$) that make the system borderline stable. Record these values.
   By varying both the initial displacement and initial velocity
   [6:u0 =?],
   [7:Du0 =?],
   find a few more initial values that make the system borderline stable. Record these values.

ANALYZING THE DATA

1. Plot the borderline stability values as a phase plane plot with displacement vertical and velocity horizontal.
2. Superpose the simple model relationship

$$\tfrac{1}{2} M\dot{v}^2 + \frac{EA}{L_o} L_o^2 \sin^4 \alpha_o \left[\bar{v}^2 + \bar{v}^3 + \tfrac{1}{4}\bar{v}^4 \right] = \mathcal{U}_o, \qquad \bar{v} \equiv v/(L_o \sin \alpha_o).$$

That is, for different values of v, solve for $\dot{v}$ and plot.
3. Do the data and the relationship agree?

4. Conjecture why the instability point does not coincide with that of the static snap-through point.

5. Conjecture why the simple model does not form a sharp point at the saddle point.

Part III. Impulse Tests

The stability boundary established in Part II can be thought of as the edge of a bowl, and the test performed there placed the system initially at or near the edge. The system can also be put or sent there by being given an initial momentum through an impulse. This is not quite the same as being given an initial velocity because the momentum is accumulated over the period of the impulse, and during this time the position changes.

The objective of the data collection and analysis is to correlate the impulse parameters of magnitude and duration with reaching the boundary of the stability region. If QED is not already running, begin by launching it from the command line in the working directory and Analyze the problem using [Gallery_ODE:5].

COLLECTING THE DATA

1. Go to the [ICs] section and set all initial conditions to zero and the time step as
   [6:u0 =0.0],
   [7:Du0 =0.0],
   [8:Dt =0.10],
   [9:#step =4000].

2. Go to the [Loads] section to impose a short-duration impulse. That is, set
   [5:P =2.0],
   [6:dT =201],
   [7:dt =1.0];
   all other loads are zero. This is actually a multiple-ping loading with periodicity of 200 s and duration 1 s – only the first ping will be of relevance here. The duration of the ping is specified by [7:dt=?].

3. [Press y] to look at the response gallery.
   [Press q g] to look at the alternate response gallery to see the displacement, velocity, and force histories as the top row, respectively. Observe that the force is a sharp impulse. Record the maximum velocity achieved.

4. Repeat for different load levels and different durations such that the impulse $\frac{1}{2} P \, dt$ is constant. That is,
   [5:P =.125, .25, .50, 1.0, 2.0, 4.0, 8.0, 10.],
   [7:dt =16., 8.0, 4.0, 2.0, 1.0, 0.5, .25, 0.2].
   Record the maximum velocities achieved and call them DataSet I.

5. With the duration fixed, adjust the load level until the orbit traces the borderline around the nearest equilibrium point. That is,
   [5:P =?],
   [7:dt =0.2].

Record the impulse. Record the complete history file `<qed.dyn>>` by [`pressing w`] while viewing the gallery.

6. Launch DiSPtool (from the command line) and [`press v`] to view the columns of data in `<<qed.dyn>>`. [`Press p`] for the parameters list and specify

[`n:file =qed.dyn`,
[`1:plot_1 =1 2`],
[`2:plot_2 =1 3`],
[`3:plot_3 =2 3`],
[`4:plot_4 =1 4`].

[`Press q R`] for complete rereading of the file. Confirm that the same information is displayed in both QED and DiSPtool. Rename `<<qed.dyn>>` for archiving.

ANALYZING THE DATA

1. For DataSet I, plot the maximum velocity against impulse duration.

    What is the trend of the data?

    Although each case has the same impulse, conjecture why the velocities (and hence the stability behavior) are different.

    Conjecture on the connection between these results and those of Part II.

2. Plot the complete trajectory history (from the renamed `<<qed.dyn>>` file) as a phase plane plot.

3. Superpose on the plot the lines

$$\dot{\bar{v}}^2 + 2\eta\bar{v}\dot{\bar{v}} + (\eta^2 + \omega_o^2)\bar{v}^2 + \omega_o^2\bar{v}^3 + \tfrac{1}{4}\omega_o^2\bar{v}^4 = V,$$

where

$$\eta = \frac{C}{M}, \qquad \omega_o^2 = \frac{K^*}{M}, \qquad K^* = \frac{EA}{L_o}2\sin^2\alpha_o$$

by solving for $\dot{v}$ in terms of v for different values of V. This is a plot of contours of a Lyapunov function.

4. Comment on what is observed.

Partial Results

Partial results for the quasi-static loading of Part I are shown in Figure 6.19(a). Partial results for the trajectories of Part III are shown in Figure 6.19(b). The Lyapunov curves are the dotted lines.

Background Readings

Some background material on the stability under motion in the large is given in References [30, 65, 68, 113, 133]. Reference [77] gives a readable introduction to the subject.

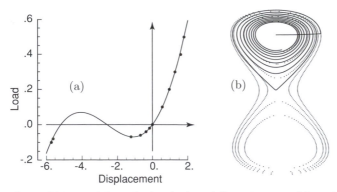

Figure 6.19. Partial results for the instability of a truss: (a) static deflections, (b) phase plane trajectories.

6.6 Dynamic Instability Under Follower Loads

As discussed in Section 6.5, an unstable structure invariably goes in motion and comes to rest only when it has found a new equilibrium position. In some cases, however, the new equilibrium state is actually a motion, called a *limit cycle*. This section explores the effect of follower loads on the stability behavior of structures and how the instability can lead to limit cycles. The cantilevered beam shown in Figure 6.20 is used to illustrate the basic concepts.

In preparation, use DiSPtool to create a smoothed ramp load-history file that increases from zero to unity over a period of 0.01 s and lasts for 0.2 s. (Section 6.5 demonstrates how this can be accomplished.) This is the primary load history, so call it <<primary.66>>. In addition, use DiSPtool to create a sine-squared pulse that begins at $t = 0.0101$ s and ends at $t = 0.0103$ s and has a resolution of $dt = 5\,\mu$s. This is the ping load history, so call it <<ping.66>>.

Part I. Follower Load on a Beam
The cantilevered beam is loaded on its end as shown in Figure 6.20. The special feature of the load is that, as the end of the beam deflects and rotates, the load also rotates. The load could be conceived as a pressure that is always normal to the end

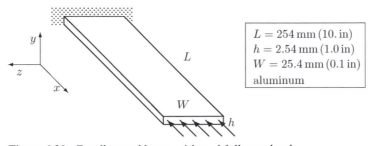

| $L = 254\,\text{mm}\,(10.\,\text{in})$ |
| $h = 2.54\,\text{mm}\,(1.0\,\text{in})$ |
| $W = 25.4\,\text{mm}\,(0.1\,\text{in})$ |
| aluminum |

Figure 6.20. Cantilevered beam with end follower load.

surface. The objective of the data collection and analysis is to establish the basic behavior of the limit cycles.

Begin by launching QED from the command line in the working directory.

CREATING THE MODEL

1. Create a $[10 \times 1.0 \times 0.1]$ aluminum beam using the [Frame:1D_beam] geometry. That is,
   ```
   [0:geometry =1],
   [1:X_length =10.0],
   [4:# X_span =1],
   [a:mat#1 =1],
   [b:thick1 =0.1],
   [c:depth1 =1.0].
   ```
2. Use BCs of [0:type=1(=AmB)] to fix one end. That is,
   ```
   [0:type =1],
   [5:end_A =000000],
   [6:others =111111],
   [7:end_B =111111].
   ```
3. Use Loads of [0:type=1(=near xyz)] to apply a horizontal force at the right boundary. That is,
   ```
   [0:type =1],
   [1:X_pos =10],
   [2:Y_pos =0],
   [4:X_load =-1];
   ```
   all other loads are zero.
4. Mesh using $<10>$ elements through the length. Set,
   ```
   [0:Global_ID =22],
   [1:X_elems =10],
   [a:damping =.08].
   [c:follower =1.0].
   ```
   [Press s] to render the mesh.
   If necessary, [press o] to set the orientation as $\phi_x = 0, \phi_y = 0, \phi_z = 0$.

COLLECTING THE DATA

1. Analyze the problem as Explicit nonlinear. Set
   ```
   [n:P(t)name =primary.66],
   [0:type =1],
   [1:scale{P} =170],
   [2:scale{Grav} =0.0],
   [5:time inc =5e-6],
   [6:#steps =20000],
   [7:snaps@ =200],
   [9:node =11].
   ```

[Press p] to set the parameters of the ping load:
[n:P(t)name =ping.66],
[1:X_pos =5],
[2:Y_pos =0],
[4:X_load =0],
[5:Y_load =100];
all other loads are zero.
[Press qs] to perform the time integration.
Explicit time integration is only conditionally stable and therefore requires a very small time step and a corresponding very large number of steps.

2. View a movie of the displaced Shapes with
[1:vars =111],
[2:xMult =1.0],
[3:snapshot =1],
[5:rate =0],
[6:size =1.5],
[7:max# =100],
[8:divs =4],
[9:slow =100].
[Press s] to process the snapshots, and [press s] again to show the movie.
Observe that the transverse deflection is only slight and that it dies down over time.

3. View the Traces of the displacement history. Set
[0:style type =1],
[1:vars =1],
[2:xMult =1],
[5:rate =0],
[6:node# =11],
[7:elem# =1],
[8:elem type =3].
Individual panels can be zoomed by [pressing z #], where the panels are numbered 1 through 6, top row first. Observe that the transverse deflection [Var:V] is highly damped.

4. Return to the Analysis section and change the primary load magnitude to
[1:scale{P} =190].
[Press s] to perform the time integration. Before exiting the Analysis section copy the <<stadyn.dyn>> file to a name such as <<dyn100.190>>.
View a movie of the deformed Shapes with
[1:vars =111],
[5:rate =0],
[6:size =1.5],
[7:max# =100].
Observe that the transverse deflection is now much larger and that it persists for the complete duration. Confirm this by looking at the Traces of the displacement history.

5. Launch DiSPtool and [View] the responses in <<dyn100.190>>; the stored columns have the sequence $[t, P_1, P_2, P_3, u, v, w, \dot{u}, \dot{v}, \dot{w}, \dots,]$. In particular, plot v against $\dot{v}$ (col 6 vs col 9) and observe that the trace limits to an ellipse by spiraling outward. The limiting trace is called a *limit cycle*.
6. Return to the Analysis section and [press p] to change the ping load magnitude to
   [4:X_load =0],
   [5:Y_load =800].
   [Press q s] to perform the time integration. Before exiting the Analysis section copy the <<stadyn.dyn>> file to a name such as <<dyn800.190>>.
7. Switch to DiSPtool and [press z#] to zoom the plot of v against $\dot{v}$. [Press i] to insert the new trace.
   To make the comparison easier to see [press e] and [press v] to shrink the display.
   Observe that the trace initially goes outside the limit cycle, but then limits to it by spiraling inward. The limit cycle is said to act as an *attractor*.
8. Return to the Analysis section and [press p] to change the ping load magnitude back to
   [4:X_load =0],
   [5:Y_load =100].
   Run the analysis for the four values of primary load:
   [1:scale{P} =190, 200, 210, 220].
   Each time, record the steady-state amplitude.
9. Return to the Mesh section and, in turn, change the damping to
   [a:damping =.08, .04, .02].
   Remesh each time and run the explicit analysis with
   [1:scale{P} =190].
   Each time, record the steady-state amplitude.

ANALYZING THE DATA
1. Plot the steady-state amplitude data against primary load.
2. Plot the steady-state amplitude data against damping.
3. A free vibration is represented by

$$v(t) = Ae^{i\mu t} = Ae^{i(\mu_R + i\mu_I)t} = Ae^{-\mu_I t}e^{i\mu_R t} = Ae^{-\mu_I t}[\cos(\mu_R t) + i \sin(\mu_R t)].$$

This indicates oscillations of frequency (μ_R) with an exponential decay in the presence of damping (μ_I).
The applied load is constant in that $P(t) = $ constant (although its orientation changes); conjecture why the time data do not exhibit a damped oscillation when the load is above critical.
4. A suggested simple model conceives of the limit cycle behavior as a forced-frequency vibration response even though the primary load is constant.
Conjecture about the basis for this model.
Does the model help explain the data?

Part II. Frequency Response of Follower Loads

Follower loads give rise to a nonconservative dynamic system. Their contribution to the system is in the form of a nonsymmetric geometric stiffness matrix; this makes vibration eigenanalysis unsuited for assessing the stiffness behavior. This exploration will access the stiffness behavior by analyzing the vibration response that is due to ping loads.

The objective of the data collection and analysis is to establish the origin of the instability. If QED is not already running, begin by launching it from the command line in the working directory.

COLLECTING THE DATA

1. Return to the Mesh section and change the damping to
   [a:damping =.02].
   Remesh.
2. Analyze the beam for its linear Vibration response using the subspace iteration algorithm. That is, set
   [0:type =1],
   [2:scale[K_G] =0],
   [8:# modes =10],
   [a:algorithm =1].
   Because the scale is zero, the analysis will give the free-vibration response of the unloaded beam.
   [Press s] to perform the eigenanalysis.
   Before exiting the analysis section, record the first two frequencies in the <<sta-dyn.out>> file.
3. Analyze the problem as Explicit nonlinear. Set
   [n:P(t)name =primary.66],
   [p:-->],
   [0:type =1],
   [1:scale{P} =0.0],
   [2:scale{Grav} =0.0],
   [5:time inc =5e-6],
   [6:#steps =40000].
   [7:snaps@ =500],
   [9:node =11].
   Note that the primary load will be zero. [Press p] to set the parameters of the ping load:
   [n:dP(t)name =ping.66],
   [1:X_pos =5],
   [2:Y_pos =0],
   [4:X_load =0],
   [5:Y_load =100];
   all other loads are zero.
   [Press qs] to perform the time integration.

4. The following exercise shows how to use DiSPtool to create an amplitude spectrum.

Launch DiSPtool (from the command line) and [press v] to view the columns of data in <<stadyn.dyn>>; the stored columns have the sequence

$[t, P_1, P_2, P_3, u, v, w, \dot{u}, \dot{v}, \dot{w}, \ldots,].$

[Press p] to access the input parameters and set

[n:file =stadyn.dyn],

[1:plot_1 =1 5],

[2:plot_2 =1 6],

[3:plot_3 =1 8],

[4:plot_4 =1 9],

[6:line #1 #2 =1 8001].

[Press q] to quit the parameters list and display the traces.

[Press s] to access the [STORE columns] menu. Choose to store a two-column file of time [col 1] and transverse displacement [col 6] by setting

[n:filename =disp.1],

[1:1st col =1],

[2:2nd col =6],

[3:3rd col =0],

[4:# pts =8001],

[5:# thin =2].

[Press s] to store the columns to disk.

Launch a second copy of DiSPtool and [press to] to go to the [Time Domain:One channel] section. [Press p] to input the stored file. That is, set

[n:filename =disp.1],

[1:dt =200e-6],

[3:#fft =4096],

[4:MA filt# =1],

[5:MA filt begin =1],

[6:plot#1#2 =1 4096],

[w:window =0].

[Press q] and then [press r] to see the frequency spectrum in the second panel. An expanded view is obtained by [pressing z2] and then [press E] multiple times. [Press >] multiple times to move the cursor to the first peak. (The speed of the cursor can be changed by [pressing F] multiple times.) Record the frequency and amplitude of the first two spectral peaks. These frequencies should be close to those of the Vibration eigenanalysis.

5. Switch back to QED and change the primary load in the Explicit nonlinear Analysis to

[1:scale{P} =40].

[Press s] to run the analysis.

Switch to the first DiSPtool, [press q] to quit the [store] menu and [press R] to force the reading of the current <<stadyn.dyn>> file. [Press s] to access the [STORE columns] menu and [press s] again to update the <<disp.1>> file.

Switch to the second DiSPtool, [press q] to quit the [expand] view and [press R] to force the reading of the updated <<disp.1>> file. [Press z2] to zoom on the amplitude spectrum and record the frequency and amplitude of the first two spectral peaks.

Observe that the lower frequency increased whereas the second frequency decreased.

6. Repeat the process for primary load levels of
   [1:scale{P} =80, 120, 160, 165, 170, 175].

ANALYZING THE DATA

1. Plot the frequency data against primary load.
   Observe that the two frequencies coalesce at the higher loads.
2. Plot the amplitude data against primary load.
   Observe that the amplitudes increase significantly with load.
3. Section 7.5 develops a simple model for a cantilever beam with an end follower force. Relating $L \to L/2$, $K \to \alpha EI/L$, $M \to \frac{1}{2}\rho A(L/2)$ then gives the linearized equations

$$
\alpha \frac{EI}{L}
\begin{bmatrix} 2 & -1 \\ -1 & 1 \end{bmatrix}
\begin{Bmatrix} \phi_1 \\ \phi_2 \end{Bmatrix}
- \frac{PL}{2}
\begin{bmatrix} 1 & -1 \\ 0 & 0 \end{bmatrix}
\begin{Bmatrix} \phi_1 \\ \phi_2 \end{Bmatrix}
+ \frac{\rho AL^3}{16}
\begin{bmatrix} 3 & 1 \\ 1 & 1 \end{bmatrix}
\begin{Bmatrix} \ddot{\phi}_1 \\ \ddot{\phi}_2 \end{Bmatrix}
= \begin{Bmatrix} 0 \\ 0 \end{Bmatrix},
$$

where $A = Wh$ and $I = Wh^3/12$. Assuming harmonic motion of the form $e^{i\mu t}$ then leads to the frequency expression

$$
\mu^2 = \frac{28}{L^4}\frac{\alpha EI}{\rho A} - \frac{4}{L^2}\frac{P}{\rho A} \pm \frac{4}{L^2}\sqrt{\left(\frac{P}{\rho A}\right)^2 - \frac{14}{L^2}\frac{\alpha EI}{\rho A}\frac{P}{\rho A} + \frac{41}{L^4}\left(\frac{\alpha EI}{\rho A}\right)^2}.
$$

Choosing $\alpha = 5.18$ makes the lowest mode coincide with the exact solution for the free vibration of a beam when $P = 0$ [30]. The special feature of this expression is that the square-root term can go to zero at the loads

$$
P = \frac{\alpha EI}{L^2}\left[7 \pm \sqrt{8}\right], \qquad P_c = 4.17\frac{\alpha EI}{L^2},
$$

and thus the frequencies coalesce.

4. Superpose, as a continuous line, a plot of $\mu/2\pi$ on the frequency plot.
   Does the simple model capture the essence of the data?
5. An implication of the simple model is that, for loads greater than the critical value, then $\mu = \mu_R \pm i\mu_I$ and the time response is

$$
v(t) = Ae^{i\mu t} = Ae^{i(\mu_R \pm i\mu_I)t} = Ae^{\pm \mu_I t}e^{i\mu_R t}.
$$

Thus at least one of the modes increases exponentially. Because μ_R is nonzero, a single-frequency oscillation occurs as the amplitude increases. That is, when the beam becomes unstable it does not lose its stiffness properties – this is the characteristic of a dynamic instability in contrast to static instability.
Do the data collected so far support the implications of the simple model?

Part III. The Role of Damping in the Motion

Another implication of the simple model of Part II is that the system behaves as if it has negative damping. It can therefore be conjectured that, if the system has some real (positive) damping, then this will affect the threshold load level for the instability. The objective of the data collection and analysis of the exploration of this section is to determine the effect of damping on the threshold load to initiate the instability.

The first task is to establish the relation between critical load and damping and, in particular, an estimate of the critical load for zero damping. This latter task cannot be done by simply setting the damping to zero because the initial transients would never die down. Instead a limiting procedure is used.

If QED is not already running, begin by launching it from the command line in the working directory.

COLLECTING THE DATA

1. Return to the Mesh section and change the damping to
   [a:damping =.12].
   Remesh.
2. Analyze the problem as Explicit nonlinear. Set
   [n:P(t)name =primary.66],
   [p:-->],
   [1:scale{P} =190],
   [2:scale{P} =0.0],
   [5:time inc =5e-6],
   [6:#steps =40000],
   [7:snaps@ =500],
   [9:node =11].
   [Press p] to set the parameters of the ping load:
   [n:dP(t)name =ping.66],
   [1:X_pos =5],
   [2:Y_pos =0],
   [4:X_load =0],
   [5:Y_load =-100];
   all other loads are zero.
   [Press qs] to perform the time integration.
3. Launch DiSPtool (from the command line) and [press v] to view the columns of data in <<stadyn.dyn>>; the stored columns have the sequence
   $[t, P_1, P_2, P_3, u, v, w, \dot{u}, \dot{v}, \dot{w}, \ldots,]$.
   [Press p] to access the input parameters and set
   [n:file =stadyn.dyn],
   [1:plot_1 =1 5],
   [2:plot_2 =1 6],
   [3:plot_3 =5 8],
   [4:plot_4 =6 9],

[6:line#1#2 =1 8001].

[Press q] to quit the parameters list and display the traces.

Observe that the motion is a damped oscillation.

Switch to **QED** and increment the primary load to

[1:scale{P} =194],

and rerun the analysis. Switch to **DiSPtool** and [press R] to view the updated responses.

Repeat the load incrementing until the responses just begin to exhibit limit cycle behavior. Record the value of load; we refer to this as the threshold load and use the symbol P_t.

4. Return to the Mesh section and change the damping to
   [a:damping =.10, .08, .06, .04, .02, .01].
   After each change, remesh and redo the analysis until the load to cause limit cycles is established. Record each of these loads.

ANALYZING THE DATA

1. Plot the threshold load data against damping.
   Observe that there is a threshold value of load below which limit cycles do not occur. Estimate this value and call it P_c.

2. The earlier simple model can be made to account for damping by adding

$$\frac{\eta A L^3}{2}\begin{bmatrix} 1 & 0 \\ 0 & 1 \end{bmatrix}\begin{Bmatrix} \dot{\phi}_1 \\ \dot{\phi}_2 \end{Bmatrix}$$

to the equations of motion. This leads to a characteristic equation that is quartic in μ with no obvious factorization. An approximation for small damping leads to

$$\mu^2 = \frac{28}{L^4}\frac{\alpha EI}{\rho A} - \frac{4}{L^2}\frac{P}{\rho A} - i\frac{4\eta A}{\rho A} \pm \frac{4}{L^2}\sqrt{\left(\frac{P}{\rho A}\right)^2} \cdots .$$

The term inside the square root is real-only and gives the critical load

$$P_c = \frac{\alpha EI}{L^2}\left[7 - \sqrt{8}\right] + \frac{2\eta^2 A^2}{\rho A}.$$

This is not the threshold load.

3. Superpose a plot of P_c on the preceding limit load plot.
   Does the simple model capture the essential behavior?

Part IV. Motion in the Large

The analyses of Parts II and III established the origin of the instability and imply that the beam will eventually have an infinite displacement. This physically unrealistic implication is due to these being linear analyses. The exploration of this section now investigates the motion in the large of the beam after the instability has been initiated.

   The specific objective of the data collection and analysis is to characterize the limit cycle and relate its size to the level of loading. If QED is not already running, begin by launching it from the command line in the working directory.

COLLECTING THE DATA

1. Return to the Mesh section and change the damping to `[a:damping =.02]`.
   Remesh.
2. Analyze the problem as Explicit nonlinear for the primary load levels `[1:scale{P} =175, 180, 190, 200, 210, 220]`.
   Each time use DiSPtool to observe the responses. In particular, estimate and record the steady-state response of v (this can be done by zooming on the panel and moving the cursor).
3. Return to the Mesh section and change the damping to `[a:damping =.12]`.
   Remesh.
   Analyze the problem as Explicit nonlinear for the primary load levels `[1:scale{P} = 200, 210, 220, 230, 240, 250]`.
   Each time use DiSPtool to observe the responses and record the steady-state response of v. Observe that at the higher load levels the limit cycle is no longer elliptical.

ANALYZING THE DATA

1. Make a log–log plot of v against $\Delta P = P - P_t$, where P_t is the threshold load for the corresponding value of damping.
   Do the data reasonably fall along a straight line?
2. When the beam becomes unstable, it behaves as if it has negative damping. When this damping exceeds the preexisting damping, then the large deflection ensues. The contribution from the follower force is nonlinear so the damping decreases and becomes positive. This decreases the amplitude back to where the damping is negative and the process repeats.
   The essential nonlinearity from Section 7.5, Model IV, comes from

$$\begin{Bmatrix} Q_1 \\ Q_2 \end{Bmatrix} = -\frac{PL}{2} \begin{Bmatrix} \sin(\phi_2 - \phi_1) \\ 0 \end{Bmatrix} \approx \frac{PL(1 - \frac{1}{6}\phi^2)}{2} \begin{bmatrix} 1 & -1 \\ 0 & 0 \end{bmatrix} \begin{Bmatrix} \phi_1 \\ \phi_2 \end{Bmatrix},$$

where ϕ is some averaged angle. Consider the load to be applied in three stages: P_o just causes the radical to be zero, $P_t = P_o + \Delta P$ is the threshold load to initiate the instability, and $P = P_o + \Delta P + \delta P$ is the current applied load. For small angles, the effective damping has the approximation

$$c^* = -\gamma_1 \delta P[1 - \frac{\gamma_2}{\delta P}\phi^2],$$

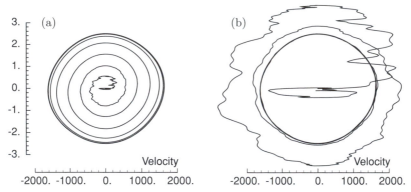

Figure 6.21. Partial results for the dynamic stability of beams; convergence to the limit cycle: (a) convergence from inside, (b) convergence from outside.

where γ_i are group constants. After the frequencies coalesce, the beam behaves as a SDoF system and has the model

$$M\ddot{\phi} - \gamma_1 \delta P[1 - \frac{\gamma_2}{\delta P}\phi^2]\dot{\phi} + K\phi = 0.$$

This is a form of the Van der Pol equation, and an analysis [30, 58] gives the solution amplitude:

$$A = \frac{2\sqrt{\delta P}}{\sqrt{\gamma_2}}.$$

A simple explanation for this solution is that damping is zero when $\phi = \sqrt{\delta P}/\sqrt{\gamma_2}$, and this amount again occurs at the extreme excursion.
Superpose, on the preceding log–log plot, lines of slope 1/2.
Do the data exhibit this 1/2 power-law behavior?
3. Conjecture about the implementation of an exploration of limit cycles under zero (or very near zero) damping.
4. Conjecture about possible monitors that, if implemented in QED, would give further insight into the origins of the dynamic instability and the formation of limit cycles.

Partial Results

Partial results for the formation of limits cycles in Part I are shown in Figure 6.21. The limit cycle acts as an attractor.

Partial results for the threshold load in Part III are shown in Figure 6.22(a). There appears to be a definite limiting threshold value of load. Partial results for the amplitude behavior in Part IV are shown in Figure 6.22(b).

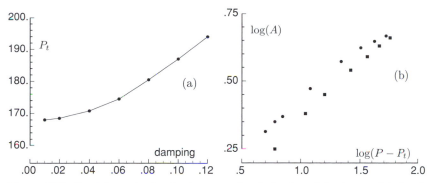

Figure 6.22. Partial results for the large motion behavior of the beam: (a) threshold load to initiate the instability, (b) amplitude response against excess load for two values of damping.

Background Readings

Aspects of the stability of systems under follower loads, that is, dynamic instability, are covered in References [133, 139], although neither of the references addresses the question of the motion in the large. Some background material on the stability under motion in the large is given in References [65, 68, 113]. Reference [77] gives a readable introduction to the subject.

7 Constructing Simple Analytical Models

When trying to understand a complex system, it is quite useful to have available some simple models – not as solutions per se but as organizational principles for seeing through the voluminous numbers produced by the FE codes. This chapter is concerned with the construction of simple analytical models; it gathers together many of the simple models used throughout the previous chapters and tries to illustrate the approach to constructing these. Although the models are approximate, by basing them on sound mechanics principles, they are more likely to capture the essential features of a problem and thus have a wider range of application. The models discussed are shown in Figure 7.1.

The term "model" is widely used in many different contexts, but here we mean a representation of a physical system that may be used to predict the behavior of the system in some desired respect. The actual physical system for which the predictions are to be made is called the *prototype*.

There are two broad classes of models: physical models and mathematical models. The physical model resembles the prototype in appearance but is usually of a different size, may involve different materials, and frequently operates under loads, temperatures, and so on, that differ from those of the prototype. The mathematical model consists of one or more equations (and, more likely nowadays, a numerical FE model) that describe the behavior of the system of interest. The equations of the mathematical model are based on certain basic laws and principles and usually involve simplifying assumptions. With the development of a valid model, it is possible to predict the immediate and future behavior of the prototype under a set of specified inputs, and to examine a priori the effect of various possible design modifications.

Usually, it is difficult to maintain exact similitude between model and prototype, and the duplication of all the details of the prototype becomes impossible. Some details of the design are obviously unimportant and may be omitted from the model confidently, but it is not obvious for other details. Whenever a model is used, it must always be borne in mind that it may not fully represent the prototype and that (physical) experimental validation is necessary.

This chapter begins by reviewing the fundamentals of solid mechanics and then introduces the powerful concept of virtual work as a means of reformulating problems in terms of stationary principles. Stationary principles lend themselves

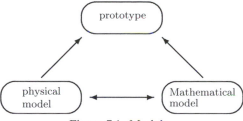

Figure 7.1. Models.

conveniently to obtaining approximate solutions and thus form a sound mechanics basis for model construction. Dimensional analysis can be used to reduce the number of variables in a problem and is therefore a good tool for correlating data as well as for the overall design of a model. Its strength is that it can be applied to highly complex phenomena requiring only a knowledge of the variables involved; its weakness is that it often has a limited predictive capability. The Ritz method is our main analytical tool used to convert the continuous systems into discrete systems. The resulting models are typically a system of (low-order) differential equations with good predictive capability. We illustrate its use on a wide range of problems. Sometimes, the role of simple models is to provide an analytical construct in which to investigate problems in a tractable way rather than describe the physical problem itself. That is, the model does not resemble the physical model but captures the essential mechanical features of a phenomenon to be investigated. These mechanical models are used to elucidate the main features of the postbuckling behavior of structures. Finally, one significant source of complexity in a problem is the modeling of the loads. Of concern here is the situation in which the given information is about the physical problem, not the loads directly, and we wish to establish a model for the actual loads. The main situations considered involve the impact of two bodies.

Many of the simple models developed are actually nonlinear and require numerical methods for their solution. Figure 5.13 shows the opening screen for QED's collection of simple models that can be interrogated in multiple ways.

7.1 Fundamentals of Solid Mechanics

The three fundamental concepts of solids mechanics are deformation and compatibility, stress and equilibrium, and the material constitutive relation. These are covered in many texts such as References [32, 38, 82]; the summary here will just extract those concepts of immediate use.

Large-Deformation Measure of Strain

Consider a common global coordinate system as shown in Figure 7.2 and associate x_i^o with the undeformed configuration and x_i with the deformed configuration. That is,

$$\text{initial position:} \quad \hat{x}^o = \sum_i x_i^o \hat{e}_i, \qquad \text{final position:} \quad \hat{x} = \sum_i x_i \hat{e}_i,$$

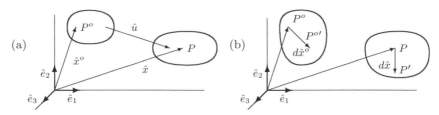

Figure 7.2. Undeformed and deformed configurations: (a) displacement, (b) deformation of neighboring points.

where the hat ($\wedge$) indicates a vector and both vectors are referred to as the common set of unit vectors $\hat{e}_i$.

A deformation is a comparison of two states and is expressed in the form

$$x_i = x_i(x_1^o, x_2^o, x_3^o, t).$$

In this Lagrangian system, all entities are expressed in terms of the initial position coordinates and time. Consider a deformation in the vicinity of the point P; that is, consider two points separated by dx_i^o in the undeformed configuration and by dx_i in the deformed configuration as shown in Figure 7.2(b). The positions of the two points are related through the Taylor series expansion

$$P: \ x_i = x_i(x_i^o),$$

$$P': \ x_i' = x_i + dx_i = x_i(x_i^o + dx_i^o) \approx x_i(x_i^o) + \sum_j \frac{\partial x_i}{\partial x_j^o} dx_j^o + \cdots + .$$

If dx_i^o is small, that is, the neighboring points are very close to each other, then

$$x_i' - x_i = dx_i \approx \sum_j \frac{\partial x_i}{\partial x_j^o} dx_j^o. \tag{7.1}$$

This describes how the separation in the deformed configuration is related to the separation in the undeformed configuration.

Strain is a measure of the stretching of the material points within a body; it is a measure of the relative displacement without rigid-body motion and is an essential ingredient for the description of the constitutive behavior of materials. The easiest way to distinguish between deformation and the local rigid-body motion is to consider the change in distance between two neighboring material particles. Let two material points before deformation have coordinates (x_i^o) and $(x_i^o + dx_i^o)$; and after deformation have the coordinates (x_i) and $(x_i + dx_i)$. The initial and final distances between these neighboring points are given by

$$dS_o^2 = \sum_i dx_i^o dx_i^o = (dx_1^o)^2 + (dx_2^o)^2 + (dx_3^o)^2,$$

$$dS^2 = \sum_i dx_i dx_i = \sum_{i,j,m} \frac{\partial x_m}{\partial x_i^o} \frac{\partial x_m}{\partial x_j^o} dx_i^o dx_j^o,$$

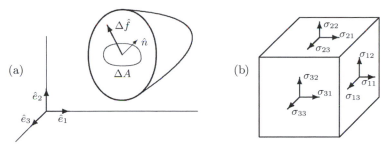

Figure 7.3. Stress concept: (a) exposed forces on an arbitrary section cut, (b) stressed cube.

respectively. Only in the event of stretching or straining is dS^2 different from dS_o^2. That is,

$$dS^2 - dS_o^2 = dS^2 - \sum_i dx_i^o dx_i^o = \sum_{i,j,m} \left[\frac{\partial x_m}{\partial x_i^o} \frac{\partial x_m}{\partial x_j^o} - \delta_{ij} \right] dx_i^o dx_j^o = \sum_{i,j} 2 E_{ij} dx_i^o dx_j^o,$$

(where δ_{ij} is the $[3 \times 3]$ unit matrix in component form) is a measure of the relative displacements. It is insensitive to rotation, as can be easily demonstrated by considering a rigid-body motion. It is also easy to observe that E_{ij} is a symmetric tensor of the second order; it is called the *Lagrangian strain tensor*.

Sometimes it is convenient to deal with displacements and displacement gradients instead of the position variables. These are obtained by using the relations

$$x_m = x_m^o + u_m, \qquad \frac{\partial x_m}{\partial x_i^o} = \frac{\partial u_m}{\partial x_i^o} + \delta_{im}.$$

The Lagrangian strain tensor E_{ij} can be written in terms of the displacement by

$$E_{ij} = \frac{1}{2} \left[\frac{\partial u_i}{\partial x_j^o} + \frac{\partial u_j}{\partial x_i^o} + \sum_m \frac{\partial u_m}{\partial x_i^o} \frac{\partial u_m}{\partial x_j^o} \right].$$

Typical expressions for E_{ij} in unabridged notations are

$$E_{11} = \frac{\partial u_1}{\partial x_1^o} + \frac{1}{2} \left[\left(\frac{\partial u_1}{\partial x_1^o} \right)^2 + \left(\frac{\partial u_2}{\partial x_1^o} \right)^2 + \left(\frac{\partial u_3}{\partial x_1^o} \right)^2 \right],$$

$$E_{22} = \frac{\partial u_2}{\partial x_2^o} + \frac{1}{2} \left[\left(\frac{\partial u_1}{\partial x_2^o} \right)^2 + \left(\frac{\partial u_2}{\partial x_2^o} \right)^2 + \left(\frac{\partial u_3}{\partial x_2^o} \right)^2 \right],$$

$$E_{12} = \frac{1}{2} \left(\frac{\partial u_1}{\partial x_2^o} + \frac{\partial u_2}{\partial x_1^o} \right) + \frac{1}{2} \left[\frac{\partial u_1}{\partial x_1^o} \frac{\partial u_1}{\partial x_2^o} + \frac{\partial u_2}{\partial x_1^o} \frac{\partial u_2}{\partial x_2^o} + \frac{\partial u_3}{\partial x_1^o} \frac{\partial u_3}{\partial x_2^o} \right]. \qquad (7.2)$$

Note the presence of the nonlinear terms in the square brackets.

Stress and Equilibrium

Consider a small surface element of area ΔA on an imagined exposed surface A in the deformed configuration as shown in Figure 7.3. There must be forces and moments acting on ΔA to make it equipollent to the effect of the rest (removed) of

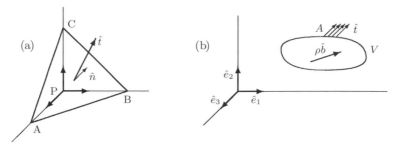

Figure 7.4. Equilibrium of volumes: (a) small tetrahedron, (b) arbitrary volume.

the material. That is, when the pieces are put back together these forces cancel each other. Let these forces be thought of as contact forces and so give rise to contact stresses (even though they are inside the body). Cauchy formalized this by introducing his concept of a traction vector.

Let $\hat{n}$ be the unit vector that is perpendicular to the surface element ΔA, and let $\Delta \hat{f}$ be the resultant force exerted from the other part of the surface element with the negative normal vector. We assume that as ΔA becomes vanishingly small, the ratio $\Delta \hat{f} / \Delta A$ approaches a definite limit $d\hat{f}/dA$. The vector obtained in the limiting process is

$$\lim_{\Delta A \to 0} \frac{\Delta \hat{f}}{\Delta A} = \frac{d\hat{f}}{dA} \equiv \hat{t}^{(\hat{n})},$$

which is called the *traction vector*. This vector represents the force per unit area acting on the surface and its limit exists because the material is assumed continuous. The superscript $\hat{n}$ is a reminder that the traction is dependent on the orientation of the area.

We simplify the notation by introducing [for the cube shown in Figure 7.3(b)]

$$\sigma_{ij} \equiv t_j^{(\hat{e}_i)},$$

where i refers to the face and j to the component. More specifically,

$$\sigma_{11} \equiv t_1^{(\hat{e}_1)}, \qquad \sigma_{13} \equiv t_3^{(\hat{e}_1)}, \qquad \sigma_{31} \equiv t_1^{(\hat{e}_3)}, \ldots.$$

The normal projections of $\hat{t}^{(\hat{n})}$ on these special faces are the normal stress components $\sigma_{11}, \sigma_{22}, \sigma_{33}$, whereas projections perpendicular to $\hat{n}$ are shear-stress components $\sigma_{12}, \sigma_{13}, \sigma_{21}, \sigma_{23}, \sigma_{31}, \sigma_{32}$.

We know that the traction vector $\hat{t}^{(\hat{n})}$ acting on an area $dA\hat{n}$ depends on the normal $\hat{n}$ of the area. The particular relation can be obtained by considering a traction on an arbitrary surface of the tetrahedron shown in Figure 7.4. On the three faces perpendicular to the coordinate directions, the components of the three traction vectors are denoted by σ_{ij}. The vector acting on the inclined surface ABC is $\hat{t}$ and the unit normal vector $\hat{n}$. The equilibrium of the tetrahedron requires that the resultant force acting on it must vanish.

This leads to the elegant relation

$$t_i = \sum_j \sigma_{ji} n_j. \qquad (7.3)$$

This compact relation says that we need know only nine numbers $[\sigma_{ij}]$ (associated with the cube shown in Figure 7.3(b) to be able to determine the traction vector on any area passing through a point. These elements are called the Cauchy stress components and form the Cauchy stress tensor. It is a second-order tensor and, as will be shown, is symmetric.

Consider an arbitrary volume V taken from the deformed body; then Newton's laws of motion become

$$\int_A t_i \, dA + \int_V \rho b_i \, dV = \int_V \rho \ddot{u}_i \, dV,$$

$$\sum_{jk} \int_A \epsilon_{ijk} x_j t_k \, dA + \sum_{jk} \int_V \epsilon_{ijk} x_j b_k \rho \, dV = \sum_{jk} \int_V \epsilon_{ijk} x_j \ddot{u}_k \rho \, dV,$$

where $\hat{t}$ is the traction on the boundary surface A, $\hat{b}$ is the body force per unit mass, and ϵ_{ijk} is the permutation symbol. These are the equations of motion in terms of t_i. We can now obtain the equations of motion in terms of the stress σ_{ij} by using $t_i = \sum_p \sigma_{pi} n_p$ and noting that, by the integral theorem,

$$\int_A t_i \, dA = \sum_p \int_A \sigma_{pi} n_p \, dA = \sum_p \int_V \frac{\partial \sigma_{pi}}{\partial x_p} dV.$$

Because the volume V is arbitrary, we conclude that the integrands must vanish and therefore

$$\sum_j \frac{\partial \sigma_{ij}}{\partial x_j} + \rho b_i = \rho \ddot{u}_i, \qquad \sigma_{ij} = \sigma_{ji}.$$

These equations of motion are written in expanded notation as, for example,

$$\frac{\partial \sigma_{11}}{\partial x_1} + \frac{\partial \sigma_{21}}{\partial x_2} + \frac{\partial \sigma_{31}}{\partial x_3} + \rho b_1 = \rho \ddot{u}_1.$$

It is worth repeating that, because of the symmetry property of the stress tensor, only six components are independent. As a result, the preceding number of independent stress components is reduced because $\sigma_{12} = \sigma_{21}, \sigma_{13} = \sigma_{31}$, and $\sigma_{23} = \sigma_{32}$. It is also worth noting that the equations of motion in terms of the Cauchy stress measure are applicable to large deformations.

Work and Strain Energy Concepts

The work done by a force at a point is the vector dot product of the force and the displacement at the point. For example, in terms of our global coordinate system,

$$dW = \hat{P} \cdot d\hat{u} = P_x du + P_y dv + P_z dw.$$

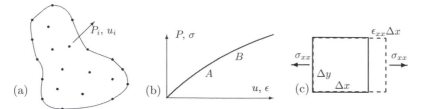

Figure 7.5. Typical elastic behavior: (a) discretized arbitrary body, (b) force-displacement behavior, (c) stressed infinitesimal element showing strain.

The systems we are interested in have multiple forces and moments, so we will generalize the preceding expression for work to

$$dW = \sum_i P_i du_i = \{P\}^T \{du\}. \tag{7.4}$$

It is understood that P_i and u_i are common (pointing in the same direction as indicated in Figure 7.5) components of generalized forces and displacements, respectively. Thus the individual contribution to the work could refer to forces and displacements $P_3 \, du_3$ or torques and twists $T_9 \, d\phi_9$, and so on.

Consider an arbitrary subvolume V of the loaded body, and let it have displacement increments du_i; then the external work increment that is due to the tractions t_i and body force b_i is

$$dW_e = \sum_i \int_A t_i du_i \, dA + \sum_i \int_V \rho b_i du_i \, dV.$$

But $t_i = \sum_j \sigma_{ij} n_j$, so substitute and use the integral theorem and equilibrium conditions to get

$$dW_e = \sum_{i,j} \int_V \sigma_{ij} d\epsilon_{ij} \, dV = d\mathcal{U}.$$

The term $\mathcal{U}$ is the strain energy so that this relation is the formal equivalence between the external work done and the internal stored energy.

We would also like to write the work expression in terms of the undeformed configuration. There are two common measures of stress in use: The Cauchy stress σ_{ij} (as just derived) is referred to the deformed configuration, whereas the Kirchhoff stress σ_{ij}^K is appropriate when Lagrangian variables are used. As shown in References [32, 38, 82], the relation between these variables is given by

$$d\epsilon_{mn} = \sum_{i,j} \frac{\partial x_i^o}{\partial x_m} \frac{\partial x_j^o}{\partial x_n} dE_{ij} , \qquad \sigma_{mn} = \sum_{i,j} \frac{\rho}{\rho^o} \frac{\partial x_m}{\partial x_i^o} \frac{\partial x_n}{\partial x_j^o} \sigma_{ij}^K.$$

The deformed and undeformed volumes are related by $dV = dV^o \rho^o / \rho$; then the work–energy relation becomes

$$dW_e = \sum_{i,j} \int_{V^o} \sigma_{ij}^K \, dE_{ij} \, dV^o = d\mathcal{U}. \tag{7.5}$$

The explicit relation between the strain energy and the components of stress and strain is

$$dU = \int_{V^o} \left[\sigma_{xx}^K dE_{xx} + \sigma_{yy}^K dE_{yy} + \sigma_{xy}^K 2 dE_{xy} + \cdots + \right] dV^o. \tag{7.6}$$

Note that there are no products such as $\sigma_{xx}^K dE_{yy}$ or $\sigma_{xx}^K dE_{xy}$ and that the shear stress multiplies twice the shear strain.

The stresses are obtained from the strains through the constitutive relation. For example, a reasonable assumption about elastic behavior is that the work done by the applied forces is transformed completely into strain (potential) energy, and this strain energy is completely recoverable. That the work is transformed into potential energy that is completely recoverable means the material system is conservative. Using material variables (Lagrangian strain and Kirchhoff stress), the increment of work done on the small volume is

$$dW_e = \int_{V^o} \left[\sum_{i,j} \sigma_{ij}^K dE_{ij} \right] dV^o.$$

The potential is composed entirely of the strain energy U; the increment of strain energy is

$$dU = dU(E_{ij}) = \int_{V^o} \left[\sum_{i,j} \frac{\partial \overline{U}}{\partial E_{ij}} dE_{ij} \right] dV^o,$$

where $\overline{U}$ is the strain energy density. From the hypothesis, we can equate dW_e and dU, and because the volume is arbitrary, the integrands must be equal; hence we have

$$\sigma_{ij}^K = \frac{\partial \overline{U}}{\partial E_{ij}}. \tag{7.7}$$

A material described by this relation is called *hyperelastic*. Note that it is valid for large deformations and for anisotropic materials.

A special case of practical interest is called linear elasticity. Organize the stresses and strains in the vector forms

$$\{\sigma^K\} \equiv \{\sigma_{11}^K, \sigma_{22}^K, \sigma_{33}^K, \sigma_{12}^K, \sigma_{23}^K, \sigma_{13}^K\}^T, \qquad \{E\} \equiv \{E_{11}, E_{22}, E_{33}, 2E_{12}, 2E_{23}, 2E_{13}\}^T;$$

then an example constitutive relation can be written in the matrix form

$$\{\sigma^K\} = [D]\{E\},$$

where $[D]$ is of size $[6 \times 6]$. Because of the symmetry of both the stress and strain, we have $[D]^T = [D]$. Special materials are reduced forms of this relation; for example, an isotropic material is described by

$$E_{11} = \frac{1}{E}[\sigma_{11}^K - v(\sigma_{22}^K + \sigma_{33}^K)], \qquad 2E_{12} = \frac{2(1+v)}{E}\sigma_{12}^K,$$

$$\sigma_{11}^K = \frac{E}{(1+v)(1-2v)}[(1-v)E_{11} + v(E_{22} + E_{33})], \qquad \sigma_{12}^K = \frac{E}{2(1+v)}2E_{12}, \tag{7.8}$$

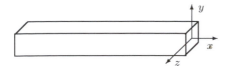

Figure 7.6. A slender structural member in local coordinates.

where E is the Young's modulus, v is the Poisson's ratio, and the relations for the other components are obtained by permuting the subscripts. This is called *Hooke's law*. A more complicated nonlinear relation for rubber materials is discussed in Section 5.3.

Energies for Particular Systems

As part of constructing simple models, we often begin by reducing the dimensionality of the body. For example, a plate has one dimension much smaller than the other two and a beam has one dimension much larger than the other two as shown in Figure 7.6. We now state the energies for these systems.

Assume the material relation is linear, and, on a local level, the deformations (strains and displacements) are small. Also consider the initial state to be the zero-strain zero-stress state, then the strain energy given in Equation (7.6) becomes

$$\mathcal{U} = \tfrac{1}{2} \int_V [\sigma_{xx}\epsilon_{xx} + \sigma_{yy}\epsilon_{yy} + \tau_{xy}\gamma_{xy} + \cdots +] dV = \tfrac{1}{2} \int_V \{\sigma\}^T \{\epsilon\} dV. \qquad (7.9)$$

This can be put in the alternative form by using Hooke's law:

$$\mathcal{U} = \tfrac{1}{2} \int_V \{\epsilon\}^T [D]\{\epsilon\} dV = \tfrac{1}{2} \int_V \{\sigma\}^T [C]\{\sigma\} dV. \qquad (7.10)$$

The work done by the external forces is

$$\mathcal{W}_e = \tfrac{1}{2}\{P\}^T\{u\} = \tfrac{1}{2}\sum_i P_i u_i.$$

In the following discussion, these relations will be particularized to the structural systems of interest by writing the distributions of stress and strain in terms of resultants.

The strain energy expression for *plane stress*, for example, reduces to

$$\mathcal{U} = \tfrac{1}{2} \int_V [\sigma_{xx}\epsilon_{xx} + \sigma_{yy}\epsilon_{yy} + \tau_{xy}\gamma_{xy}] dV$$

$$= \tfrac{1}{2} \int_A \frac{Et}{(1 - v^2)} [\epsilon_{xx}^2 + \epsilon_{yy}^2 + 2v\epsilon_{xx}\epsilon_{yy}] dx dy + \tfrac{1}{2} \int_A Gt\gamma_{xy}^2 \, dx dy,$$

where A is the area of the plate and t is the thickness. This is the expression used in the analysis of plates.

The simplest of all elastic structural members is the spring. Let F be the force in the spring and u the resulting displacement; then $F = Ku$ and

$$\text{spring:} \qquad \mathcal{U}_E = \tfrac{1}{2}\frac{F^2}{K} = \tfrac{1}{2}Ku^2.$$

For the rod–truss member, there is only an axial stress present, and it is uniformly distributed on the cross section. Let F be the resultant force; then $\sigma_{xx} = F/A = E\epsilon_{xx}$ and

$$\text{axial:} \qquad \mathcal{U}_E = \frac{1}{2}\int_0^L \frac{F^2}{EA}dx = \frac{1}{2}\int_0^L EA[\frac{du}{dx}]^2\,dx.$$

For the beam–frame member in bending, there is only an axial stress, but it is distributed linearly on the cross section in such a way that there is no resultant axial force. Let M be the resultant moment; then $\sigma_{xx} = -My/I = E\epsilon_{xx}$ and

$$\text{bending:} \qquad \mathcal{U}_E = \frac{1}{2}\int_0^L \frac{M^2}{EI}dx = \frac{1}{2}\int_0^L EI[\frac{d^2v}{dx^2}]^2\,dx,$$

where I is the second moment of area. For a circular shaft in torsion, the shear stress is linearly distributed on the radius, giving $\tau = M_x r/J = G\gamma$ and

$$\text{torsion:} \qquad \mathcal{U}_E = \frac{1}{2}\int_0^L \frac{M_x^2}{GJ}dx = \frac{1}{2}\int_0^L GJ[\frac{d\phi}{dx}]^2\,dx,$$

where ϕ is the twist per unit length, M_x is the resultant axial torque, G is the shear modulus, and J is the polar moment of area.

Note that all strain energy expressions are of the form

$$\mathcal{U}_E = \frac{1}{2}\int_L \frac{\text{load}^2}{\text{stiffness}}dx = \frac{1}{2}\int_L \text{stiffness} \times \text{deformation}^2\,dx.$$

Different particular structures will have their appropriate resultant load, deformation, and stiffness per unit length.

The elastic stiffness is based on the elastic modulus material property; however, the total stiffness of a structure or structural component also depends on the level of stress. This is a second-order effect often neglected in linear analyses but is of profound significance in nonlinear and stability analyses. To see how stress can affect the stiffness, consider the strain energy in a member (beam–frame or rod–truss) slightly deflected off its axis. The Lagrangian strain in a beam bending in the $x - y$ plane is

$$E_{xx} = \frac{\partial u}{\partial x^o} + \frac{1}{2}\left[(\frac{\partial u}{\partial x^o})^2 + (\frac{\partial v}{\partial x^o})^2\right].$$

The kinematics of thin-beam theory gives

$$u(x^o, y^o) \approx u(x^o) - y^o\frac{\partial v}{\partial x^o}, \qquad v(x^o, y^o) \approx v(x^o).$$

Substitute these into the strain–displacement relation to get

$$E_{xx} = \frac{\partial u}{\partial x^o} + \frac{1}{2}\left[(\frac{\partial u}{\partial x^o})^2 + (\frac{\partial v}{\partial x^o})^2\right] - y^o\frac{\partial^2 v}{\partial x^{o2}} + \frac{1}{2}\left[(-y^o\frac{\partial^2 v}{\partial x^{o2}})^2\right]$$

$$\approx \frac{\partial u}{\partial x^o} - y^o\frac{\partial^2 v}{\partial x^{o2}} + \frac{1}{2}(\frac{\partial v}{\partial x^o})^2.$$

The approximation is based on the assumption that the transverse deflection is significantly larger than that of the axial displacement. We recognize the leading term as that of the linear theory of a rod–truss; the second as that of the linear flexure theory of a beam–frame; the third term, which is nonlinear, comes from the large deflection of the member. This third term remains even if we set $y^o = 0$ in order to recover the rod–truss behavior. This axial strain gives rise to an axial stress and, assuming the linear relationship $\sigma_{xx}^K = E E_{xx}$, gives the strain energy as

$$\mathcal{U} = \tfrac{1}{2} \int_{V^o} \sigma_{xx}^K E_{xx} \, dV^o = \tfrac{1}{2} \int_{V^o} E E_{xx}^2 \, dA \, dx^o.$$

After substitution and integration we are left with the three terms

$$\mathcal{U} = \tfrac{1}{2} \int_L E I_{zz} \Big(\frac{\partial^2 v}{\partial x^{o2}}\Big)^2 dx^o + \tfrac{1}{2} \int_L E A \Big(\frac{\partial u}{\partial x^o}\Big)^2 dx^o + \tfrac{1}{2} \int_L \bar{F}_o \Big(\frac{\partial v}{\partial x^o}\Big)^2 dx^o. \quad (7.11)$$

In this, we introduced the term

$$\bar{F}_o = E A \frac{\partial u}{\partial x^o},$$

which corresponds to an axial load. Strictly speaking, this is deformation dependent, but in the construction of simple models it is assumed to be preexisting in the structure in the form of a prestress. Thus the energy expression comprises the contribution for the elastic behavior (axial and bending) plus a term associated with the axial loading and transverse deflection; this third term is referred to as the *geometric strain energy* and is designated by $\mathcal{U}_G$.

7.2 Stationary Principles in Mechanics

In principle, the use of compatibility, equilibrium, and material behavior will always yield a solution. In practice, however, their direct use on complex problems is invariably too awkward to effect a solution. For this reason, these fundamental principles are usually first manipulated into more useful forms for application. This section reviews the use of stationary principles (sometimes called *variational principles*) in formulating mechanics problems.

The fundamental principle on which the stationary principles are based is that of virtual work. This principle is essentially an alternative statement of equilibrium; what makes it effective in formulating problems (and hence constructing simple models) is that it does not use free-body equilibrium diagrams and therefore is suitable for approximations. Good background material is developed in References [72, 74, 112, 136].

Principle of Virtual Work

References [27, 30, 32] give general derivations of the principle of virtual work; here we will be content with a simpler motivational derivation.

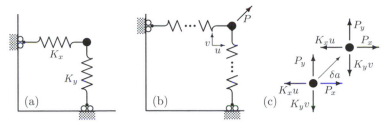

Figure 7.7. Three stages of loading: (a) no load, unstrained, (b) loaded, elastic straining, (c) free bodies of the particle under virtual displacement, loads unchanged.

Consider a particle in a state of equilibrium under the action of a set of forces as shown in Figure 7.7(b). The particle is attached to linear springs that are free to move on rollers as indicated, thus always keeping their original orientation. For generality, let there also be acting on the particle a body force $\hat{F}^b$ (such as gravity). The particle achieves a static equilibrium position under the displacements u and v, also as shown in Figure 7.7(b). The two equilibrium conditions are

$$P_x + F_x^b - K_x u = 0, \qquad P_y + F_y^b - k_y v = 0.$$

These are for the free body of the particle separated from the springs as shown in the lower left of Figure 7.7(c).

Now imagine that the particle is displaced a small amount δa and the forces move with it, but with no change in the magnitude or the direction of any of the forces. It is important to understand that we are not speaking of the actual displacements of the particle caused by the action of the applied forces; we are speaking of an imagined displacement in which the forces are imagined to behave as stated. For this reason, the displacement δa is referred to as a *virtual displacement* to distinguish it from the actual real displacements. Throughout this section, the symbol δ before any quantity will indicate that it is a virtual (or imagined) variation of the quantity.

The work done on the particle during the virtual displacement is the sum of all the components of force in the direction of δa, times δa. This is called the *virtual work*. Let the virtual displacement have components $\delta u, \delta v$ in the coordinate directions; we have for the virtual work

$$\delta W = [P_x + F_x^b - K_x u]\delta u + [P_y + F_y^b - K_y v]\delta v. \qquad (7.12)$$

By virtue of the fact that the particle is already in equilibrium (that is, the terms in brackets are zero) we conclude that the total virtual work is zero:

$$\delta W = 0.$$

In Equation (7.12), we identify three types of virtual work: $P_x \delta u$ is the work of the externally applied load, $F_x^b \delta u$ is the work of the body forces, and $K_x u \delta u$ is the internal work done by the spring. We lump the first two types together and call them external work δW_e. The internal work we call strain energy δU because

$$K_y u\, \delta u = \delta[\tfrac{1}{2} K_y u^2] = \delta U.$$

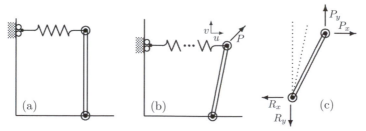

Figure 7.8. Three stages of loading: (a) no load, unstrained, (b) loaded, elastic straining, (c) free body after virtual displacement, loads unchanged.

(Note that δ, as an operator, can be treated as similar to a derivative.) Thus our virtual work statement becomes

$$\delta \mathcal{W} = \delta \mathcal{W}_e - \delta \mathcal{U} = 0 \qquad \text{or} \qquad \delta \mathcal{W}_e = \delta \mathcal{U}. \qquad (7.13)$$

This almost resembles a statement of conservation of energy; that is, the work done by the external forces is equal to the energy stored by the elastic body. But, as we will see, because of the δ operator, the virtual work principle is profoundly richer than the conservation of energy.

Thus, if a particle is in equilibrium under the action of a set of forces, the total virtual work done by the forces during a virtual displacement is zero. Note that the converse statement is not true; just because the virtual work is zero we cannot conclude that the body is in equilibrium. (This is easily demonstrated for a particle acted on by a vertical gravitational force but given a virtual displacement in the horizontal direction.) It is correct to say that if the virtual work is zero *for every possible* virtual displacement then the body is in equilibrium, but this would be an impractical approach to testing for equilibrium.

We can simplify matters considerably by introducing the idea of *independent* virtual displacements; an arbitrary displacement of a point can always be written as the sum of three independent displacements. This leads to the much more useful statement that if the virtual work is zero *for every independent* virtual displacement then the body is in equilibrium. With this seemingly simple result we have managed to interpret Equation (7.12) as a condition that defines equilibrium rather than a result of equilibrium. Furthermore, some unknown forces that would appear in a free-body diagram will not appear in the virtual work statement if they do no work. For example, the reactions at the pinned boundary of Figure 7.8 appear in the free body but do no work because the real and virtual displacements are constrained to be always zero.

These simple examples identify the important features for the general statement of virtual work. First note, with reference to Figures 7.7 and 7.8, that there are three states of the particle:

A, an original state in which there are no loads and the springs are unstrained,
B, a final equilibrium state after the loads are applied and the springs have been stretched, and

C, an imaginary state in which a virtual displacement is present but geometric constraints are not violated.

We emphasize that the true displacement of the structure is the difference in displacement between States A and B, and that the virtual displacement is the difference between States B and C.

To generalize this idea, suppose the deformation of the structure can be described in terms of N independent parameters $u_1, u_2, \ldots, u_N$, (they could be the displacements of the junctions of a multispring system, for example) then, by the chain rule for differentiation, the virtual strain energy can be written as

$$\delta \mathcal{U} = \delta \mathcal{U}(u_1, u_2, \ldots, u_N) = \frac{\partial \mathcal{U}}{\partial u_1}\delta u_1 + \frac{\partial \mathcal{U}}{\partial u_2}\delta u_2 + \cdots + = \sum_i^N \frac{\partial \mathcal{U}}{\partial u_i}\delta u_i.$$

The virtual work of the applied loads can be written as

$$\delta W_e = P_1\delta u_1 + P_2\delta u_2 + \cdots + = \sum_i^N P_i\delta u_i,$$

where P_i are the forces acting parallel to the displacements u_i (some of which could be zero). These lead to the principle of virtual work being rewritten as

$$\delta W = \delta \mathcal{U} - \delta W_e = \sum_i^N \left[\frac{\partial \mathcal{U}}{\partial u_i} - P_i\right]\delta u_i = 0.$$

Because this must be true for each independent variation δu_i, we conclude that

$$\frac{\partial \mathcal{U}}{\partial u_i} - P_i = 0, \qquad i = 1, 2, \ldots, N. \tag{7.14}$$

In this way, the single virtual work statement becomes N simultaneous equations.

The virtual work form of equilibrium for a general nonlinear body may be stated [using Equation (7.5)] as [5, 30]

$$\delta W = \delta W_e - \delta \mathcal{U} = \delta W_e - \sum_{p,q} \int_{V^o} \sigma_{pq}^K \delta E_{pq} dV^o = 0. \tag{7.15}$$

This virtual work form is completely general, but there are further developments that are more convenient to use in some circumstances. Some of these developments are now summarized.

Conservative Systems

A system is *conservative* if the work done in moving the system around a closed path is zero. An elastic body (even if nonlinear) is an example of such a system; elastic-plastic behavior is nonconservative. *Follower loads*, such as a jet nozzle reaction and pressure on a deforming body, are also nonconservative.

Reconsider the virtual strain energy contribution to the total virtual work written as

$$-\delta \mathcal{U}(u_1, u_2, \ldots, u_N) = -\sum_i^N \frac{\partial \mathcal{U}}{\partial u_i}\delta u_i \equiv -\sum_i^N F_i\delta u_i = \delta W_i.$$

Thus the derivatives

$$F_i \equiv -\frac{\partial \mathcal{U}}{\partial u_i}$$

have the interpretation of forces and $\mathcal{U}$ has the interpretation of a potential. Indeed, all conservative loads can be derived from a *potential* according to

$$P_i = -\frac{\partial \mathcal{V}}{\partial u_i}, \qquad \mathcal{V} = \mathcal{V}(u_1, u_2, \ldots, u_N),$$

where $\mathcal{V}$ is a possibly nonlinear function of all of the DoFs. The external virtual work term now becomes

$$\delta \mathcal{W}_e = -\sum_i P_i \delta u_i = -\sum_i \frac{\partial \mathcal{V}}{\partial u_i} \delta u_i = -\delta \mathcal{V}.$$

The principle of virtual work can be rewritten as

$$\delta \mathcal{W} = -\delta \mathcal{V} - \delta \mathcal{U} = 0 \qquad \text{or} \qquad \delta \Pi \equiv \delta[\mathcal{U} + \mathcal{V}] = 0. \tag{7.16}$$

The term inside the brackets is called the *total potential energy*. This relation is called the *principle of stationary potential energy*. We may now restate the principle of virtual work:

> *For a conservative system to be in equilibrium, the first-order variation in the total potential energy must vanish for every independent admissible virtual displacement.*

Another way of stating this is that, among all the displacement states of a conservative system that satisfy compatibility and the boundary constraints, those that also satisfy equilibrium make the total potential energy stationary.

For discrete systems,

$$\delta \Pi \equiv \delta[\mathcal{U} + \mathcal{V}] = \sum_i^N \frac{\partial}{\partial u_i}\Big[\mathcal{U} + \mathcal{V}\Big]\delta u_i = 0.$$

Because the δu_i can be varied arbitrarily, the principle of stationary potential energy in terms of generalized coordinates leads to

$$\mathcal{F}_i = \frac{\partial}{\partial u_i}\Big[\mathcal{U} + \mathcal{V}\Big] = 0 \qquad \text{for} \quad i = 1, 2, \ldots, N. \tag{7.17}$$

The expression $\mathcal{F}_i = 0$ is our statement of static equilibrium for conservative systems. This is a system of N equations for the N unknown DoFs. Thus, as pointed out before, in comparison with the conservation of energy theorem, this is much richer, because instead of just one equation, it leads to as many equations as there are unknown DoFs.

Dynamical Systems

D'Alembert's principle converts a dynamic problem into an equivalent problem of static equilibrium by treating the inertia as a body force. That is, the total body force comprises $\rho^o b_i \rightarrow (\rho^o b_i - \rho^o \ddot{u}_i)$ where $-\rho^o \ddot{u}_i dV^o$ is the inertia force of an infinitesimal volume. The virtual work of the body force is

$$\delta \mathcal{W}^b = \sum_i \int_{V^o} \rho^o b_i \delta u_i \, dV - \sum_i \int_{V^o} \rho^o \ddot{u}_i \delta u_i \, dV^o.$$

In writing this relation, we suppose that the performance of the virtual displacement consumes no time; that is, the real motion of the system is stopped while the virtual displacement is performed. Consequently, the time variable is conceived to remain constant while the virtual displacement is executed.

With the introduction of the concept of *kinetic energy* defined as

$$\mathcal{T} \equiv \tfrac{1}{2} \sum_i \int_{V^o} \rho^o \dot{u}_i \dot{u}_i dV^o \qquad \text{such that} \qquad \delta \mathcal{T} \equiv \sum_i \int_{V^o} \rho^o \dot{u}_i \, \delta \dot{u}_i \, dV^o,$$

the principle of virtual work becomes

$$\delta \mathcal{W} = \delta \mathcal{W}^s + \delta \mathcal{W}^b - \delta \mathcal{U} + \delta \mathcal{T} - \frac{d}{dt} \sum_i \int_{V^o} (\dot{u}_i \delta u_i) \rho^o dV^o = 0.$$

Hamilton disposed of the last term in this relation by integrating the equation over time and stipulating that the configuration has no variations at the extreme times. In the special case in which the applied loads, both body forces and surface tractions, can be derived from a scalar potential function $\mathcal{V}$, the variations become complete variations and we can write

$$\delta \int_{t_1}^{t_2} [\mathcal{T} - (\mathcal{U} + \mathcal{V})] dt = 0. \tag{7.18}$$

This equation is Hamilton's principle [52].

Hamilton's principle provides a complete formulation of a dynamical problem; however, to obtain solutions to some problems, the Hamilton integral formulation must be converted into one or more differential equations of motion. For a computer solution, these must be further reduced to equations using discrete unknowns.

Accept that we can write a function as $u = u(u_1, u_2, \ldots, u_N)$, where u_i are the generalized coordinates. The time derivative of such a function is

$$\dot{u} = \sum_{j=1}^{N} \frac{\partial u}{\partial u_j} \dot{u}_j.$$

Consequently, we see that the kinetic energy is a function of the following form:

$$\mathcal{T} = \mathcal{T}(u_1, u_2, \ldots, u_N; \dot{u}_1, \dot{u}_2, \ldots, \dot{u}_N).$$

The variation in the kinetic energy is given by

$$\delta \mathcal{T} = \sum_{j=1}^{N} \frac{\partial \mathcal{T}}{\partial u_j} \delta u_j + \sum_{j=1}^{N} \frac{\partial \mathcal{T}}{\partial \dot{u}_j} \delta \dot{u}_j.$$

Figure 7.9. Possible instantaneous shapes for a moving fixed–fixed cable. The end conditions cannot change.

We can use integration by parts on the second term to obtain

$$\int_{t_1}^{t_2} \delta \, \mathcal{T} \, dt = \int_{t_1}^{t_2} \sum_{j=1}^{N} \left\{ \frac{\partial \mathcal{T}}{\partial u_j} - \frac{d}{dt} \frac{\partial \mathcal{T}}{\partial \dot{u}_j} \right\} \delta u_j \, dt,$$

where we used the fact that the variations at the extreme times are zero.

The total potential of the conservative forces is a function of the form

$$\mathcal{U} + \mathcal{V} = \Pi = \Pi(u_1, u_2, \ldots, u_N), \qquad \delta \Pi = \sum_{j=1}^{N} \frac{\partial \Pi}{\partial u_j} \delta u_j.$$

Additionally, we have that the virtual work of the nonconservative forces is given by

$$\delta \, \mathcal{W}^d = \sum_{j=1}^{N} Q_j \delta u_j.$$

Hamilton's extended principle (in which the variation is inside the integral) now takes the form

$$\int_{t_1}^{t_2} \sum_{j=1}^{N} \left\{ -\frac{d}{dt} \left(\frac{\partial \mathcal{T}}{\partial \dot{u}_j} \right) + \frac{\partial \mathcal{T}}{\partial u_j} - \frac{\partial (\mathcal{U} + \mathcal{V})}{\partial u_j} + Q_j \right\} \delta u_j \, dt = 0.$$

Because the virtual displacements δu_j are independent and arbitrary, and because the time limits are arbitrary, then each integrand is zero. This leads to the *Lagrange's equation of motion*:

$$\mathcal{F}_i \equiv \frac{d}{dt} \left(\frac{\partial \mathcal{T}}{\partial \dot{u}_i} \right) - \frac{\partial \mathcal{T}}{\partial u_i} + \frac{\partial}{\partial u_i} \left(\mathcal{U} + \mathcal{V} \right) - Q_i = 0 \qquad (7.19)$$

for $i = 1, 2, \ldots, N$. The expression $\mathcal{F}_i = 0$ is our statement of (dynamic) equilibrium. It is apparent from Lagrange's equation that, if the system is not in motion, then we recover the principle of stationary potential energy expressed in terms of generalized coordinates.

Ritz Method for Continuous Systems

Consider the simple cable shown in Figure 7.9. At instant t_1, we could describe the transverse deflection with the polynomial

$$v(x, t_1) = a_0 + a_1 x + a_2 x^2 + \cdots + .$$

At a later instant t_2, it has a different shape and hence could be described by the polynomial

$$v(x, t_2) = b_0 + b_1 x + b_2 x^2 + \cdots + .$$

In fact, at each time we have a different shape and therefore can think of the polynomial coefficients changing in time, that is,

$$v(x, t) = a_0(t) + a_1(t)x + a_2(t)x^2 + \cdots + .$$

What is noticed, however, is that at no time is the polynomial representation allowed to violate the fixed boundary conditions at the ends. That is, the geometric boundary conditions cannot be violated.

To generalize this idea, assume an expansion for the displacements in the form

$$u(x, y, z; t) = \sum_{i}^{N} a_i(t) g_i(x, y, z),$$

where the $g_i(x, y, z)$ are linearly independent trial functions and individually satisfy the geometric boundary conditions. The coefficients a_i are the generalized coordinates; here they are considered functions of time. To ensure that the characteristics of the system are taken into consideration, the trial functions must satisfy the essential (geometric) boundary conditions of the problem. Let the preceding expression be represented by matrix expressions in the form

$$u(x, y, z; t) = \{g(x, y, z)\}^T \{a(t)\}, \qquad \dot{u}(x, y, z; t) = \{g(x, y, z)\}^T \{\dot{a}(t)\}.$$

Note that the functions $\{g(x, y, z)\}$ do not change in time. The strain is given by

$$\{E(x)\} = \{\partial g(x)\}^T \{a\},$$

where the notation $\{\partial g(x)\}$ indicates the appropriate spatial derivatives of the Ritz functions (in the case of linear rods it is $\partial/\partial x$). The strain energy (in the linear case) is expressed as

$$\mathcal{U} = \tfrac{1}{2} \int_{V^o} E E_{xx}^2 \, dV^o = \tfrac{1}{2} \int_{V^o} \{a\}^T \{\partial g\} E \{\partial g\}^T \{a\} \, dV^o = \tfrac{1}{2} \{a\}^T [K] \{a\},$$

where

$$[K] \equiv \int_{V^o} \{\partial g\}^T E \{\partial g\} \, dV^o = [N \times N]$$

is the stiffness matrix. The kinetic energy is

$$\mathcal{T} = \tfrac{1}{2} \int_{V^o} \dot{u}^2 \, dV^o = \tfrac{1}{2} \int_{V^o} \rho^o \{\dot{a}\}^T [g]^T [g] \{\dot{a}\} \, dV^o = \tfrac{1}{2} \{\dot{a}\}^T [M] \{\dot{a}\},$$

where

$$[M] \equiv \int_{V^o} \rho^o [g]^T [g] \, dV^o$$

is the mass matrix. The potential of the applied loads is

$$\mathcal{V} = -\sum_i P_i u(x_i) = -\sum_i P_i \{g_j(x_i)\}^T \{a\} = -\{P\}[g]^T \{a\} = -\{\tilde{P}\} \{a\}.$$

The notation $g_j(x_i)$ means that each Ritz function is evaluated at the position where the load P_i is applied. Similar expressions are obtained for other loads associated with the other components of displacement. Because, for a given structure, all functions within these integrals are known functions of the coordinates, the indicated integrations can be carried out either explicitly or numerically.

The Lagrange's equation of motion are

$$\mathcal{F}_i = \frac{d}{dt}\left(\frac{\partial \mathcal{T}}{\partial \dot{a}_i}\right) - \frac{\partial \mathcal{T}}{\partial a_i} + \frac{\partial(\mathcal{U}+\mathcal{V})}{\partial a_i} - Q_i = 0 \qquad \text{for } i = 1, 2, \ldots, N,$$

which then leads to

$$[K]\{a\} + [C]\{\dot{a}\} + [M]\{\ddot{a}\} = \{\tilde{P}\} + \{Q\}.$$

The Ritz method has thus lead to a system of linear equations to determine the generalized coordinates. The properties of the continuous system are represented by the matrices $[K]$, $[C]$, $[M]$, which are obtained by quadratures. Examples of the use of the Ritz approach to constructing simple analytical models are given in Section 7.4.

Finite Element Formulation

The Ritz method is formalized for computers in the form of the FE method. Introduce the notation

$$\delta E_{kl} \rightarrow \{6 \times 1\} = \{\delta E\} = \{\delta E_{11},\ \delta E_{22},\ \delta E_{33},\ 2\delta E_{12},\ 2\delta E_{23},\ 2\delta E_{13}\}^T,$$

$$\sigma_{kl}^K \rightarrow \{6 \times 1\} = \{\sigma^K\} = \{\sigma_{11}^K,\ \sigma_{22}^K,\ \sigma_{33}^K,\ \sigma_{12}^K,\ \sigma_{23}^K,\ \sigma_{13}^K\}^T,$$

so that the integrand of the internal virtual work can be written as

$$\{\sigma^K\}^T\{\delta E\}.$$

The variation of the Lagrangian strain is given by

$$2\delta E_{ij} = \frac{\partial \delta u_i}{\partial x_j^o} + \frac{\partial \delta u_j}{\partial x_i^o} + \sum_k \left[\frac{\partial u_k}{\partial x_i^o}\frac{\partial \delta u_k}{\partial x_j^o} + \frac{\partial \delta u_k}{\partial x_i^o}\frac{\partial u_k}{\partial x_j^o}\right].$$

Let the displacements in a small region of volume V_m^o be represented by

$$u_i(x_1^o, x_2^o, x_3^o) = \sum_k h_k(x_1^o, x_2^o, x_3^o)u_{ik} = [h]\{u_i\} \quad \text{or} \quad \{u\} = [H]\{u\},$$

where $h_k(x_1^o, x_2^o, x_3^o)$ are known shape (or interpolation) functions and u_{ik} are unknown nodal values. All relevant quantities can now be written in terms of both of these. For example, the derivatives and variations of derivatives are given by

$$\frac{\partial u_i}{\partial x_j^o} = \sum_k \frac{\partial h_k}{\partial x_j^o}u_{ik} = [\partial h]\{u\}, \qquad \delta\frac{\partial u_i}{\partial x_j^o} = \sum_k \frac{\partial h_k}{\partial x_j^o}\delta u_{ik} = [\partial h]\{\delta u\}.$$

Hence the variation of strain can be written symbolically as

$$\{\delta E\} = [B_E]\{\delta u\},$$

where $[B_E]$ contains various spatial derivatives of h_k as well as the nodal displacement values $\{u\}$.

The internal virtual work becomes

$$\sum_{i,j} \int_{V^o} \sigma_{ij}^K \delta E_{ij} dV^o = \int_{V^o} \{\delta E\}^T \{\sigma^K\} dV^o$$

$$\Rightarrow \quad \sum_m \{\delta u\}_m^T \Big\{ \int_{V_m^o} [B_E]^T \{\sigma^K\} dV_m^o \Big\} = \sum_m \{\delta u\}_m^T \{F\}_m = \{\delta u\}^T \{F\},$$

where $\{F\}$ is the assemblage of all the element forces $\{F\}_m$ given by

$$\{F\} = \sum_m \int_{V_m^o} [B_E]^T \{\sigma^K\} dV_m^o. \tag{7.20}$$

The force vector $\{F\}$ is interpreted as the set of nodal forces that are (virtual) work equivalent to the actual distributed stresses and strains. The integral is done for each element, and the assemblage process forms the vector $\{F\}$ of size $\{N \times 1\}$.

Equation (7.20) is the fundamental relation in the FE analysis of nonlinear mechanics. It is the nonlinear equivalent of the linear stiffness relation

$$\{F\} = [K]\{u\}, \qquad [K] = \sum_m [k]_m.$$

In nonlinear mechanics, the stiffness has a more refined meaning than implied in linear analyses, as seen shortly.

The virtual work of the applied body forces (without inertia) and surface tractions leads to

$$\sum_i \int_{V^o} \rho f_i \delta u_i \, dV^o + \sum_i \int_{A^o} t_i \delta u_i \, dA^o$$

$$\Rightarrow \quad \{P\} = \Big\{ \int_{V^o} [H]^T \{f\} dV^o \Big\} + \Big\{ \int_{A^o} [H]^T \{t\} dA^o \Big\}.$$

The virtual work of the inertia loads leads to

$$-\int_V (\rho^o \ddot{u}_i + \eta^o \dot{u}_i)\delta u_i dV \quad \Rightarrow \quad -[M]\{\ddot{u}\} - [C]\{\dot{u}\}$$

$$\text{or} \quad -\sum_m \Big[\int_{V_m^o} \rho [H]^T [H] dV^o \Big]\{\ddot{u}\} - \sum_m \Big[\int_{V_m^o} \eta [H]^T [H] dV^o \Big]\{\dot{u}\}.$$

Assemblage is done as in the linear case, and these give the equations of motion as

$$[M]\{\ddot{u}\} + [C]\{\dot{u}\} = \{P\} - \{F\}. \tag{7.21}$$

This is our governing equation of motion for a general nonlinear system discretized in the form of FEs. Note that the mass and damping matrices are determined initially, and the load distribution is also determined initially (although its history will change); only the body-stress term $\{F\}$ needs to be updated.

We can view the FE method as an application of the Ritz method in which, instead of the trial functions spanning the complete domain, the individual functions span only subdomains (the FEs) of the complete region. In order that a FE solution should be a Ritz analysis, it must satisfy the essential boundary conditions. However, in the selection of the displacement functions, no special attention need be given to the natural boundary conditions because these conditions are imposed with the load vector and are satisfied approximately in the Ritz solution. The accuracy with which these natural boundary conditions are satisfied depends on the specific trial functions used and on the number of elements used to model the problem.

The variation of the nodal forces leads to

$$\{\delta F\} = [\frac{\partial F}{\partial u}]\{\delta u\} = [K_T]\{\delta u\},$$

where $[K_T]$ is the *tangent* (or *total*) stiffness of the system. This is the matrix we wish to establish in explicit form. Substituting for $\{F\}$ in terms of the stress leads to

$$\{\delta F\} = \sum_m \int_{V_m^o} \left[[B_E]^T\{\delta\sigma^K\} + [\delta B_E]^T\{\sigma^K\} \right] dV_m^o.$$

The stresses are a function of the strains so that the first term becomes

$$[B_E]^T\{\delta\sigma^K\} = [B_E]^T[\frac{\partial\sigma_i^K}{\partial E_j}]\{\delta E\} = [B_E]^T[D][B_E]\{\delta u\}.$$

The matrix $[D]$ is the tangent modulus for the material. The strain operator is considered a function of the displacement gradients so that the second term becomes

$$[\delta B_E]^T\{\sigma^K\} = \{\delta u,_x\}^T[\frac{\partial B_E}{\partial u,_x}]^T\{\sigma^K\} = \{\delta u\}^T[B_D]^T[\sigma^K]\{B_D\}.$$

We therefore have for the total stiffness

$$[K_T] = \sum_m \int_{V_m^o} \left[[B_E]^T[D][B_E] + [B_D]^T[\sigma^K]\{B_D\} \right] dV_m^o = [K_E] + [K_G].$$

The integral is over the element volume V_m^o, and the summation is associated with the assemblage of the collection of elements. We see that the total stiffness relation comprises two parts: One is related to the tangent modulus properties of the material; the other is related to the current value of stress. The first matrix is often called the elastic stiffness because in the linear case it is primarily a function of the elastic material properties. The second matrix is called the *initial stress matrix* because in the linear case it depends on the stress and not on the material properties. The combination of both matrices is sometimes referred as the *tangent stiffness*. For nonlinear problems, both matrices depend on the stress and current deformation with the first distinctly related to the material tangent modulus $\partial\sigma_{ij}^K/\partial E_{pq}$ and the second distinctly related to the changing geometry of the element (through δB_E). It would seem appropriate to call the first matrix the tangent stiffness and the second the *geometric stiffness*; we refer to the combination as the *total stiffness*. All matrices are symmetric.

References [32–34] give explicit forms for the matrices appearing in this FE formulation.

7.3 Models, Similitude, and Dimensional Analysis

Buckingham's Π (pi) theorem states [73, 114, 135] "If an equation is dimensionally homogeneous, it can be reduced to a relationship among a complete set of dimensionless products." A corollary can be stated as "The number of independent dimensionless products required is equal to the number of variables involved minus the number of primary dimensions involved." This theorem can be used to reduce the number of variables in a problem, but it does not yield a complete solution nor does it reveal the complete character of a problem. We illustrate its use on a few typical problems and show how, to some degree, it can yield simple models.

The use of dimensional analysis, both as a tool for correlating data and as a basis for the design of models, is well established and thoroughly documented in the literature – a classic work in the area is Reference [73]. A strength of dimensional analysis is that it can be applied to highly complex phenomena requiring only a knowledge of the variables involved, and yet deliver some important insights. However, it is essential to have a good understanding of the fundamental physical laws underlying the phenomenon involved because omission of one or more important variables generally will not be detected by dimensional analysis, and errors in a model design are not likely to be detected until prototype behavior is found to be not as predicted. The variables are written in terms of fundamental (or primary) units such as length (L), force (F), time (T), mass (M), and so on.

Model I. Static Loading

Consider the stress in a statically loaded structure possibly undergoing large deflections and strains (but no plasticity). Let the geometry be represented by G, the material properties by the modulus E, and the loads by P. Note that all geometric quantities such as width, hole diameter, or thickness have a definite relation to each other and therefore only one characteristic length is necessary. A similar comment can be made about the load. The stress at some position x is therefore assumed to be a function of the quantities as

$$\sigma = f(x, E, G, P).$$

There are only two fundamental units present, force and length, but there are five quantities; hence we can form $5 - 2 = 3$ nondimensional groups. For example, using E and G as the base quantities (because they contain force and length), we can form

$$\Pi_1 = E^a G^b \sigma, \qquad \Pi_2 = E^c G^d x, \qquad \Pi_3 = E^e G^f P,$$

where $a, b, \ldots, f$ are yet to be determined. We write these groups in terms of the primary dimensions F and L as

$$\Pi_1 = \left(\frac{F}{L^2}\right)^a L^b \left(\frac{F}{L^2}\right), \qquad \Pi_2 = \left(\frac{F}{L^2}\right)^c L^d L, \qquad \Pi_1 = \left(\frac{F}{L^2}\right)^e L^f F.$$

Ensuring that they are nondimensional requires that the exponents sum to zero for each primary dimension. Thus,

$$a + 1 = 0, \quad -2a + b - 2 = 0; \quad c = 0, -2c + d + 1 = 0; \quad e + 1 = 0, \quad -2e + f = 0$$

leads to $a = -1, b = 0, c = 0, d = -1, e = -1, f = -2$, giving the relations

$$\Pi_1 = \frac{\sigma}{E}, \quad \Pi_2 = \frac{x}{G}, \quad \Pi_3 = \frac{P}{EG^2}, \quad \text{or} \quad \frac{\sigma}{E} = f\left(\frac{x}{G}, \frac{P}{EG^2}\right).$$

We see how the number of effective variables is reduced.

For complete similarity between model and prototype, we require that all independent dimensionless products be the same. That is,

$$\Pi_1 = \frac{\sigma_m}{E_m} = \frac{\sigma_p}{E_p}, \quad \Pi_2 = \frac{x_m}{G_m} = \frac{x_p}{G_p}, \quad \Pi_3 = \frac{P_m}{E_m G_m^2} = \frac{P_p}{E_p G_p^2}.$$

It is interesting that the load scaling is a combination of material and geometry. Furthermore, if model and prototype are of the same material, but the former is a half-scale model then the load must be reduced by a factor of four to give the same stresses in both models.

Suppose we know (or assume) that the structure behaves linearly; then it depends linearly on the load giving

$$\frac{\sigma}{E} = \frac{P}{EG^2} f^*\left(\frac{x}{G}\right) \quad \text{or} \quad \sigma = \frac{P}{G^2} f\left(\frac{x}{G}\right).$$

We conclude that the stress does not depend on the Young's modulus. This seems to imply that stress does not depend on the material properties, which is erroneous. An isotropic elastic material depends on two parameters, the modulus E and Poisson's ratio ν. If ν is included in the original expression for stress, then it would form its own Π term (because it is nondimensional to begin with), and the linear expression for stress would become

$$\sigma = \frac{P}{G^2} f\left(\frac{x}{G}, \nu\right).$$

This highlights the need for the inclusion of appropriate quantities, which is something not always obvious a priori. Also, because ν is nondimensional, it cannot be adjusted between model and prototype. That is, we must have the same Poisson's ratio in model and prototype, and unless the same materials are used for both, it is difficult to satisfy. In certain types of experimental model studies – for example, stress analysis using plastic photoelastic models [31] – Poisson's ratio is usually not the same for model and prototype, and the effect of Poisson's ratio on the stress distribution must be assessed. Reference [135] discusses this in more detail and also has a discussion of the additional restrictions imposed when gravity loading is considered.

Model II. Vibration

Consider a vibration problem in which the frequency is assumed to be a function of geometry, modulus, and mass density ρ:

$$\omega = f(G, E, \rho).$$

There are three fundamental units present, force, length, and time, but there are four quantities; hence we can form only $4 - 3 = 1$ nondimensional group. For example, using E, G, and ρ as the base quantities, we can form

$$\Pi_1 = E^a G^b \rho^c \omega \qquad \text{or} \qquad \Pi_1 = \left(\frac{F}{L^2}\right)^a L^b \left(\frac{M}{L^3}\right)^c \frac{1}{T},$$

where M is the mass primary unit.

There is a problem here because there are too many primary units; SI units use $[MLT]$ with force as a derived quantity calculated as mass times acceleration with units $[ML/T^2]$; common units use $[FLT]$ with mass as a derived quantity calculated as force divided by acceleration with units $[FT^2/L]$. Explicitly, the mass density is derived as the weight density divided by gravity $((W/V)/g)$.

Using the common units, we have

$$\Pi_1 = \left(\frac{F}{L^2}\right)^a L^b \left(\frac{F}{L^3}\frac{T^2}{L}\right)^c \frac{1}{T} \qquad \Rightarrow \qquad a + c = 0, \quad -2a + b - 4c = 0, \quad 2c - 1 = 0.$$

This gives

$$\Pi_1 = G\sqrt{\frac{\rho}{E}}\,\omega \qquad \text{or} \qquad 1 = f^*\left(G\sqrt{\frac{\rho}{E}}\,\omega\right),$$

which can be rewritten as

$$\omega = \frac{1}{G}\sqrt{\frac{E}{\rho}} \times \text{constant}.$$

Suppose the structure is beamlike; then we would expect a dependence on the cross-sectional moment of inertia $I = ab^3/12$. Working as previously, we would find a second nondimensional term $\Pi_2 = I/G^4$, giving

$$0 = f^*\left(G\sqrt{\frac{\rho}{E}}\,\omega, \frac{I}{G^4}\right) \qquad \text{or} \qquad \omega = \frac{1}{G}\sqrt{\frac{E}{\rho}}\,f_2\left(\frac{I}{G^4}\right).$$

This did not lead to the hoped-for result that frequency depends on $\sqrt{I}$. Dimensional analysis is not sensitive enough to distinguish between multiple quantities with similar units. Furthermore, dimensional analysis does not distinguish reciprocal relations; thus I/G^4 and G^4/I are equally valid.

The interpretation of $1/G$ in the preceding discussion is that if the overall structure is increased uniformly, then the frequency will decrease. It gives no information about what would happen if a single dimension (thickness, hole diameter, and so on) is changed keeping all other dimensions the same.

The similarity condition between model and prototype is

$$\omega_m G_m \sqrt{\frac{\rho_m}{E_m}} = \omega_p G_p \sqrt{\frac{\rho_p}{E_p}}.$$

A 1/10-scale aluminum model of a steel structure would give $\omega_m = 10\omega_p$ because the ratio E/ρ is the same for both materials. An acrylic model, on the other hand, gives $\omega_m \approx 3\omega_p$ because $E_p/\rho_p \approx 10 E_m/\rho_m$. These considerations ignored the effect of damping, which is difficult to satisfy between model and prototype. Experimental studies are described in References [131, 132].

Model III. Nonlinear Materials

It is very difficult to do a true similitude study for elastic-plastic and other non-linear material behavior. To get a sense of the nondimensional quantities involved, consider the stress at a typical point in a statically loaded structure

$$\sigma = f(G, E, \sigma_Y, P),$$

where σ_Y is the yield stress for elastic-plastic behavior. There are only two fundamental units present, force and length, but there are five quantities; hence we can form $5 - 2 = 3$ nondimensional groups. For example, using σ_Y and G as the base quantities (because they contain force and length), we can form

$$\Pi_1 = \sigma_Y^a G^b E, \qquad \Pi_2 = \sigma_Y^c G^d \sigma, \qquad \Pi_3 = \sigma_Y^e G^f P.$$

Proceeding as done earlier leads to (with slight rearrangement)

$$\Pi_1 = \frac{E}{\sigma_Y}, \quad \Pi_2 = \frac{\sigma}{\sigma_Y}, \quad \Pi_3 = \frac{P}{\sigma_Y G^2} \quad \text{or} \quad \frac{\sigma}{\sigma_Y} = f\left(\frac{E}{\sigma_Y}, \frac{P}{EG^2}\right).$$

This has identified σ_Y as a significant nondimensionalizing factor.

A comparison between different nonlinear constitutive relations can be made by use of the Ramberg–Osgood parameters, which are suggested by the preceding relations, that is, by defining

$$\bar{\sigma} = \frac{\sigma}{\sigma_Y}, \qquad \bar{\epsilon} \equiv \frac{\epsilon E}{\sigma_Y}.$$

Note that strain is nondimensional to begin with and can always be added to any other nondimensional quantity. The stress–strain diagrams of the similar materials should coincide and plot as, for instance,

$$\bar{\epsilon} = \bar{\sigma} + \frac{3}{7}(\bar{\sigma})^m,$$

as shown in Figure 7.10. The exponent m determines the amount of work hardening. This is good only for proportional loading and no unloading occurs.

A very interesting use of this similarity law is given in the experimental studies of References [13, 14], where polymers are used to simulate elastic-plastic rolling behavior of aluminum at elevated temperatures.

Figure 7.10. Ramberg–Osgood replotting of elastic-plastic constitutive behavior.

Model IV. Impact

Consider the dynamic problem of a relatively concentrated mass M_c traveling with velocity V impacting a structure characterized by its geometry G, material E, and density ρ. The stress at a typical location is then a function of

$$\sigma = f(G, E, \rho, M_c, V).$$

There are three fundamental units present, force, length, and time, and there are six quantities; hence we can form $6 - 3 = 3$ nondimensional groups. Using M_c, V, G as the base quantities, the nondimensional groups can be shown to be

$$\Pi_1 = \frac{\sigma G^3}{M_c V^2}, \qquad \Pi_2 = \frac{E G^3}{M_c V^2}, \qquad \Pi_3 = \frac{\rho G^3}{M_c}.$$

As pointed out in Reference [114], these groupings are not unique, and an equally valid set could be

$$\Pi_1 = \frac{\sigma}{\rho V^2}, \qquad \Pi_2 = \frac{E}{\rho V^2}, \qquad \Pi_3 = \frac{M_c}{\rho G^3}.$$

Multiplication or division between any of the Π groups produces equally valid groups.

The stress relation can be written as

$$\sigma = \frac{M_c V^2}{G^3} f\left(\frac{E G^3}{M_c V^2}, \frac{\rho G^3}{M_c} \right).$$

If the model and prototype are made of the same material and the impacting velocity is the same, then the similarity conditions requires

$$\frac{G_m^3}{M_{cm}} = \frac{G_p^3}{M_{cp}}.$$

Thus to have the stresses be the same in both models requires that the impacting mass scale as the length cubed. Suppose the event is the collision between two automobiles or two ships and all models are of the same material and geometrically similar, then the scaled model and prototype are expected to have the stress (and hence damage), provided the velocities are the same.

This analysis ignored a number of pertinent features that are relevant in actual impact events. These features include strain-rate effects, local plasticity, wave-propagation effects, cracking, and local buckling, to name a few. Reference [114] has a very good discussion of some experimental results for impact loading scale-model testing. Section 7.6 considers some alternative impact models.

Model V. Using Governing Equations

Sometimes the relevant governing equation is available for a problem, and this can be used to further refine the dimensional analysis. The procedure is illustrated with a few examples.

Consider the prediction of strains in an elastic structure under the influence of a dynamic loading over some part of the surface of the structure. For simplicity, let the structure be represented by the SDoF system

$$Ku + C\dot{u} + M\ddot{u} = P(t),$$

with the initial conditions $u(0) = u_o$, $\dot{u}(0) = v_o$. (This can be considered as a typical mode from a complex structure with multiple modes.) Introduce the dimensionless variables

$$\bar{u} = \frac{u}{G}, \qquad \bar{t} = \frac{t}{\tau}, \qquad P(t) = P_o f(t), \qquad \bar{P}_o = \frac{P_o}{P_*},$$

where G is a characteristic length, possibly even the initial displacement u_o, for example, and τ is a characteristic time that we now determine. Making the substitution to the new variables results in

$$[\frac{K\tau^2}{M}]\bar{u} + [\frac{C\tau}{M}]\bar{u}' + \bar{u}'' = [\frac{\tau^2 P_o}{MG}]f(t),$$

where the prime indicates differentiation with respect to $\bar{t}$. We would like the terms in square brackets to be nondimensional; hence we choose

$$\tau = \sqrt{\frac{M}{K}},$$

giving us

$$\bar{u} + [\frac{C}{\sqrt{KM}}]\bar{u}' + \bar{u}'' = [\frac{P_o}{KG}]f(t)$$

and the initial conditions $\bar{u} = 1$, $\bar{u}' = v_o\sqrt{M/K}/G$. Note that $\sqrt{K/M}$ is the natural frequency (or the modal frequency in the complex case) of the system and therefore the characteristic time scale is related to this frequency.

To make our example explicit, suppose we wish to compare the results from a prototype and model

$$u_p, \ t_p, \ K_p, \ C_p, \ M_p, \ P_p, \ G_p \Longleftrightarrow u_m, \ t_m, \ K_m, \ C_m, \ M_m, \ P_m, \ G_m.$$

The responses

$$\frac{u_p}{G_p} = \frac{u_m}{G_m}$$

are the same at times

$$\frac{t_p}{\sqrt{M_p/K_p}} = \frac{t_m}{\sqrt{M_m/K_m}}$$

if the damping and initial velocities are specified by

$$\frac{C_p}{\sqrt{M_p K_p}} = \frac{C_m}{\sqrt{M_m K_m}}, \qquad \frac{v_p}{G_p}\sqrt{\frac{M_p}{K_p}} = \frac{v_m}{G_m}\sqrt{\frac{M_m}{K_m}}.$$

The condition on the damping is difficult to satisfy unless the model and prototype materials are the same.

We have one additional requirement: The force histories in nondimensional times must be equal,

$$f_p\left(\frac{t}{\sqrt{M_p/K_p}}\right) = f_m\left(\frac{t}{\sqrt{M_m/K_m}}\right).$$

The force scale is established from the condition

$$\frac{P_o}{K_p L_p} = \frac{P_{om}}{K_m L_m},$$

which indicates that if the same material is used in model and prototype, the pressures at scaled locations and times must be the same.

If the same materials are used in model and prototype, the time scale will equal the length scale. Because the length scale is generally greater than unity, it follows that when modeling with the same materials, corresponding times in the model will be shorter than for the prototype. Thus, in a problem of this type, the model and prototype loading pressure–time relationship, when expressed in terms of P/P_o and the dimensionless time variable, must be identical, whereas in terms of real time, the model and prototype pressure–time relationship must be different. This is difficult to achieve in practice if, say, the loading is from an impact or blast.

Consider the more general dynamics problems of a 3D linear elastic body governed by Navier's equations [28]:

$$(\lambda + \mu) \sum_k \frac{\partial^2 u_k}{\partial x_k \partial x_i} + \mu \sum_k \frac{\partial^2 u_i}{\partial x_k^2} + \rho f_i = \rho \ddot{u}_i + \eta \dot{u}_i,$$

where λ and μ are the Lamé constants and f_i are the applied body forces. Introduce nondimensional length and time as

$$\bar{u}_i = \frac{u_i}{G}, \qquad \bar{t} = \frac{t}{\tau}, \qquad \tau \equiv G\sqrt{\frac{\rho}{\mu}} = \frac{G}{c_S},$$

where c_S is the shear wave speed. The governing equation becomes

$$\left[\frac{\lambda + \mu}{\mu}\right] \sum_k \frac{\partial^2 \bar{u}_k}{\partial \bar{x}_k \partial \bar{x}_i} + \sum_k \frac{\partial^2 \bar{u}_i}{\partial \bar{x}_k^2} + \frac{G}{c_S^2} f_i = \bar{u}_i'' + [\eta G c_S]\bar{u}_i',$$

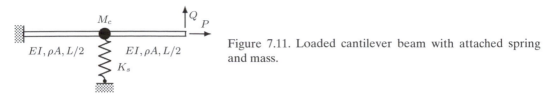

Figure 7.11. Loaded cantilever beam with attached spring and mass.

where the primes indicate derivatives with respect to time $\bar{t}$. The first bracketed term is

$$\frac{\lambda + \mu}{\mu} = \frac{1 + 3\nu}{1 - 2\nu},$$

where ν is the Poisson's ratio and is therefore nondimensional. This indicates that, for similitude, the model and prototype must have the same Poisson's ratio.

7.4 Some Simple Models With the Ritz Method

This section is a collection of simple models describing various aspects of structural mechanics. The Ritz method is the main analytical tool used to convert the continuous systems into discrete systems, and the emphasis is on the principles used in the construction process.

Model I. Dynamics of a Cantilevered Beam

Consider the cantilevered beam shown in Figure 7.11 with a spring of stiffness α and concentrated mass of M_c attached halfway along its length. We wish to determine the mass and stiffness associated with this structure for a vibration in the vertical direction only.

In this example, the essential boundary conditions at $x = 0$ are

$$v(0, t) = 0, \qquad \phi(0, t) = 0$$

for all time. Choose an expansion for the transverse displacement as

$$v(x, t) = a_0(t) + a_1(t)\left[\frac{x}{L}\right] + a_2(t)\left[\frac{x}{L}\right]^2 + a_3(t)\left[\frac{x}{L}\right]^3 + \cdots + .$$

The essential boundary conditions require that $a_0 = 0$, $a_1 = 0$, leading to

$$v(x, t) = a_2(t)\left[\frac{x}{L}\right]^2 + a_3(t)\left[\frac{x}{L}\right]^3 + \cdots + .$$

Each of the additional polynomial terms would be a Ritz function. We consider a one-term solution and a two-term solution.

For the one-term solution,

$$v(x, t) = a_2(t)\left[\frac{x}{L}\right]^2,$$

and the total strain energy is given by

$$\mathcal{U} = \frac{1}{2}\int_0^L EI[\frac{\partial^2 v}{\partial x^2}]^2 dx + \frac{1}{2}K_s v(L/2)^2 + \frac{1}{2}\int_0^L \bar{F}_o[\frac{\partial v}{\partial x}]^2 dx$$

$$= \frac{1}{2}a_2^2 \int_0^L EI[\frac{2}{L^2}]^2 dx + \frac{1}{2}K_s a_2^2 (\frac{1}{2})^4 + \frac{1}{2}a_2^2 \int_0^L P[\frac{2x}{L^2}]^2 dx$$

$$= \frac{1}{2}a_2^2[\frac{4EI}{L^3} + \frac{K_s}{16} + \frac{4P}{3L}].$$

The total kinetic energy of the beam is readily found to be

$$\mathcal{T} = \int_0^L \frac{1}{2}\rho A\dot{v}^2 dx + \frac{1}{2}M_c\dot{v}^2\Big|_{L/2}$$

$$= \frac{1}{2}\dot{a}_2^2 \int_0^L \rho A(\frac{x}{L})^4 dx + \frac{1}{2}M_c\dot{a}_2^2(\frac{1}{2})^4 = \frac{1}{2}\dot{a}_2^2[\frac{\rho AL}{5} + \frac{M_c}{16}].$$

The potential of the applied load is

$$\mathcal{V} = -Qv(L) = -Qa_2.$$

Note that the effect of the axial load P is already accounted for in the strain energy expression. Substituting these into Lagrange's equation leads to the equation of motion:

$$\mathcal{F}_{a_2} = [\frac{\rho AL}{5} + \frac{M_c}{16}]\ddot{a}_2 + [\frac{4EI}{L^3} + \frac{K_s}{16} + \frac{4P}{3L}]a_2 - Q = 0.$$

The first bracketed term can be interpreted as an effective mass, and the second as that of an effective stiffness. Observe that the axial load contributes to the stiffness.

A more accurate model of the deformed vibration shape has a third-order displacement term. That is,

$$v(x,t) = a_2(t)g_2(x) + a_3(t)g_3(x), \qquad g_2(x) \equiv [\frac{x}{L}]^2, \qquad g_3(x) \equiv [\frac{x}{L}]^3.$$

The total strain energy is given by

$$\mathcal{U} = \frac{1}{2}\int_0^L EI[a_2g_2'' + a_3g_3'']^2 dx + \frac{1}{2}K_s[a_2g_2 + a_3g_3]^2\Big|_{L/2} + \frac{1}{2}\int_0^L \bar{F}_o[a_2g_2' + a_3g_3']^2 dx$$

$$= \frac{1}{2}[\frac{4EI}{L^3} + \frac{K_s}{16} + \frac{4P}{3L}]a_2^2 + [\frac{6EI}{L^3} + \frac{K_s}{32} + \frac{3P}{2L}]a_2a_3 + \frac{1}{2}[\frac{12EI}{L^3} + \frac{K_s}{64} + \frac{9P}{5L}]a_3^2.$$

The total kinetic energy of the beam is

$$\mathcal{T} = \frac{1}{2}\int_0^L \rho A[\dot{a}_2g_2 + \dot{a}_3g_3]^2 dx + \frac{1}{2}M_c[\dot{a}_2g_2 + \dot{a}_3g_3]^2\Big|_{L/2}$$

$$= \frac{1}{2}[\frac{\rho AL}{5} + \frac{M_c}{16}]\dot{a}_2^2 + [\frac{\rho AL}{6} + \frac{M_c}{32}]\dot{a}_2\dot{a}_3 + \frac{1}{2}[\frac{\rho AL}{7} + \frac{M_c}{64}]\dot{a}_3^2.$$

The potential of the applied load is

$$\mathcal{V} = -Qv(L) = -Q[a_2g_2 + a_3g_3]\big|_L = -Q[a_2 + a_3].$$

Substituting these into Lagrange's equation leads to

$$\mathcal{F}_{a_2} = \left[\frac{\rho A L}{5} + \frac{M_c}{16}\right]\ddot{a}_2 + \left[\frac{\rho A L}{3} + \frac{M_c}{16}\right]\ddot{a}_3 + \left[\frac{4EI}{L^3} + \frac{K_s}{16} + \frac{4P}{3L}\right]a_2$$

$$+ \left[\frac{12EI}{L^3} + \frac{K_s}{16} + \frac{3P}{L}\right]a_3 - Q = 0,$$

$$\mathcal{F}_{a_3} = \left[\frac{\rho A L}{3} + \frac{M_c}{16}\right]\ddot{a}_2 + \left[\frac{\rho A L}{7} + \frac{M_c}{64}\right]\ddot{a}_3 + \left[\frac{12EI}{L^3} + \frac{K_s}{16} + \frac{3P}{L}\right]a_2$$

$$+ \left[\frac{12EI}{L^3} + \frac{K_s}{64} + \frac{9P}{5L}\right]a_3 - Q = 0.$$

These are arranged in matrix form as

$$\left[\frac{\rho A L}{210}\begin{bmatrix} 42 & 35 \\ 35 & 30 \end{bmatrix} + \frac{M_c}{64}\begin{bmatrix} 4 & 2 \\ 2 & 1 \end{bmatrix}\right]\begin{Bmatrix} \ddot{a}_2 \\ \ddot{a}_3 \end{Bmatrix}$$

$$+ \left[\frac{EI}{L^3}\begin{bmatrix} 4 & 6 \\ 6 & 12 \end{bmatrix} + \frac{K_s}{64}\begin{bmatrix} 4 & 2 \\ 2 & 1 \end{bmatrix} + \frac{P}{15L}\begin{bmatrix} 20 & 45 \\ 45 & 27 \end{bmatrix}\right]\begin{Bmatrix} a_2 \\ a_3 \end{Bmatrix} = \begin{Bmatrix} Q \\ Q \end{Bmatrix}.$$

Observe that the K_{11} stiffness and M_{11} mass are the same as already obtained for the SDoF case. The vibration problem requires solving a $[2 \times 2]$ eigenvalue problem.

Model II. Vibration of Rectangular Plates

For rectangular plates, we take the strain and kinetic energies as, respectively,

$$\mathcal{U} = \tfrac{1}{2}\int_A D[\nabla^2 w]^2\, dxdy, \qquad \mathcal{T} = \tfrac{1}{2}\int_A \rho h\, \dot{w}^2\, dxdy, \qquad \nabla^2 \equiv \frac{\partial^2}{\partial x^2} + \frac{\partial^2}{\partial y^2},$$

where $D = Eh^3/12(1 - \nu^2)$ is the plate flexural rigidity and the deflection $w(x, y; t)$ is a function of two space variables.

The first step is to obtain appropriate Ritz functions. For 1D systems, the simplest approach (as just shown) is to assume a polynomial and then impose the geometric BCs. This becomes somewhat unwieldy in the 2D case. Instead, it is sometimes simpler to use a product function. For rectangular problems, for example, we assume

$$w(x, y; t) = a(t)g_x(x)g_y(y),$$

where $g_x(x)$ satisfies the boundary conditions on the x edges and $g_y(y)$ satisfies the boundary conditions on the y edges. This can then be generalized for an arbitrary number of modes by

$$w(x, y, t) = g_x(x)g_y(y)[a_{00}(t) + a_{10}(t)x + a_{20}(t)x^2 + a_{01}(t)y + a_{02}(t)y^2$$

$$+ a_{11}(t)xy + \cdots +].$$

Note that the coefficients (which are the generalized coordinates) are functions of time.

Consider, for example, a plate simply supported on all edges. Represent the x behavior as

$$w = a_o + a_1 x + a_2 x^2.$$

The geometric edge conditions are such that $w = 0$ at each end; this leads to

$$w = a_2[-ax + x^2].$$

The natural BC of zero moment need not be considered. A similar function is obtained for the y behavior. Put both of them together to get

$$w(x, y, t) = a_2(t)[\frac{x}{a} - \frac{x^2}{a^2}][\frac{y}{b} - \frac{y^2}{b^2}].$$

Noting that

$$\nabla^2 w = \frac{\partial^2 w}{\partial x^2} + \frac{\partial^2 w}{\partial y^2} = a_2(t)[-\frac{2}{a^2}][\frac{y}{b} - \frac{y^2}{b^2}] + [\frac{x}{a} - \frac{x^2}{a^2}][-\frac{2}{b^2}],$$

we compute the strain energy term as

$$\mathcal{U} = \tfrac{1}{2} D a_2^2 \frac{8ab}{60} \left\{ \frac{1}{a^4} + \frac{5}{3a^2 b^2} + \frac{1}{b^4} \right\}.$$

The kinetic energy computes to

$$\mathcal{T} = \tfrac{1}{2} \rho h \dot{a}_2^2 \frac{ab}{900}.$$

This gives the stiffness and mass as, respectively,

$$K = D \frac{8ab}{60} \left\{ \frac{1}{a^4} + \frac{5}{3a^2 b^2} + \frac{1}{b^4} \right\}, \qquad M = \rho h \frac{ab}{900}.$$

The corresponding frequency for harmonic motion is

$$\omega = \sqrt{\frac{D}{\rho h}} \sqrt{120} \sqrt{\frac{1}{a^4} + \frac{1.67}{a^2 b^2} + \frac{1}{b^4}}, \qquad \omega_{\text{exact}} = \sqrt{\frac{D}{\rho h}} \pi^2 \sqrt{\frac{1}{a^4} + \frac{2}{a^2 b^2} + \frac{1}{b^4}}.$$

The comparison with the exact result [30] is basically $\sqrt{120}$ versus π^2, or 10.95 versus 9.97. This is a reasonable approximation for a one-term solution.

As an interesting side point, suppose the preceding strain energy expression for the plate was not used and instead the plate was conceived as two beamlike behaviors such that

$$\mathcal{U} \approx \tfrac{1}{2} \int_a E I_{yy} [\frac{\partial^2 w}{\partial x^2}]^2 dx + \tfrac{1}{2} \int_b E I_{xx} [\frac{\partial^2 w}{\partial y^2}]^2 dy, \qquad I_{yy} = \tfrac{1}{12} bh^3, \quad I_{xx} = \tfrac{1}{12} ah^3.$$

The final expressions for the stiffness and frequency become

$$K = D^* \frac{8ab}{60} \left\{ \frac{1}{a^4} + \frac{1}{b^4} \right\}, \qquad \omega = \sqrt{\frac{D^*}{\rho h}} \sqrt{120} \sqrt{\frac{1}{a^4} + \frac{1}{b^4}}, \qquad D^* = \tfrac{1}{12} E h^3.$$

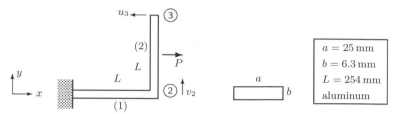

Figure 7.12. Two-member frame.

Although these expressions are not as accurate as the previous ones, it nonetheless does a good job in capturing the behavior of the essential parameters. This example is a reminder that the energy expressions need not be true or exact in order to construct a good simple model.

Model III. Compound Systems

The previous examples illustrate the use of Ritz functions that spanned the whole domain. This is not essential indeed for complex structures comprising multiple- (and different) part types, it is advantageous to use piecewise Ritz functions. The approach is demonstrated by establishing the equations of motion for the simple frame shown in Figure 7.12. The load acts halfway along the vertical member.

This is an example in which it would be very difficult to establish a single Ritz function to cover the whole domain. Instead, we consider the frame as comprising two subdomains and write separate Ritz functions for each. Furthermore, the frame has both bending and axial behaviors, but we assume that the vibrations are dominated by the flexible behavior.

Let the horizontal and vertical members be considered as separate domains as indicated. For the first domain, assume that the action is dominated by the flexural behavior and hence we have

$$u(x) = 0, \qquad v(x) = a_0 + a_1 x + a_2 x^2 + \cdots + .$$

Assume the second member is also dominated by the flexible behavior, but in addition it also has a rigid-body motion imposed on it by the motion of the tip of Member 1. Thus,

$$u(y) = b_0 + b_1 y + b_2 y^2 + \cdots + , \qquad v(y) = \text{constant}.$$

There are two sets of geometric constraints for the problem. The first requires zero displacement and rotation at the fixed end. The second requires continuity of displacement and rotation at the joint connecting the two members. Let v_2 be the vertical deflection at Node 2 and u_3 be the horizontal deflection at Node 3; then a set of displacements consistent with the constraints is

Member 1: $\qquad u(x) = 0, \qquad v(x) = v_2[\frac{x}{L}]^2,$

Member 2: $\qquad u(y) = v_2 2[\frac{y}{L} - \frac{y^2}{L^2}] + u_3[\frac{y}{L}]^2, \qquad v(y) = v_2.$

Note that the rotation at the joint is given by the alternative forms

$$\phi = \frac{\partial v}{\partial x}^{(1)} = 2v_2[\frac{x}{L^2}]\Big|_{x=L} = \frac{2v_2}{L}, \qquad \phi = \frac{\partial u}{\partial y}^{(2)} = \left[v_22[\frac{1}{L} - 2\frac{y}{L^2}] + u_32[\frac{y}{L^2}]\right]\Big|_{y=0} = \frac{2v_2}{L},$$

indicating the continuity of rotation. Thus a positive vertical deflection at the joint, results in a positive rotation of the joint, which in turn results in a right-to-left deflection of the vertical tip (a positive u_3).

The strain and kinetic energies are integrals over the complete body; when the body is segmented, then we simply segment the integrals. The strain energy becomes

$$\mathcal{U} = \frac{1}{2}\int_{(1)} EI[\frac{\partial^2 v}{\partial x^2}]^2 dx + \frac{1}{2}\int_{(2)} EI[\frac{\partial^2 u}{\partial y^2}]^2 dx = \frac{1}{2}EIv_2^2\frac{4}{L^3} + \frac{1}{2}EI[-v_22 + u_3]^2\frac{4}{L^3}.$$

The kinetic energy is

$$\mathcal{T} = \frac{1}{2}\int_{(1)} \rho A \dot{v}^2 dx + \frac{1}{2}\int_{(2)} \rho A \dot{u}^2 dx + \frac{1}{2}\int_{(2)} \rho A \dot{v}^2 dx$$

$$= \frac{1}{2}\rho A L \dot{v}_2^2\frac{1}{3} + \frac{1}{2}\rho A L[\dot{v}_2^2\frac{2}{15} - \dot{v}_2\dot{u}_3\frac{1}{5} + \dot{u}_3^2\frac{1}{5}] + \frac{1}{2}\rho A L \dot{v}_2^2.$$

Finally, the potential of the applied load is

$$\mathcal{V} = -[-P][-u(y = L/2)] = -P[v_22 + u_3]\frac{1}{4}.$$

Substituting these into Lagrange's equations then leads to the equations of motion:

$$\frac{EI}{L^3}\begin{bmatrix} 20 & -8 \\ -8 & 4 \end{bmatrix}\begin{Bmatrix} v_2 \\ u_3 \end{Bmatrix} + \frac{\rho A L}{30}\begin{bmatrix} 34 & -3 \\ -3 & 6 \end{bmatrix}\begin{Bmatrix} \ddot{v}_2 \\ \ddot{u}_3 \end{Bmatrix} = \frac{P}{4}\begin{Bmatrix} -2 \\ 1 \end{Bmatrix}.$$

Thus the single applied load gives rise to components associated with each DoF.

Figure 7.13 shows the free-vibration mode shapes as predicted by the simple model and a comparison with the FE-generated mode shapes. An obvious point is that the simple model is incapable of predicting the behavior of the higher vibration modes.

This example illustrates the basic features of the FE approach to structural dynamics.

Model IV. Large Deflections of an Elastica
An *elastica* is a slender member that supports both axial and bending loads; however, it does not experience any axial stretching. This is a good (approximate) model to describe the large-deflection behavior of slender beams. Some exact solutions can be found in References [30, 75].

Consider the plane deflection of the elastica shown in Figure 7.14; let s be the distance along the elastica. Hence we have

$$\frac{dx}{ds} = \cos\phi, \qquad \frac{dy}{ds} = \sin\phi,$$

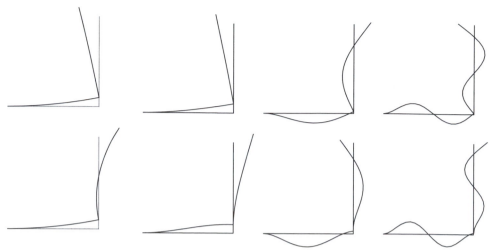

Figure 7.13. Free-vibration mode shapes. Leftmost are results from the simple model, the remainder are FE-generated mode shapes.

where ϕ is the slope. A point originally at position $x^o = s$, $y^o = 0$, moves to a location x, y a distance s along the elastica because the elastica is inextensible. Hence, the two displacements are given by

$$u = x - x^o = x - s, \qquad v = y - y^o = y - 0.$$

We can put this in differential form as

$$u_{,s} = \frac{du}{ds} = \frac{dx}{ds} - 1 = \cos\phi - 1, \qquad v_{,s} = \frac{dv}{ds} = \frac{dy}{ds} = \sin\phi,$$

where the subscript comma indicates differentiation. It is clear that these equations satisfy the constraint $[1 + u_{,s}]^2 + v_{,s}^2 = 1$ that says that there is no axial stretching of the beam (although there are axial forces). Furthermore, it says that the transverse and axial deflections are related to each other. We use this connection to develop some approximate forms for the elastica equations.

The total potential for a beam with tip loads P_u and P_v resting on an elastic foundation is

$$\Pi = \frac{1}{2} \int E\epsilon_{ss}^2 \, dA \, ds - P_u u\big|_{(s=L)} - P_v v\big|_{(s=L)} + \frac{1}{2} \int K_v v^2 \, ds.$$

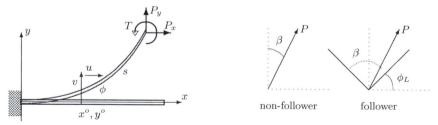

Figure 7.14. An elastica with tip loads.

The large-deflection strain in the beam is axial:

$$\epsilon_{ss} = u,_s + \tfrac{1}{2}v,_s^2 - y\phi,_s = -\tfrac{1}{2}u,_s^2 - y\phi,_s , \qquad \epsilon_{ss}^2 = \tfrac{1}{4}u,_s^4 - u,_s^2 y\phi,_s + y^2\phi,_s^2 .$$

Substitute into the total potential and integrate on the cross section to get

$$\Pi = \tfrac{1}{2}\int EA\tfrac{1}{4}u,_s^4 \, ds + \tfrac{1}{2}\int EI\phi,_s^2 \, ds - \int P_u u,_s \, ds - P_v v_L + \tfrac{1}{2}\int K_v v^2 \, ds.$$

Note that the end shortening needed to be calculated as the integral of $u,_s$ over the length of the beam. We now use the approximations

$$u,_s = \sqrt{1 - v,_s^2} - 1 \approx -\tfrac{1}{2}v,_s^2 - \tfrac{1}{8}v,_s^4 - \cdots -,$$

$$\phi,_s = \frac{d\phi}{ds} = \frac{1}{\cos\phi}\frac{d^2v}{ds^2} = \frac{1}{\sqrt{1 - v,_s^2}}\frac{d^2v}{ds^2} \approx v,_{ss}[1 + \tfrac{1}{2}v,_s^2 + \tfrac{1}{8}v,_s^4 + \cdots +].$$

Substitute these into the potential and order according to the significance of the terms (with the higher-order nonlinear terms being less significant) to get

$$\Pi = \tfrac{1}{2}\int \left[EIv,_{ss}^2 + P_u v,_s^2 + K_v v^2\right] ds - P_v v_L + \tfrac{1}{2}\int \left[EIv,_{ss}^2 v,_s^2 + \tfrac{1}{4}P_u v,_s^4\right] ds + \cdots +.$$

It is interesting that the EA terms do not appear to this order of approximation.

Consider a cantilever beam with a tip transverse load. The linear solution for the problem is $v(x) = (P_v/EIL)[3Lx^2 - x^3]$. So, in the style of the Ritz method, we assume the deflected shape to be given by

$$v(s) = [\tfrac{3}{2}(s/L)^2 - \tfrac{1}{2}(s/L)^3]v_1,$$

where v_1 is the tip deflection. Substitute into the total potential and integrate over the length L to get

$$\Pi = \tfrac{3}{2}EI\frac{v_1^2}{L^3} - P_v v_1 + \tfrac{81}{80}EI\frac{v_1^4}{L^5} + \cdots +.$$

Equilibrium is given by

$$\mathcal{F}_{v_1} = \frac{\partial\Pi}{\partial v_1} = 3EI\frac{v_1}{L^3} - P_v + \tfrac{81}{20}EI\frac{v_1^3}{L^5} + \cdots + \cdots = 0.$$

We can view this as an approximate expansion (in terms of v_1) of the true equilibrium equation. The first two terms are the linear solution.

The load deflection and stiffness expressions are

$$P_v = \frac{3EI}{L^3}\left[1 + \tfrac{81}{60}\frac{v_1^2}{L^2}\right]v_1, \qquad u_L = \left[0 - \tfrac{3}{5}\frac{v_1^2}{L^2} - \tfrac{1}{4}\frac{v_1^4}{L^4}\right]L, \qquad K_T = \frac{3EI}{L^3}\left[1 + \tfrac{81}{20}\frac{v_1^2}{L^2}\right].$$

The deflections are shown plotted in Figure 5.6, where they are compared with a FE analysis. Clearly the solution has a limited range, but it is a substantial improvement over the linear solution; in particular, it gives a reasonable estimate for the axial displacement.

As a second example, consider a simply supported beam that can buckle under axial compression. The beam rests on an elastic foundation. We assume the deflected (buckled) shape to be given by $v(s) = v_1 \sin(n\pi s/L)$; this satisfies the kinematical end conditions, and n specifies the number of half-waves. Substitute into the total potential and integrate over the length L to get

$$\Pi = \tfrac{1}{2}\big[EI(\tfrac{n\pi}{L})^4\tfrac{L}{2} - P(\tfrac{n\pi}{L})^2\tfrac{L}{2} + K_v\tfrac{L}{2}\big]v_1^2$$

$$+ \tfrac{1}{2}\big[EI(\tfrac{n\pi}{L})^6\tfrac{L}{8} - \tfrac{1}{4}P(\tfrac{n\pi}{L})^4\tfrac{3L}{8}\big]v_1^4 + \cdots + .$$

Equilibrium is given by

$$\mathcal{F}_{v_1} = \frac{\partial \Pi}{\partial v_1} = \big[EI(\tfrac{n\pi}{L})^4\tfrac{L}{2} - P(\tfrac{n\pi}{L})^2\tfrac{L}{2} + K_v\tfrac{L}{2}\big]v_1$$

$$+ \big[EI(\tfrac{n\pi}{L})^6\tfrac{L}{8} - \tfrac{1}{4}P(\tfrac{n\pi}{L})^4\tfrac{3L}{8}\big]2v_1^3 + \cdots + = 0.$$

Consider the first-term approximation

$$\big[EI(\tfrac{n\pi}{L})^4 - P(\tfrac{n\pi}{L})^2 + K_v\big]v_1 = 0, \qquad \frac{\partial^2 \Pi}{\partial v_1^2} = \big[EI(\tfrac{n\pi}{L})^4 - P(\tfrac{n\pi}{L})^2 + K_v\big].$$

This has two solutions

$$v_1 = 0 \qquad \text{and} \qquad P = EI(\tfrac{n\pi}{L})^2 + K_v(\tfrac{L}{n\pi})^2 = P_c, \qquad v_1 = ?$$

The first of these is the unbuckled position of the beam. The second gives the Euler buckling loads [29, 30, 123], and the stability criterion shows it is neutrally stable because the second derivative of Π is zero. This first approximation gives no information about the postbuckled deflection.

Consider the two-term expansion (replacing EI terms with P_c and K_v terms) to get

$$\Big[[P_c - P] + \tfrac{1}{2}(\tfrac{n\pi}{L})^2[P_c - K_v(\tfrac{L}{n\pi})^2 - \tfrac{3}{4}P]v_1^2\Big]v_1 = 0.$$

Again there are two solutions:

$$v_1 = 0 \qquad \text{and} \qquad \frac{P}{P_c} = \frac{1 + \tfrac{1}{2}(n\pi v_1/L)^2[1 - (K_v/P_c(L/n\pi)^2]}{1 + \tfrac{1}{2}\tfrac{3}{4}(n\pi v_1/L)^2}. \qquad (7.22)$$

This second approximation allowed determination of the relation between the load and the deflection in the postbuckling region. The end shortening is calculated by

$$\Delta = \int_L u_{,s}\,ds \approx -\int_L [\tfrac{1}{2}v_{,s}^2 + \tfrac{1}{8}v_{,s}^4]\,ds = -[(n\pi v_1/L)^2 + \tfrac{3}{16}(n\pi v_1/L)^4]\tfrac{1}{4}L.$$

For $P > P_c$, the stability criterion shows it to be stable.

Model V. Reduced Dimensional Modeling

The previous examples used the Ritz method to replace a continuous system with a discrete system; another powerful use of the Ritz method is in replacing one continuous system with another continuous system but of reduced dimensionality. This

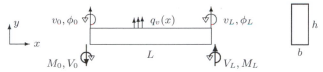

Figure 7.15. Rectangular block with in-plane distributed and end loads.

is at the base of all simple structural models, for example, replacing a 3D bar with a 1D beam model or a 1D rod model. The approach is often referred to as the *semidirect method* [27] because it replaces the true 3D deformation state with an assumed simpler deformation state. The approach will be illustrated in developing a beam theory that takes the shear deformation and rotational inertia into account. This approach is especially useful in constructing higher-order waveguide theories for wave propagation [28].

Consider a rectangular block of length L, thickness h, and width b, as shown in Figure 7.15. Assume that the loads are in the $x - y$ plane only; then if b is small compared with the other dimensions, the block can also be assumed to be in a state of plane stress where σ_{zz}, σ_{xz}, σ_{yz} are zero.

We begin by expanding the displacement fields in a Taylor series about the midplane displacements $\bar{u}(x, 0)$ and $\bar{v}(x, 0)$ as

$$\bar{u}(x, y) \approx \bar{u}(x, 0) + y\frac{\partial \bar{u}}{\partial y}\big|_{y=0} + \cdots + = u(x) - y\phi(x) + \cdots +,$$

$$\bar{v}(x, y) \approx \bar{v}(x, 0) + y\frac{\partial \bar{v}}{\partial y}\big|_{y=0} + \cdots + = v(x) + y\psi(x) + \cdots +,$$

where we have used the notation

$$u(x) = \bar{u}(x, 0), \quad v(x) = \bar{v}(x, 0), \quad \phi(x) = -\frac{\partial \bar{u}}{\partial y}\big|_{y=0}, \quad \psi(x) = \frac{\partial \bar{v}}{\partial y}\big|_{y=0}.$$

Let us be primarily interested in flexural deformations, then set $\bar{u}(x, 0) = 0$. Furthermore, for illustration purposes, we retain only one term in each expansion and obtain the approximate displacement fields as

$$\bar{u}(x, y) \approx -y\phi(x), \qquad \bar{v}(x, y) \approx v(x).$$

This says that the deformation is governed by two independent functions, $v(x)$ and $\phi(x)$, that depend on only the position along the centerline. We make the assumption that these kinematic representations do not change under dynamic conditions.

The strains corresponding to the preceding deformations are

$$\epsilon_{xx} = \frac{\partial \bar{u}}{\partial x} = -y\frac{\partial \phi}{\partial x}, \qquad \epsilon_{yy} = \frac{\partial \bar{v}}{\partial x} = 0, \qquad \gamma_{xy} = \frac{\partial \bar{u}}{\partial y} + \frac{\partial \bar{v}}{\partial x} = \left[-\phi + \frac{\partial v}{\partial x}\right].$$

For a slender beam undergoing flexural deformation we would expect that $\sigma_{yy} \ll \sigma_{xx}$, so that we can set $\sigma_{yy} = 0$. The plane-stress system then becomes

$$\sigma_{xx} = -yE\frac{\partial \phi}{\partial x}, \qquad \sigma_{xy} = G\left[-\phi + \frac{\partial v}{\partial x}\right], \qquad \sigma_{yy} = 0.$$

These stresses are obviously not in equilibrium [for example, σ_{yy} on the top boundary must be equal to $q_v(x)$ whereas σ_{xy} should be zero], but recall that equilibrium will actually be imposed through the use of the variational principle. The strain energy is then

$$\mathcal{U} = \tfrac{1}{2} \int_V [\sigma_{xx}\epsilon_{xx} + \sigma_{xy}\gamma_{xy}]\, dV = \tfrac{1}{2} \int_V [E\epsilon_{xx}^2 + G\gamma_{xy}^2]\, dV.$$

Substitute for the strains to get the total strain energy as

$$\mathcal{U} = \tfrac{1}{2} \int_o^L \int_{-h/2}^{h/2} \left[Ey^2 [\frac{\partial \phi}{\partial x}]^2 + G[\phi - \frac{\partial v}{\partial x}]^2 \right] b\, dy\, dx$$

$$= \tfrac{1}{2} \int_o^L \left[EI[\frac{\partial \phi}{\partial x}]^2 + GA[\phi - \frac{\partial v}{\partial x}]^2 \right] dx,$$

where A is the area and $I = bh^3/12$ is the second moment of area. The total kinetic energy is

$$T = \tfrac{1}{2} \int_V \rho[\dot{u}^2 + \dot{v}^2]\, dV = \tfrac{1}{2} \int_0^L \int_A \rho[y^2\dot{\phi}^2 + \dot{v}^2]\, dA\, dx = \tfrac{1}{2} \int_0^L [\rho A\dot{v}^2 + \rho I\dot{\phi}^2]\, dx.$$

If the applied surface tractions and loads are as shown in Figure 7.15, then the potential of these loads is

$$\mathcal{V} = -\int_o^L q_v(x)v\, dx - M_L\phi_L + M_0\phi_0 - V_L v_L + V_0 v_0$$

$$= -\int_o^L q_v(x)v\, dx - M\phi \Big|_0^L - Vv \Big|_0^L.$$

Hamilton's principle for the simplified block may now be stated as

$$\delta \int_{t_1}^{t_2} \left\{ \int_0^L \left[\tfrac{1}{2}[\rho A\dot{v}^2 + \rho I\dot{\phi}^2] - \tfrac{1}{2}EI[\frac{\partial \phi}{\partial x}]^2 - \tfrac{1}{2}GA[-\phi + \frac{\partial v}{\partial x}]^2 + q_v v \right] dx \right.$$

$$\left. + M\phi \Big|_0^L + Vv \Big|_0^L \right\} dt = 0.$$

There are two variables, $v(x, t)$ and $\phi(x, t)$, that are subject to variation.

Taking the variation inside the integrals and using integration by parts, we get

$$\int_{t_1}^{t_2} \left\{ \int_0^L \left[GA[-\phi + \frac{\partial v}{\partial x}] + EI\frac{\partial^2 \phi}{\partial x^2} - \rho I\ddot{\phi} \right] \delta\phi\, dx \right.$$

$$+ \int_0^L \left[GA\frac{\partial}{\partial x}[-\phi + \frac{\partial v}{\partial x}] - \rho A\ddot{v} + q_v \right] \delta v\, dx$$

$$\left. + \left[EI\frac{\partial \phi}{\partial x} - M \right]\delta\phi \Big|_0^L + \left[GA[-\phi + \frac{\partial v}{\partial x}] - V \right]\delta v \Big|_0^L \right\} dt = 0. \qquad (7.23)$$

The facts that the variations δv and $\delta \phi$ can be varied separately and arbitrarily, and that the limits on the integrals are also arbitrary, lead us to conclude from the integral terms in square brackets that

$$GA\frac{\partial}{\partial x}\left[\frac{\partial v}{\partial x}-\phi\right]=\rho A\ddot{v}-q_v,$$

$$EI\frac{\partial^2\phi}{\partial x^2}+GA\left[\frac{\partial v}{\partial x}-\phi\right]=\rho I\ddot{\phi}. \tag{7.24}$$

This is a set of coupled partial differential equations. The other terms in squares brackets in Equation (7.23) must also be zero; these lead to the associated BCs (at each end of the beam) specified in terms of any pair of conditions selected from the following groups:

$$\left\{v=\text{specified}\quad\text{or}\quad GA[\frac{\partial v}{\partial x}-\phi]=V\right\},\quad\left\{\phi=\text{specified}\quad\text{or}\quad EI\frac{\partial\phi}{\partial x}=M\right\}. \tag{7.25}$$

The first condition in each case is referred to as the *geometric* BC whereas the second condition is referred to as the *natural* BC. Collectively, these equations are called the *Timoshenko equations* of motion for a deep beam. In comparison with the elementary Bernoulli–Euler beam theory, this theory accounts for the shear deformation as well as for the rotational inertia.

Because we started with an approximation for the deformation distributions, this gave us an approximate potential function from which we derived a governing differential equation and a set of boundary conditions most consistent with that approximation. We can imagine, therefore, proposing a different potential and deriving natural BCs and governing equations that are different but nonetheless consistent. This is a key point: We use the derivation of the strain and kinetic energies to establish the form of the potential, but then we are free to add adjustable parameters because the variational principle will correctly implement these parameters.

Rather than associate the parameters with the deformation, it is a little more convenient to modify the potential itself. Reasoning that the shear is represented less accurately in our modeling, we modify the energy terms associated with it as

$$\mathcal{U}=\frac{1}{2}\int_o^L\left[EI(\frac{\partial\phi}{\partial x})^2+GAK_1[\phi-\frac{\partial v}{\partial x}]^2\right]dx,\qquad \mathcal{T}=\frac{1}{2}\int_0^L\left[\rho A\dot{v}^2+\rho I K_2\dot{\phi}^2\right]dx,$$

where the adjustable parameters K_1 and K_2 have been introduced. The modified differential equations become

$$GAK_1\frac{\partial}{\partial x}\left[\frac{\partial v}{\partial x}-\phi\right]=\rho A\ddot{v}-q_v,$$

$$EI\frac{\partial^2\phi}{\partial x^2}+GAK_1\left[\frac{\partial v}{\partial x}-\phi\right]=\rho I K_2\ddot{\phi}, \tag{7.26}$$

and the associated modified boundary conditions are

$$\left\{v=\text{specified}\quad\text{or}\quad GAK_1[\frac{\partial v}{\partial x}-\phi]=V\right\},\quad\left\{\phi=\text{specified}\quad\text{or}\quad EI\frac{\partial\phi}{\partial x}=M\right\}.$$

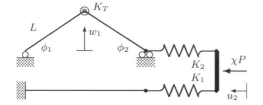

Figure 7.16. 2DoF simple model for the buckling of a plate or beam.

The coefficients can be evaluated many ways. The group constant GAK_1 is essentially the shear stiffness. Although isotropic elastic materials have a definite relation between G and E, here they are allowed to be distinct. Indeed, to recover the elementary theory, it is necessary to set $GAK_1 = \infty$; this would imply that there is no shear deformation even though there are shear forces present. Another special case of interest is that in which rotational effects are neglected (i.e., $\rho I K_2 = 0$). Without information to the contrary, values of $K_1 = \pi^2/12$ and $K_2 = 1.0$ can be chosen; see Reference [28] for more discussions and references concerning the adjustable parameters.

7.5 Mechanical Models for Postbuckling

Sometimes, the role of simple models is to provide an analytical construct in which to investigate problems in a tractable way rather than describe the physical problem itself. That is, the model (which we refer to as a *mechanical model*) does not resemble the physical model but captures the essential mechanisms of a phenomenon to be investigated. A very simple example would be the replacement of an impacted cantilever beam with a single spring–mass system. Unlike the previous section, no attempt will be made here to relate the values of mass and stiffness to their physical counterparts; it is pointed out, however, that this can often be done by "pegging" the results to the phenomenon of interest (vibration frequency or buckling load, for example).

   The postbuckling behavior of structures is quite complicated, and we illustrate the use of some simple mechanical models to elucidate its main features.

Model I. Elastic Postbuckling

The essential deformations for the buckling of beams and plates are an in-plane compression and a nonlinear out-of-plane flexure. The simple model shown in Figure 7.16 is made of rigid links, each of length L with a torsional spring K_T representing the flexural stiffness. It also has axial springs associated with the shortening u_2. As we will see, the spring K_2 introduces a nonlinear coupling into the system even for small deflections w_1 and u_2.

   With the small-deflection assumption, the angles are given by

$$\phi_1 \approx \frac{w_1}{L}, \qquad \phi_2 \approx \frac{w_1}{L},$$

and the twist of the spring is $\phi = \phi_1 + \phi_2 = 2w_1/L$. The end shortening of the links is computed as

$$\Delta = L[2 - \cos\phi_1 - \cos\phi_2] \approx L[2 - (1 - \tfrac{1}{2}\phi_1^2 + \cdots +) - (1 - \tfrac{1}{2}\phi_1^2 + \cdots +)]$$

$$= L\phi_1^2 \approx w_1^2/L.$$

Introduce new generalized coordinates defined as

$$x_1 \equiv \frac{w_1}{L} \quad \text{or} \quad x_2 \equiv \frac{u_2}{L}.$$

The strain energies for the torsion and axial springs become, respectively,

$$U_T = \tfrac{1}{2}K_T[\phi]^2 = \tfrac{1}{2}K_T[2\phi_1]^2 = \tfrac{1}{2}K_T[2\frac{w_1}{L}]^2 = \tfrac{1}{2}K_T 4x_1^2,$$

$$U_S = \tfrac{1}{2}K_1 u_2^2 + \tfrac{1}{2}K_2[u_2 - \Delta]^2 = \tfrac{1}{2}[K_1 + K_2]L^2 x_2^2 - \tfrac{1}{2}K_2 L^2[2x_1^2 x_2 - x_1^4].$$

We see the nonlinear contribution of the K_2 spring in the last term. What is also interesting to observe is that this spring couples the in-plane and out-of-plane deflections.

The total potential for the problem can be written as

$$\Pi = U + V = \tfrac{1}{2}K_T 4x_1^2 + \tfrac{1}{2}(K_1 + K_2)L^2 x_2^2 - \tfrac{1}{2}K_2 L^2[2x_1^2 x_2 - x_1^4] - \chi P L x_2. \quad (7.27)$$

This is the general structure of the potential function. This model can exhibit a bifurcation and is postbuckle stable, as expected physically from Figure 7.16 with linear springs. The equilibrium equations are

$$\mathcal{F}_1 = \frac{\partial \Pi}{\partial x_1} = \left[4K_T - 2K_2 L^2[x_2 - x_1^2] \right]x_1 = 0,$$

$$\mathcal{F}_2 = \frac{\partial \Pi}{\partial x_2} = (K_1 + K_2)L^2 x_2 - K_2 L^2 x_1^2 - \chi P L = 0.$$

There are two *equilibrium paths* – an equilibrium path is a collection of equilibrium states that are changed by a controlling variable, in this case the load χP. The first path is obtained by setting $x_1 = 0$; the second is obtained by setting the first bracketed terms to zero. This results in

$$\text{I: } x_1 = 0, \qquad\qquad x_2 = \frac{\chi P}{(K_1 + K_2)L};$$

$$\text{II: } x_1 = \sqrt{\frac{\chi P - 2K_1 L\gamma}{K_1 L}} \quad x_2 = \frac{\chi P - 2K_T/L}{K_1 L}, \qquad \gamma \equiv \frac{(K_1 + K_2)K_T}{K_1 K_2 L^2}.$$

The first path is simply the linear end shortening of the springs before buckling occurs. For the second path, buckling occurs when $\chi P = 2K_1 L\gamma$.

We simplify some of the coefficients so that the essential mechanisms become even clearer. Let

$$K_1 = \tfrac{1}{2}K, \qquad K_2 = \tfrac{1}{2}K, \qquad 4K_T = KL^2, \qquad \chi PL/KL^2 \longrightarrow \chi.$$

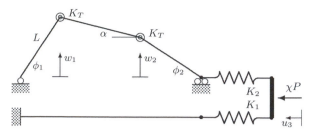

Figure 7.17. 3DoF simple model for a plate.

The total potential becomes

$$\Pi \quad \Rightarrow \quad \tfrac{1}{2}x_1^2 + \tfrac{1}{2}x_2^2 - \tfrac{1}{2}x_1^2 x_2 + \tfrac{1}{4}x_1^4 - \chi.$$

The equilibrium equations are

$$\mathcal{F}_1 = \frac{\partial \Pi}{\partial x_1} = \big[1 + x_1^2 - x_2\big]x_1 = 0,$$

$$\mathcal{F}_2 = \frac{\partial \Pi}{\partial x_2} = -\tfrac{1}{2}x_1^2 + x_2 - \chi = 0.$$

There are two equilibrium paths. The first is obtained by setting $x_1 = 0$; the second is obtained by setting the first bracketed terms to zero. These result in

$$\text{I: } x_1 = 0, \quad x_2 = \chi,$$
$$\text{II: } x_1 = \sqrt{2[\chi - 1]}, \quad x_2 = 2\chi - 1.$$

Note that these paths do not exist for all positive loads; for example, the second path exists only for $\chi \geq 1$.

Model II. Secondary Buckling and Mode Jumping

The phenomenon of mode jumping is intimately associated with the interaction of two buckling modes. Our objective here is to develop some simple models through which we can explore some of its aspects.

Because mode jumping is associated with the interaction of two modes, we need at least two flexural DoFs. The simple model shown in Figure 7.17 (which is a modified version of that introduced by Stein [117]) has sufficient DoFs. Consider the torsional springs first. With the small-deflection assumption, the angles are given by

$$\phi_1 \approx \frac{w_1}{L}, \qquad \phi_2 \approx \frac{w_2}{L}, \qquad \alpha \approx \frac{w_1 - w_2}{L}.$$

The twist of the springs are then

$$\phi_1 = \phi_1 + \alpha \approx \frac{2w_1 - w_2}{L}, \qquad \phi_2 = \phi_2 - \alpha \approx \frac{2w_2 - w_1}{L}.$$

The strain energy is

$$\mathcal{U}_T = \tfrac{1}{2}K_T\big[\frac{2w_1 - w_2}{L}\big]^2 + \tfrac{1}{2}K_T\big[\frac{2w_2 - w_1}{L}\big]^2 = \frac{K_T}{2L^2}\big[5w_1^2 - 8w_1 w_2 + 5w_2^2\big].$$

Introduce new generalized coordinates defined as

$$x_1 \equiv \tfrac{1}{2}(w_1 + w_2), \quad x_2 \equiv \tfrac{1}{2}(w_1 - w_2) \quad \text{or} \quad w_1 = x_1 + x_2, \quad w_2 = x_1 - x_2,$$

and $x_3 \equiv u_3$, where x_1 and x_2 are the amplitudes of the symmetric and antisymmetric deformation modes. After substitution, we get

$$\mathcal{U}_T = \frac{K_T}{L^2}\big[x_1^2 + 9x_2^2\big].$$

This result is interesting for two reasons. First, the energies of the two modes are uncoupled, and we can utilize this to add springs without affecting the coupling. Second, the antisymmetric mode has a larger coefficient, and hence we expect the symmetric mode to occur first.

The end shortening of the links is computed as

$$\Delta = L[3 - \cos\phi_1 - \cos\alpha - \cos\phi_2] \approx \frac{1}{2L}\big[w_1^2 + [w_1 - w_2]^2 + w_2^2\big] = \frac{1}{L}[x_1^2 + 3x_2^2].$$

The strain energy of the axial springs is therefore

$$\begin{aligned}
\mathcal{U}_S &= \tfrac{1}{2}K_1 u_3^2 + \tfrac{1}{2}K_2[u_3 - \Delta]^2 \\
&= \tfrac{1}{2}[K_1 + K_2]x_3^2 - \tfrac{1}{2}K_2[x_1^2 + 3x_2^2]x_3/L + \tfrac{1}{2}K_2[x_1^2 + 3x_2^2]^2/L^2.
\end{aligned}$$

We see the nonlinear contribution of the K_2 spring in the second and third terms.

We can write the total potential for the problem as

$$\begin{aligned}
\Pi = \mathcal{U} + \mathcal{V} = {}&\frac{K_T}{L^2}\big[x_1^2 + 9x_2^2\big] + \tfrac{1}{2}\big[K_1 + K_2\big]x_3^2 \\
&- \frac{K_2}{L}\big[x_1^2 + 3x_2^2\big]x_3 + \frac{K_2}{2L^2}\big[x_1^2 + 3x_2^2\big]^2 - \chi P x_3.
\end{aligned} \tag{7.28}$$

This is the general structure of the potential function. This model can exhibit a bifurcation, but it cannot exhibit a mode jump even though it has sufficient DoFs. We can utilize the uncoupling of the modes to modify the potential in various ways. Consider a nonlinear axial and a nonlinear torsional spring attached at the center of the middle link; then we just add to the potentials

$$\mathcal{U}_A = \tfrac{1}{2}\alpha_1 x_1^2 + \tfrac{1}{4}\alpha_2 x_1^4 + \cdots +, \qquad \mathcal{U}_T = \tfrac{1}{2}\beta_1 x_2^2 + \tfrac{1}{4}\beta_2 x_2^4 + \cdots +, \tag{7.29}$$

respectively. Note that both α_2 and β_2 can be either positive or negative. This gives us a mechanism to change the parameters of the system without affecting the mechanics of the problem.

Allman [2], as part of a discussion of mode jumping in plates, introduced an idealized problem with chosen parameters that make the manipulations less cumbersome. The potential energy (modified slightly) of the system is

$$\Pi = K\big[\tfrac{1}{2}(x_1^2 + 4x_2^2 + x_3^2) - \tfrac{1}{2}(x_1^2 + 3x_2^2)x_3 + \tfrac{1}{4}(x_1^4 + 5x_1^2 x_2^2 + 6x_2^4)\big] - \chi P x_3.$$

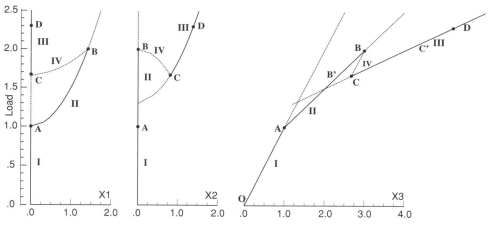

Figure 7.18. Equilibrium paths for Allman's problem.

We can have a combination of Equations (7.28) and (7.29) coincide with Allman's by making the following associations:

$$K_1 = \tfrac{2}{5}K, \quad P = K, \quad \alpha_1 = 4K - \frac{2K_T}{L^2}, \quad \alpha_2 = \frac{47}{72};$$

$$K_2 = \tfrac{3}{5}K, \quad L = \frac{6}{5}, \quad \beta_1 = K - 18K_T, \quad \beta_2 = -\frac{13}{2}.$$

The main conclusion of this example is that, in terms of the mechanical model with nonlinear springs, to have mode jumping we need a softening mechanism associated with the antisymmetric mode in the large postbuckling region; that is, β_2 must be negative.

The equilibrium equations are

$$\mathcal{F}_1 = \frac{\partial \Pi}{\partial x_1} = x_1\left[1 + x_1^2 + \tfrac{5}{2}x_2^2 - x_3\right] = 0,$$

$$\mathcal{F}_2 = \frac{\partial \Pi}{\partial x_2} = x_2\left[4 + \tfrac{5}{2}x_1^2 + 6x_2^2 - 3x_3\right] = 0,$$

$$\mathcal{F}_3 = \frac{\partial \Pi}{\partial x_3} = \left[-\tfrac{1}{2}(x_1^2 + 3x_2^2) + x_3\right] - \chi = 0.$$

There are four equilibrium paths. We obtain the first three by setting $x_1 = 0$, $x_2 = 0$; $x_2 = 0$; and $x_1 = 0$; respectively. We obtain the fourth path by setting the first two bracketed terms to zero. This results in

$$\text{I: } x_1 = 0, \quad x_2 = 0, \quad x_3 = \chi;$$

$$\text{II: } x_1 = \sqrt{2}\sqrt{\chi - 1}, \quad x_2 = 0, \quad x_3 = 2\chi - 1;$$

$$\text{III: } x_1 = 0, \quad x_2 = \sqrt{(6\chi - 8)/3}, \quad x_3 = 4\chi - 4;$$

$$\text{IV: } x_1 = \sqrt{6\chi - 10}, \quad x_2 = \sqrt{4 - 2\chi}, \quad x_3 = \chi + 1.$$

These four equilibrium paths are shown plotted in Figure 7.18. Note that these paths do not exist for all positive loads; for example, the second path exists only for $\chi \geq 1$.

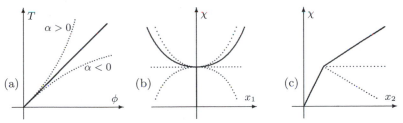

Figure 7.19. Simple model for elastic-plastic response: (a) nonlinear torsional spring behavior, (b) transverse deflection, (c) axial shortening.

We are also interested in distinguishing the stable and unstable portions of the paths. We get this stability information by looking at the spectral properties of the tangent stiffness matrix – Reference [30] shows how this is done. In Figure 7.18, the unstable equilibrium path segments are indicated as dashed lines. From this figure, equilibrium state B would result in a mode jump because for a small increase in load the current path is unstable and the nearest stable path is at C′.

Model III. Elastic-Plastic Postbuckling
The simple model shown in Figure 7.16 will be adapted to the plastic buckling case.

The effect of the plasticity will be introduced by means of a nonlinear torsion spring stiffness. Specifically, we modify the potentials by replacing $\mathcal{U}_T$ with

$$\mathcal{U}_T = \tfrac{1}{2} K_T[\phi^2 + \tfrac{1}{2}\alpha\phi^4 + \cdots +], \qquad T = \frac{\partial \mathcal{U}}{\partial \phi} = K_T[1 + \alpha\phi^2 + \cdots +]\phi.$$

Note that α can be either positive or negative, as shown in Figure 7.19(a). When α is negative, the torque–twist behavior resembles that of a beam in elastic-plastic bending. Also note that α has a contribution $\phi^4 \to x_1^4$ similar to the noncoupling nonlinear term arising from K_2.

The total potential for the problem can now be written as

$$\Pi = \mathcal{U} + \mathcal{V}$$
$$= \tfrac{1}{2}K_T[4x_1^2 + \tfrac{1}{2}\alpha 32x_1^4] + \tfrac{1}{2}(K_1 + K_2)L^2 x_2^2 - \tfrac{1}{2}K_2 L^2[2x_1^2 x_2 - x_1^4] - \chi PLx_2.$$

The equilibrium equations are

$$\mathcal{F}_1 = \frac{\partial \Pi}{\partial x_1} = \big[\,4K_T[1 + \alpha 8x_1^2] - 2K_2 L^2[x_2 - x_1^2]\big]x_1 = 0,$$

$$\mathcal{F}_2 = \frac{\partial \Pi}{\partial x_2} = (K_1 + K_2)L^2 x_2 - K_2 L^2 x_1^2 - \chi PL = 0.$$

There are two equilibrium paths. We obtain the first by setting $x_1 = 0$; we obtain the second by setting the first bracketed terms to zero. This results in

$$\text{I: } x_1 = 0, \quad x_2 = \frac{\chi P}{(K_1 + K_2)L};$$

$$\text{II: } x_1 = \sqrt{\frac{\chi P - 2K_1 L\gamma}{K_1 L[1 + \alpha 16\gamma]}}, \quad x_2 = \frac{\chi P[1 + \alpha 16\gamma] - 2K_T/L}{K_1 L[1 + \alpha 16\gamma]}, \quad \gamma \equiv \frac{(K_1 + K_2)K_T}{K_1 K_2 L^2}.$$

The first path is simply the linear end shortening of the springs before buckling occurs. For the second path, buckling occurs when

$$\chi P = 2K_1 L\gamma$$

and is independent of the torsional nonlinearity α. Where α plays a role is in the postbuckling behavior, as shown in Figures 7.19(b) and 7.19(c)): When it is positive, the postbuckle slope is steeper; when it is negative, the postbuckle slope is shallower. Indeed, when $\alpha = -1/(16\gamma)$, the postbuckle slope is flat, indicating a neutral stability. Of interest to us here is that, if α is sufficiently negative, that is, $\alpha < -1/(16\gamma)$, then the postbuckling is unstable.

As done for the previous cases, we simplify some of the coefficients so as to make the essential mechanism clearer. Let

$$K_1 = \tfrac{1}{2}K, \qquad K_2 = \tfrac{1}{2}K, \qquad 4K_T = KL^2, \qquad \chi PL/KL^2 \longrightarrow \chi.$$

The total potential becomes

$$\Pi \quad \Rightarrow \quad \tfrac{1}{2}x_1^2 + \tfrac{1}{2}x_2^2 - \tfrac{1}{2}x_1^2 x_2 + \tfrac{1}{4}(1+\alpha)x_1^4 - \chi.$$

The equilibrium equations are

$$\mathcal{F}_1 = \frac{\partial \Pi}{\partial x_1} = \left[1 + (1+\alpha)x_1^2 - x_2\right]x_1 = 0,$$

$$\mathcal{F}_2 = \frac{\partial \Pi}{\partial x_2} = -\tfrac{1}{2}x_1^2 + x_2\right] - \chi = 0.$$

There are two equilibrium paths. We obtain the first by setting $x_1 = 0$; we obtain the second by setting the first bracketed terms to zero. This results in

$$\text{I:} \quad x_1 = 0, \quad x_2 = \chi;$$

$$\text{II:} \quad x_1 = \sqrt{\frac{\chi - 1}{\tfrac{1}{2} + \alpha}}, \quad x_2 = \frac{(1+\alpha)\chi - \tfrac{1}{2}}{\tfrac{1}{2} + \alpha}.$$

Note that, for $\alpha > -\tfrac{1}{2}$, the second path exists only for $\chi \geq 1$. A negative value of α such that $\tfrac{1}{2} + \alpha < 0$ gives a transverse deflection

$$x_1 = \sqrt{\frac{1 - \chi}{|\tfrac{1}{2} + \alpha|}}.$$

This is a path with increasing displacement for decreasing load. That is, this path is unstable.

A negative value of α is to be associated with a relatively flat stress–strain curve as observed for elastic-plastic materials. We therefore anticipate that a postbuckling analysis of elastic-plastic beams should predict an unstable postbuckling behavior if there is not much work hardening. This is completely unlike the prediction based on a linear elastic analysis.

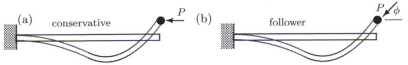

Figure 7.20. Small displacements of a cantilevered beam with two types of end loads: (a) load remains horizontal and is conservative, (b) load remains tangential and is nonconservative.

Model IV. Effect of Follower Forces

Consider a column or beam (as shown in Figure 7.20) with a quasi-static axial load P that is treated as a parameter. There are two situations that can prevail. The first is that in which the load P always remains horizontal. In this case P is a conservative load, and hence we can write a potential for it. The second is that in which it remains tangential to the tip; although the magnitude remains essentially constant, the fact that it changes its orientation means it is a changing force, and hence we cannot write a potential for it.

Consider the mechanical model of rigid bars and torsional springs shown in Figure 7.21; the two rigid bars are of length L; two torsional springs are of stiffness K_1 and K_2, and two masses M_1 and M_2. The two DoFs are the rotations of the bars. This is similar to the model used in References [30, 139] to discuss dynamic instability.

The strain energy is confined to the two springs and is given by

$$ \mathcal{U} = \tfrac{1}{2} K_1 \phi_1^2 + \tfrac{1}{2} K_2 [\phi_2 - \phi_1]^2. $$

The kinetic energy is confined to just the masses. The positions of these masses are

$$ x_2 = L \cos \phi_1, \qquad y_2 = L \sin \phi_1, $$
$$ x_3 = L \cos \phi_1 + L \cos \phi_2, \qquad y_3 = L \sin \phi_1 + L \sin \phi_2. $$

This leads to the velocities

$$ \dot{x}_2 = -L \sin \phi_1 \dot{\phi}_1, \qquad \dot{y}_2 = L \cos \phi_1 \dot{\phi}_1; $$
$$ \dot{x}_3 = -L \sin \phi_1 \dot{\phi}_1 - L \sin \phi_2 \dot{\phi}_2, \qquad \dot{y}_3 = L \cos \phi_1 \dot{\phi}_1 + L \cos \phi_2 \dot{\phi}_2. $$

The kinetic energy is therefore

$$ \mathcal{T} = \tfrac{1}{2} M_2 [\dot{x}_2^2 + \dot{y}_2^2] + \tfrac{1}{2} M_3 [\dot{x}_3^2 + \dot{y}_3^2] $$
$$ = \tfrac{1}{2} M_2 L^2 \dot{\phi}_1^2 + \tfrac{1}{2} M_3 L^2 [\dot{\phi}_1^2 + \dot{\phi}_2^2 + 2 \cos(\phi_2 - \phi_1) \dot{\phi}_1 \dot{\phi}_2]. $$

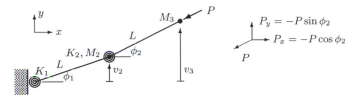

Figure 7.21. Discretized beam and the resolved components of the applied follower load.

Lagrange's equations reduce to

$$\frac{d}{dt}\left(M_2 L^2 \dot{\phi}_1 + M_3 L^2 [\dot{\phi}_1 + \cos(\phi_2 - \phi_1)\dot{\phi}_2]\right)$$

$$- M_3 L^2 \sin(\phi_2 - \phi_1)\dot{\phi}_1 \dot{\phi}_2 + K_1 \phi_1 - K_2[\phi_2 - \phi_1] - Q_1 = 0,$$

$$\frac{d}{dt}\left(M_3 L^2 [\dot{\phi}_2 + \cos(\phi_2 - \phi_1)\dot{\phi}_1]\right) + M_3 L^2 \sin(\phi_2 - \phi_1)\dot{\phi}_1 \dot{\phi}_2 + K_2[\phi_2 - \phi_1] - Q_1 = 0.$$

Let $K_2 = K_1 = K$, $M_2 = 2M_3 = 2M$; then the equations of motion can be expressed as

$$K \begin{bmatrix} 2 & -1 \\ -1 & 1 \end{bmatrix} \begin{Bmatrix} \phi_1 \\ \phi_2 \end{Bmatrix} + ML^2 \begin{bmatrix} 3 & \cos \Delta \\ \cos \Delta & 1 \end{bmatrix} \begin{Bmatrix} \ddot{\phi}_1 \\ \ddot{\phi}_2 \end{Bmatrix} + ML^2 \sin(\Delta)\dot{\phi}_1 \dot{\phi}_2 \begin{Bmatrix} -1 \\ 1 \end{Bmatrix}$$

$$= \begin{Bmatrix} Q_1 \\ Q_2 \end{Bmatrix},$$

with $\Delta \equiv \phi_2 - \phi_1$. We wish to concentrate on the effect of the follower forces; hence we assume that the nonlinearities in this system can be neglected so that the system reduces to

$$K \begin{bmatrix} 2 & -1 \\ -1 & 1 \end{bmatrix} \begin{Bmatrix} \phi_1 \\ \phi_2 \end{Bmatrix} + ML^2 \begin{bmatrix} 3 & 1 \\ 1 & 1 \end{bmatrix} \begin{Bmatrix} \ddot{\phi}_1 \\ \ddot{\phi}_1 \end{Bmatrix} = \begin{Bmatrix} Q_1 \\ Q_2 \end{Bmatrix}.$$

It remains now to consider the applied loads.

We use virtual work to establish the equivalent loads. The two components of applied force are

$$P_x = -P \cos \phi_2, \qquad P_y = -P \sin \phi_2.$$

The virtual horizontal displacement of the tip is

$$\delta u_3 = \delta x_3 = \delta[L \cos \phi_1 + L \cos \phi_2] = -L \sin \phi_1 \delta \phi_1 - L \sin \phi_2 \delta \phi_2.$$

The virtual work of the horizontal component is then

$$\delta W_x = P_x \delta u_3 = (-P \cos \phi_2)(-L \sin \phi_1 \delta \phi_1 - L \sin \phi_2 \delta \phi_2)$$

$$= PL[\sin \phi_1 \cos \phi_2 \delta \phi_1 + \sin \phi_2 \cos \phi_2 \delta \phi_2].$$

The virtual vertical displacement of the tip is $\delta v_3 = \delta y_3$, so that the virtual work of the vertical component is

$$\delta W_y = P_y \delta v_3 = (-P \sin \phi_2)(L \cos \phi_1 \, \delta \phi_1 + L \cos \phi_2 \, \delta \phi_2)$$

$$= PL[-\sin \phi_2 \cos \phi_1 \delta \phi_1 - \sin \phi_2 \cos \phi_2 \delta \phi_2].$$

The total virtual work done by both components is

$$\delta W = \delta W_x + \delta W_y = PL[\sin \phi_1 \cos \phi_2 - \sin \phi_2 \cos \phi_1] \delta \phi_1 = -PL \sin(\phi_2 - \phi_1)\delta \phi_1.$$

The generalized forces associated with the DoFs ϕ_1 and ϕ_2 give the virtual work

$$\delta W = Q_1 \delta \phi_1 + Q_2 \delta \phi_2.$$

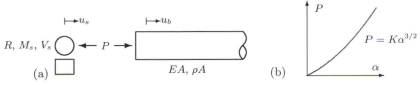

Figure 7.22. Impact of a rod with a spherical ball or flat-ended short cylinder: (a) notation and free-body diagram, (b) Hertzian contact law.

We now make the equivalence of the virtual work of the actual system to the virtual work of the generalized system, from which we conclude that $Q_2 = 0$. The system of equations then becomes

$$K \begin{bmatrix} 2 & -1 \\ -1 & 1 \end{bmatrix} \begin{Bmatrix} \phi_1 \\ \phi_2 \end{Bmatrix} + ML^2 \begin{bmatrix} 3 & 1 \\ 1 & 1 \end{bmatrix} \begin{Bmatrix} \ddot{\phi}_1 \\ \ddot{\phi}_2 \end{Bmatrix} = -PL \begin{Bmatrix} \sin(\phi_2 - \phi_1) \\ 0 \end{Bmatrix}.$$

The applied loads are deformation dependent. In the special case of the small-angle approximation, this reduces to

$$K \begin{bmatrix} 2 & -1 \\ -1 & 1 \end{bmatrix} \begin{Bmatrix} \phi_1 \\ \phi_2 \end{Bmatrix} - PL \begin{bmatrix} 1 & -1 \\ 0 & 0 \end{bmatrix} \begin{Bmatrix} \phi_1 \\ \phi_2 \end{Bmatrix} + ML^2 \begin{bmatrix} 3 & 1 \\ 1 & 1 \end{bmatrix} \begin{Bmatrix} \ddot{\phi}_1 \\ \ddot{\phi}_2 \end{Bmatrix} = \begin{Bmatrix} 0 \\ 0 \end{Bmatrix}.$$

Note that the contribution of the load term is that of a nonsymmetric stiffness matrix, that is, a nonsymmetric $[K_G]$. This contribution can give rise to a dynamic instability if P is sufficiently large. Additional aspects of this model are developed in the explorations of Section 6.6.

7.6 Simple Models for Loadings

One significant source of complexity in a problem is the modeling of the loads. All FE codes have a number of standard types of loads such as point loads with defined histories, sinusoidal loads, gravity loads, and so on. Of concern here is the situation in which the a priori information is about the physical problem, not the loads directly, and we wish to establish the actual loads.

Typically, the construction of the model involves concatenating multiple models. For example, a sphere impacting a rod involves the rigid-body equation of motion of the sphere, a contact model between the sphere and the rod, and a wave-propagation model for the rod. The main examples illustrated are for impact and blast loadings.

Model I. Elastic Impact of Rods

Consider a long rod impacted by a spherical ball or by a small flat-ended cylinder traveling at velocity V_s, as shown in Figure 7.22. We are interested in the difference made by the two contact conditions.

This analysis is based on Reference [105]. Let u_s be the motion of the center of mass of the ball and u_b be the motion of a representative point along the rod.

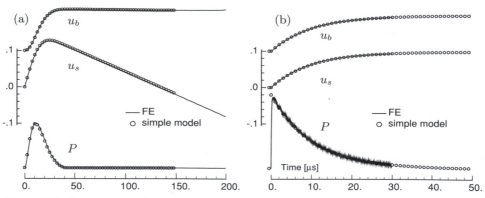

Figure 7.23. Responses for impacted rod with $V_s = 250\,\text{m/s}$ (10000 in/s): (a) Hertzian contact, (b) flush contact.

The equation of motion of the impactor is

$$M_s \ddot{u}_s = -P.$$

Assume that the impact generates a longitudinal plane wave that travels in the rod; then the velocity of our representative point is related to the force by [28]

$$\dot{u}_b = -\frac{c_o \sigma}{E} = \frac{c_o}{EA}P \quad \text{or} \quad \frac{EA}{c_o}\dot{u}_b = P,$$

with $c_o = \sqrt{E/\rho}$ being the longitudinal wave speed. This is the equation of motion for a SDoF system with damping but no mass or stiffness. Assume that this relation is valid even near the impact site. Note that both equations of motion are linear; typically in impact-type problems it is the contact conditions that will give rise to nonlinearities and thus complicate the problem.

Let the contact between the ball and the flat end of the rod be described by the Hertzian contact law; that is, the force and relative indentation are given by [46, 56, 110]

$$P = K\alpha^{3/2}, \quad \alpha = u_s - u_b, \quad K = \frac{4}{3}\sqrt{R_s}\left[\frac{k_s k_b}{k_s + k_b}\right], \quad k_i \equiv \frac{E_i}{1 - \nu_i^2}.$$

This allows u_s and u_b to be obtained from the following differential equations:

$$\ddot{u}_s = -\frac{K}{M_s}[u_s - u_b]^{3/2}, \quad \dot{u}_b = \frac{c_o K}{EA}[u_s - u_b]^{3/2},$$

with the initial conditions that, at $t = 0$,

$$u_s = 0, \quad \dot{u}_s = V_s, \quad u_b = 0.$$

These equations can be programmed to allow calculation of the resulting histories. The responses are shown in Figure 7.23(a). The force history is nonsymmetric – two impacting spheres produce a symmetric force history. Note that the ball has a rebound velocity.

The preceding differential equations can be combined into a single equation for the indentation α as

$$M_s\ddot{\alpha} + C\alpha^{1/2}\dot{\alpha} + K\alpha^{3/2} = 0, \qquad C \equiv \frac{3c_o M_s K}{2EA},$$

with the initial conditions that, at $t = 0$, $\alpha = 0$, $\dot{\alpha} = V_s$. Note that this has the form for the free vibration of a nonlinear spring–mass system with a nonlinear viscous damper – the apparent damping arises because the long rod is a conduit of energy out of the contact region. On the assumption that the coefficient C is small, a first integration of the differential equation leads to

$$P_{\max} = K^{2/5}\left[\tfrac{5}{4}M_s V_s^2\right]^{3/5}, \qquad T = \pi\sqrt{\frac{M_s}{K}}\left[\frac{K}{M_s V_s^2}\right]^{1/10},$$

where T is the contact time. The conclusion is that the maximum force is almost directly proportional to velocity whereas the contact time is fairly insensitive to the velocity.

By way of contrast, consider the case in which the impactor is flat ended so that the contact is flush. The equations of motion are similar to the preceding ones, but we impose the condition that the contacting surfaces have the same motion. That is,

$$\ddot{u}_s = \ddot{u}_b \qquad \text{or} \qquad -\frac{1}{M_s}P = \frac{c_o}{EA}\frac{dP}{dt}.$$

This differential equation for $P(t)$ can be integrated to give

$$P(t) = \frac{V_s EA}{c_o}e^{-\beta t}, \qquad u_s(t) = u_b(t) = \frac{V_s}{\beta}[1 - e^{-\beta t}], \qquad \beta \equiv \frac{EA}{c_o M_s}.$$

These responses are shown in Figure 7.23(b). What is interesting is that the maximum force occurs at $t = 0$, and it does not depend on the mass of the impactor. The reason is that the actual BC is that of an imposed velocity, and this, of course, is independent of the mass. Where the mass has a big effect is on the time decay through the parameter β – the larger the mass, the flatter the force trace, and the longer the duration of the force. Thus two very long, flat-ended rods, impacting collinearly, would produce a force history that is a step function that exists for a very long time.

Model II. Impact of a Rod on a Half-Space
The focus of the previous model was on the response in the stationary rod that is due to a relatively small impactor; here we consider a relatively long rod to be in motion impacting a very large body.

A simple model for blunt impact is shown in Figure 7.24(a), where the block is treated as an expanding tapered rod. The striker is treated as a uniform rod and the force is related to the particle velocity by [28]

$$P = \frac{E_1 A}{c_1}[V_o - \dot{v}], \qquad c_1 = \sqrt{E_1/\rho_1},$$

where the contact area is $A = \pi a^2$ and $v(t)$ is the motion of the interface. The exact solution for waves propagating in a tapered rod involves Bessel functions [28] but,

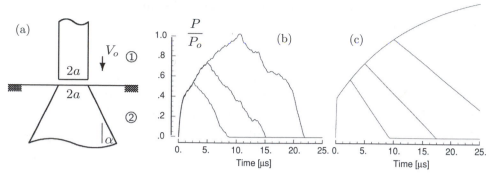

Figure 7.24. Rod impacting a half-space: (a) blunt contact modeled as a uniform rod impacting a tapered rod, (b) FE results for different striker lengths, (c) simple model results.

as an approximation, the force–deformation relation can be written as

$$P = \frac{E_2 A}{c_2}\dot{v} + \frac{E_2 A}{2a}\tan(\alpha)v \qquad \text{or} \qquad \frac{E_2 A}{c_2}\dot{v} + \frac{E_2 A}{2a}\tan(\alpha)v = P,$$

with $c_2 = \sqrt{E_2/\rho_2}$. This is the equation of motion for a SDoF spring–damper system with no mass. The effective angle of the taper, α, is unknown and is treated as an adjustable parameter. The larger this angle, the stiffer the block appears. Eliminating P and solving for $v(t)$ leads to

$$v = \frac{\beta V_o}{\gamma}[1 - e^{-\gamma t}], \quad \dot{v} = \beta V_o e^{-\gamma t}, \quad \beta \equiv \frac{1}{[1 + E_2 c_1/E_1 c_2]}, \quad \gamma \equiv \frac{(E_2 A/2a)}{(E_1 A/c_1)}\beta \tan(\alpha),$$

and the expression for the load history is

$$P(t) = V_o \frac{E_1 A}{c_1}\left[1 - \beta e^{-\gamma t}\right].$$

At first contact, the common velocity becomes βV_o; if the rods were of the same material, the common velocity would be $\frac{1}{2}V_o$. On further loading, the increasing stiffness causes the exponential behavior. For very long rods, the force would increase and asymptote to $P_{\max} = V_o E_1 A/c_1$, which is the value obtained if the striker had impacted a rigid stop.

Unloading occurs when the wave reflects off the free end of the rod and returns to the contact site. The unloading wave causes a velocity change given approximately by

$$\Delta \dot{v} = V_o - \beta V_o.$$

The time to unload is therefore $v_{\max}/\Delta\dot{v}$. These behaviors are shown in Figure 7.24(c), where α was chosen to match the rising part of the curves in Figure 7.24(b). Clearly, the simple model captures the basic behaviors of the FE results.

Model III. Elastic Impact of Plates and Beams

Consider the situation in which the impacted body is extended in two directions but relatively short in the impacted direction. We assume that this can be modeled as a plate in flexure. A transverse point force applied to an infinite plate of thickness h causes a velocity at the load site of [28]

$$\dot{w}(t) = \frac{1}{8\sqrt{\rho h D}} P(t) \quad \text{or} \quad 8\sqrt{\rho h D}\, \dot{w}(t) = P(t),$$

with $D = Eh^3/12(1 - \nu^2)$. This is the equation of motion for a SDoF system with damping but no mass or stiffness.

Consider the impact by a sphere; then, similarly to the rod impact of the first case, we get the equations of motion

$$\ddot{w}_s = -\frac{K}{M_s}[w_s - w_p]^{3/2}, \qquad \dot{w}_b = \frac{K}{8\sqrt{\rho h D}}[w_s - w_p]^{3/2},$$

with the initial conditions that, at $t = 0$,

$$w_s = 0, \qquad \dot{w}_s = V_s, \qquad w_p = 0.$$

Because these equations are of the same form as for a rod, then the two problems give the same results if

$$\frac{c_o}{EA} \Leftrightarrow \frac{1}{8\sqrt{\rho h D}} \quad \text{or} \quad h \Leftrightarrow R\left[\frac{3(1 - \nu^2)\pi^2}{16}\right]^{1/4} \approx 1.14\,R.$$

Now consider the impacted body to be extended in only one direction and assume that it behaves like a beam in flexure. The response of an infinite beam to a transverse impact has the rather complicated hereditary relation [28]

$$\dot{v}(t) = \frac{2B}{h}\sqrt{\frac{EI}{\rho A}}\int_0^t \frac{P(\tau)}{\sqrt{t - \tau}}\,d\tau, \qquad B = \frac{h}{4EI\sqrt{2\pi}}\left[\frac{EI}{\rho A}\right]^{1/4}.$$

This does not have a simple interpretation in terms of a SDoF spring–mass system. Because of the hereditary integral, the motion of the beam will persist long after the force has gone to zero.

Model IV. Impact with Plasticity

The contact between two bodies invariably causes very large local stresses. If the materials are not hardened, then these stresses cause local plasticity. Here we consider the effect of this plasticity on the contacting force history.

The basic assumption for the simple model is that one of the materials behaves rigid perfectly plastic and the other is rigid. That is, there is no elastic contribution during the loading. The geometry of the blunt impact model is shown in Figure 7.25(a). Assuming no volume change, the volume at an arbitrary time when the striker would have displaced an amount v_1 is

$$v(t) = LA = L\pi R^2 = [L - v_1]\pi a^2 = \text{constant}.$$

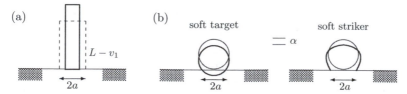

Figure 7.25. Simple models for plastic impact: (a) blunt striker, (b) a hard striker with soft target, and a soft striker with hard target will have the same result.

The contacting force is related to the average stress (which is assumed to be at the yield level) as

$$P = \sigma_Y \pi a^2 = \sigma_Y \frac{L\pi R^2}{L - v_1} = \sigma_Y \frac{LA}{L - v_1}.$$

The equation of motion is

$$M_1 \ddot{v}_1 = -P = -\sigma_Y \frac{AL}{L - v_1} \approx -\sigma_Y A\left[1 + \frac{v_1}{L}\right] \quad \text{or} \quad \ddot{v}_1 + \frac{\sigma_Y A}{M_1 L}v_1 = -\frac{\sigma_Y A}{M_1}.$$

The approximation is based on the assumption that the shortening v_1 is much smaller than the original length L of the rod. The resulting equation of motion is a linear inhomogeneous equation that can be solved directly. Imposing the initial conditions of $v_1 = 0$ and $\dot{v}_1 = V_o$ gives the solution

$$v_1(t) = (1/\omega)V_o \sin \omega t - L[1 - \cos \omega t], \qquad P(t) = M_1 \omega^2 [(V_o/\omega) \sin \omega t + L \cos \omega t],$$

where $\omega \equiv \sqrt{\sigma_Y A / M_1 L}$. This equation indicates that the contact force is independent of the elastic moduli of either material (as per assumption) and that it depends on only the yield stress and original dimensions of the striker. As in the elastic case, it gives an initial jump in force, this time of amount $M_1 \omega^2 L = \sigma_Y A$. This is followed by a slight increase in force.

Figure 7.26(a) shows the force history for two velocities. After an initial overshoot that occurs within the first microsecond or so, the results show that the contact

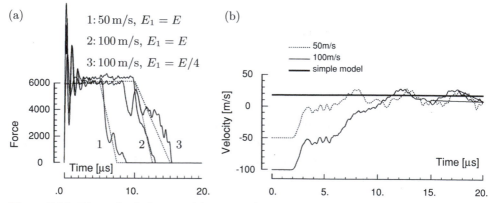

Figure 7.26. Blunt plastic impact: (a) Comparison of FE and simple model force histories, (b) velocity at free end of striker showing rebound.

force is not sensitive to velocity but the contact duration is directly proportional to velocity. Neither has an obvious dependence on the elastic properties but the time to unload does. Figure 7.26(b) shows that there is a significant loss of momentum because the rebound velocity is barely above zero.

The contact time during which plasticity dominates ends when $\dot{v}_1 = 0$, giving

$$T = \frac{1}{\omega}\tan^{-1}\left(\frac{V_o}{L\omega}\right) \approx \frac{V_o}{L\omega^2} = \frac{M_s V_o}{\sigma_Y A} = \frac{M_s V_o}{P_{max}} = \frac{\rho_s V_o L}{\sigma_Y}.$$

This shows that the plastic contact duration is directly proportional to the original velocity and mass of the striker. When elastic recovery begins in the striker, the force distribution varies from P_{max} to 0 over the length of the rod. It takes a time of L/c_1 for the unloading wave to travel from one end to the other, and therefore the elastic unloading time depends on the modulus and length of the striker. The rebound velocity can be estimated from the momentum change that occurs during the elastic unloading to give

$$\tfrac{1}{2}P_{max}L/c_1 = \rho_1 AL\,\dot{v}_R \qquad \text{or} \qquad \dot{v}_R = \sigma_Y/2c_1\rho_1.$$

These results are shown plotted in Figure 7.26(a) and capture the essence of the results as discussed earlier. The rebound behavior is shown in Figure 7.26(b).

Figure 7.25(b) shows two cases of round plastic impact: a hard striker with soft target, and a soft striker with hard target. In both cases, the indentation law obtained is the same; we need only substitute the yield stress σ_Y of the softer material. Assuming a short striker of radius R_1, the relative indentation α and the contact length $2a$ are related by

$$\alpha = v_1 - v_2 = \frac{a^2}{2R_1}.$$

The contacting force is related to the average stress (which is assumed to be at the yield level) as

$$P = \sigma_Y \pi a^2.$$

Combine both of these together under the assumption that the target does not move to get the contact law

$$P = \sigma_Y \pi 2R_1 v_1.$$

This is a linear contact law. The equation of motion of the striker can now be written as

$$M_1\ddot{v}_1 = -P = -\sigma_Y 2\pi R_1 v_1 \qquad \text{or} \qquad \ddot{v}_1 + \frac{\sigma_Y 2\pi R_1}{M_1}v_1 = 0.$$

The resulting equation of motion is similar to the blunt impact and has the solution

$$v_1(t) = (1/\omega)V_o \sin\omega t, \qquad P(t) = M_1\omega^2(V_o/\omega)\sin\omega t,$$

where $\omega \equiv \sqrt{\sigma_Y 2\pi R_1/M_1}$.

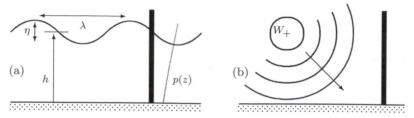

Figure 7.27. Fluid loading: (a) water waves, (b) blast loading.

The contact ends when $\dot{v}_1 = 0$, giving

$$\omega T = \pi/2 \quad \text{or} \quad T = \sqrt{\frac{\pi}{8} \frac{M_1}{\sigma_Y R_1}} = \pi R_1 \sqrt{\frac{\rho_s}{6\sigma_Y}}.$$

This shows that the plastic contact duration does not depend on the original velocity of the striker. An interesting interpretation of the last form is to see $\sqrt{\sigma_Y/\rho_s} = c_p$ as a wave speed; then the contact time is the time for the wave to travel the dimension of the striker. Keep in mind, however, that because $\sigma_Y \approx E/200$, then $c_p \approx c_o/15$, which is an order of magnitude slower than the corresponding elastic wave speed.

References [9, 15] have more examples of simple models involving plastic deformations.

Model V. Water Waves and Blast Loading

A structure standing in deep still water experiences a pressure distribution proportional to the depth as $p(z) = -\rho g z$, where ρ is the density of water. When the water is disturbed in the form of water waves, this pressure distribution changes and has an oscillatory component. Reference [119] gives a discussion of the effects of water waves on submerged structures. It is a rather complicated subject, but it shows that the pressure distribution (at a particular frequency ω) with respect to depth (z) [see Figure 7.27(a)] is

$$p(z) = \tfrac{1}{2}\rho g \eta \frac{gh}{\omega} \frac{\cosh\left(k[h+z]\right)}{\sinh(kh)} \cos(kx - \omega t) - \rho g z, \qquad \omega^2 = gk \tanh(kh).$$

The wave number k is related to the wavelength λ by $k = 2\pi/\lambda$ and has a transcendental relationship (called the spectrum relation) to the frequency. This specifies the pressure for a wave of given frequency ω and amplitude η. In truth, a real water wave would have a spectrum of both, and therefore some statistical representations are usually used.

For the deep-water situation ($h/\lambda \gg 1$), $\tanh(kh) \to 1$, giving the useful approximation

$$k \approx \frac{\omega^2}{g}.$$

On the other hand, for the shallow-water situation $(h/\lambda \ll 1)$, $\tanh(kh) \to kh$, giving the approximation

$$k \approx \frac{\omega}{\sqrt{gh}}.$$

For intermediate situations, the spectrum relation must be solved numerically.

A related loading situation is that of blast loading resulting from an explosion, for example. References [17, 50, 85] discuss blast loadings in both air and water; again, the complete analysis is rather complicated, but as an illustration, the pressure loading in water has the empirical relation

$$P(t) = P_m e^{-\alpha t}, \qquad P_m = 17570 \left[\frac{W^{1/3}}{R} \right]^{1.13}, \qquad \alpha = \frac{740}{W^{1/3}},$$

where P_m is the maximum pulse pressure in pounds per square inch, W is the equivalent charge weight of the explosive in pounds of TNT, and R is the distance from the source in inches. The pressure wave is assumed to propagate spherically with the speed of sound.

Once the wave reaches the structure, there can be a complex interaction between the structure and the fluid, and this is treated in texts such as References [41, 59, 91] that deal with this structure–fluid interaction problem. An assumption often made is that the structure is relatively stiff and large in comparison with the wave extent and therefore the pressure acting on the structure is twice that of the incident wave.

The other extreme is when the structure is small. In the context of waves, a small body is one that does not significantly modify the incident wave field. If the characteristic length (diameter, for example), is about 5% of the wavelength the assumption is reasonable. Examples of these structures are piles, pipelines, oil platform supports, and moorings. By way of simple modeling, consider a horizontal wave of velocity V_o incident upon a vertical cylinder with a small diameter. A velocity potential for ideal flow around the cylinder is [89]

$$\phi = V_o \left[1 + \frac{R}{r} \right] r \cos \theta.$$

The normal and tangential components of velocity on the surface are

$$\dot{u}_r = 0, \qquad \dot{u}_\theta = 2V_o \sin \theta.$$

The fluid pressure is

$$p = -\rho \frac{\partial \phi}{\partial t} - \tfrac{1}{2}\rho [\dot{u}_r^2 + \dot{u}_\theta^2] - \rho g z + C_B.$$

This is obtained from the Bernoulli equation, and C_B is called the Bernoulli constant. Integrating the pressure over the cylinder gives the fluid-induced force acting on the cylinder. The force per unit length of cylinder is

$$f_{mx} = -\int_{-\pi}^{+\pi} p(R, \theta, t) \cos \theta \, R \, d\theta.$$

Table 7.1. *Inertia and drag coefficients for*
common shapes. The cross-section $\square^*$ has
smoothed edges

$\Longrightarrow$	$\bigcirc$	$\square$	$\square^*$	$\triangleright$	$\triangleleft$	$\diamond$
C_m	2	2.5	2.5	2.3	2.3	2.2
C_d	2	2.0	0.6	2.0	1.3	1.5

Only the first pressure term integrates to nonzero. Noting that, on the surface,

$$\rho \frac{\partial \phi}{\partial t} = 2R \frac{dV_o}{dt} \cos \theta,$$

then the integral becomes

$$f_{mx} = 2\rho\pi R^2 \frac{dV_o}{dt} = 2\rho\pi R^2 \dot{V}_o.$$

This force is proportional to the acceleration of the fluid and therefore is referred to as an inertia (or mass) force.

A similar analysis can be done for other shaped bodies, and it is concluded that they all can be represented by

$$f_{mx} = C_m \rho \pi R^2 \dot{V}_o,$$

where C_m is called the inertia (or mass) coefficient. Typical values of this coefficient are shown in Table 7.1.

Model VI. Structures in Fluid Flows

Consider fluid flow past structures such are piles, pipelines, oil platform supports, and moorings. The flow induces two types of drag: skin drag from the friction between the fluid and surface, and form drag from the pressure differential across the body when the flow separates. We establish simple models for these drag forces. The structures referred to are stationary with moving fluids, but the simple analysis to follow is also applicable to moving structures such as aircraft or ships.

We use dimensional analysis to arrive at the simple model. Assume that the resultant force per unit length of cylinder ($q, [F/L]$) depends on cylinder radius ($R, [L]$), fluid velocity ($V_o, [L/T]$), fluid density ($\rho, [FT^2/L^4]$), and fluid viscosity ($\mu, [FT/L^2]$). That is,

$$q = f(R, V_o, \rho, \mu).$$

There are three fundamental units present, force, length, and time, and there are five quantities; hence we can form $5 - 3 = 2$ nondimensional groups. Using R, V_o, and ρ as the base quantities (because they contain force, length, and time), we can form

$$\Pi_1 = R^a V_o^b \rho^c q, \qquad \Pi_2 = R^d V_o^e \rho^f \mu.$$

We write these in terms of the primary dimensions $[F, L, T]$ as

$$\Pi_1 = L^a \left(\frac{L}{T}\right)^b \left(\frac{FT^2}{L^4}\right)^c \left(\frac{F}{L}\right), \qquad \Pi_2 = L^d \left(\frac{L}{T}\right)^e \left(\frac{FT^2}{L^4}\right)^f \left(\frac{FT}{L^2}\right).$$

Ensuring they are nondimensional requires that the exponents sum to zero for each primary dimension. This leads to $a = -1, b = -2, c = -1, d = -1, e = -1, f = -1$, giving the relations

$$\Pi_1 = \frac{q}{RV_o^2\rho}, \qquad \Pi_2 = \frac{\mu}{RV_o\rho} \qquad \text{or} \qquad q = RV_o^2\rho f\left(\frac{\mu}{RV_o\rho}\right).$$

The nondimensional factor inside the parenthesis is the reciprocal of the *Reynold's number*.

Thus the drag force is proportional to the velocity squared. We can write this in empirical form with a coefficient that depend on Reynold's number, shape, and surface roughness as

$$q = \rho C_d \rho R |V_o| V_o,$$

where C_d is called the drag force coefficient. Typical values for this coefficient are shown in Table 7.1.

The total force acting on the cylinder is the combination of drag and inertia forces. That is,

$$f_x = \rho C_d \rho R |V_o| V_o + C_m \rho \pi R^2 \dot{V}_o.$$

If the object also moves in the fluid, then relative velocities and accelerations are used.

7.7 QED's Gallery of ODEs

Although the simple analytical models are simpler than their full counterpart, they nonetheless often require numerical methods for their solution. This is especially true for the nonlinear dynamic models. QED has implemented many of these as part of the Analysis section. Each is integrated with the Runge–Kutta fourth-order scheme [101].

The equations behind each of the models are summarized here; most of them are treated analytically in Reference [58]. The systems can be set in motion through the applied loads or through specifying initial conditions in the form of initial displacement or initial velocity. Multiple load types can be superposed in the form

$$P(t) = a_1 t + a_2 \sin \omega t + a_3 \sum_m \Delta(t - t_m) + a_4 \sum_n H(t - t_n),$$

where $\Delta(t - t_m)$ is a ping load at time t_m and $H(t - t_n)$ is a step load at time t_n.

Model I. Duffing

This is a generalization of the Duffing equation in which the nonlinearity is in the associated stiffness term:

$$M\ddot{u} + C\dot{u} + K[1 + \alpha u^{\beta}]u = P(t).$$

Because β is adjustable, this can model both hardening and softening systems.

Model II. MDoF

This is a parameterized 2DoF linear system:

$$\begin{bmatrix} m_{11} & 0 \\ 0 & m_{22} \end{bmatrix} \begin{Bmatrix} \ddot{u}_1 \\ \ddot{u}_2 \end{Bmatrix} + \begin{bmatrix} c_{11} & 0 \\ 0 & c_{22} \end{bmatrix} \begin{Bmatrix} \dot{u}_1 \\ \dot{u}_2 \end{Bmatrix} + \begin{bmatrix} k_{11} & k_{12} \\ k_{21} & k_{22} \end{bmatrix} \begin{Bmatrix} u_1 \\ u_2 \end{Bmatrix} = \begin{Bmatrix} P_1 \\ P_2 \end{Bmatrix} f(t).$$

Because the stiffness matrix can be made nonsymmetric, this can be used to model both static and dynamic instabilities.

Model III. Pendulum

This implements a pendulum attached to a spring

$$M\ddot{u} + C\dot{u} + K[1 + \beta cos(u)]\sin(u) = P(t), \qquad K \longrightarrow Mg/L, \qquad \beta = K_s/Mg.$$

When $P(t)$ is sinusoidal, this system can be used to create chaotic dynamics [4].

Model IV. Van der Pol

This classic equation has nonlinear damping in the form

$$M\ddot{u} + C\dot{u}[1 + \alpha u^2] + Ku = P(t).$$

Making C and α negative gives rise to limit cycles.

Model V. Truss

This is a two-member symmetric truss that can have large deflections:

$$M\ddot{v} + C\dot{v} + 2K\left[\sin\alpha_o + \frac{v}{L_o}\right]\left[L - 1\right]\frac{L_o}{L} = P(t), \qquad L = \sqrt{1 + 2\frac{v}{L_o}\sin\alpha_o + (\frac{v}{L_o})^2}.$$

The stiffness is related to the geometry and modulus as $K \longrightarrow EA/L_o$. When the loads are large enough, this system will exhibit a snap-through buckling.

Model VI. Parametric

This is a variation on the Mathieu equation:

$$M\ddot{u} + C\dot{u} + K[\alpha + \beta \cos \omega t]u = P(t).$$

The parametric excitation, $\beta \cos \omega t$, is set separately from the applied loading.

Model VII. Flutter

Flutter is a dynamic instability and therefore requires at least two DoFs. This implements the normalized version,

$$
\begin{bmatrix} 1 & 0 \\ 0 & s^2 \end{bmatrix} \begin{Bmatrix} v \\ \phi \end{Bmatrix} - \bar{V}^2 \begin{bmatrix} 0 & 1 \\ 0 & e \end{bmatrix} \begin{Bmatrix} v \\ \phi \end{Bmatrix} + \gamma \bar{V} \begin{bmatrix} 1 & 0 \\ 0 & e \end{bmatrix} \begin{Bmatrix} v' \\ \phi' \end{Bmatrix} + \begin{bmatrix} 1 & \delta r \\ \delta r & r^2 \end{bmatrix} \begin{Bmatrix} v'' \\ \phi'' \end{Bmatrix} = \begin{Bmatrix} 0 \\ 0 \end{Bmatrix},
$$

where the prime indicates differentiation with respect to normalized time $\tau = t\sqrt{K_B/M_B}$. This is a version of the MDoF model but the parameters have explicit physical meaning.

Model VIII. Euler

The Euler equations describe the rigid-body rotations of a block:

$$
I_{xx}\dot{\omega}_x + (I_{zz} - I_{yy})\omega_y\omega_z = T_x(t),
$$

$$
I_{yy}\dot{\omega}_y + (I_{xx} - I_{zz})\omega_z\omega_x = 0,
$$

$$
I_{zz}\dot{\omega}_z + (I_{yy} - I_{xx})\omega_x\omega_y = 0,
$$

where I_{xx}, I_{yy}, I_{zz} are the principal moments of inertia.

Model IX. Impact

These models are discussed in Section 7.6.

References

[1] **Achenbach, J. D.**, *Wave Propagation in Elastic Solids*, North-Holland, New York, 1973.

[2] **Allman, D. J.**, "On the general theory of the stability of equilibrium of discrete conservative systems," *Aeronautical Journal* **27**, 29–35, 1989.

[3] **Argyris, J. H. and Mlejnek, H.-P.**, *Dynamics of Structures*, Elsevier, Amsterdam, 1991.

[4] **Baker, G. L. and Gollub, J. P.**, *Chaotic Dynamics: An Introduction*, Cambridge University Press, Cambridge, 1990.

[5] **Bathe, K.-J.**, *Finite Element Procedures in Engineering Analysis*, Prentice-Hall, Englewood Cliffs, NJ, 1982, 2nd ed., 1996.

[6] **Batoz, J.-L., Bathe, K.-J., and Ho, L.-W.**, "A study of three-node triangular plate bending elements," *International Journal for Numerical Methods in Engineering* **15**, 1771–1812, 1980.

[7] **Belytschko, T.**, "An overview of semidiscretization and time integration procedures," in *Computational Methods for Transient Analysis*, T. Belytschko and T. J. R. Hughes, editors, pp. 1–65, Elsevier, Amsterdam, 1983.

[8] **Bergan, P. G. and Felippa, C. A.**, "A triangular membrane element with rotational degrees of freedom," *Computer Methods in Applied Mechanics and Engineering* **50**, 25–69, 1985.

[9] **Birkhoff, G., MacDougall, D. P., Pugh E. M., and Taylor, G.**, "Explosives with lined cavities," *Journal of Applied Physics* **19**, 563–582, 1948.

[10] **Bishop, R. E. D. and Johnson, D. C.**, *Mechanics of Vibrations*, Cambridge University Press, Cambridge, 1960.

[11] **Boresi, A. P. and Chong, K. P.**, *Elasticity in Engineering Mechanics*, Elsevier, New York, 1987.

[12] **Brigham, E. O.**, *The Fast Fourier Transform*, Prentice-Hall, Englewood Cliffs, NJ, 1973.

[13] **Burger, C. P., Oyinlola, A. K., and Scott, T. E.**, "Full-field strain distributions in hot rolled billets by simulation," *Transactions of the ASME, Met. Engineering Quarterly* **20**, 26–29, 1976.

[14] **Burger, C. P.**, "Nonlinear photomechanics," *Experimental Mechanics* **20**, 381–389, 1980.

[15] **Choi, S-W. and Doyle, J. F.**, "A computational framework for the separation of material systems," in *Recent Advances in Structural Dynamics*, University of Southampton, Southampton, UK, CD #23, 2003.

[16] **Clough, R. W. and Penzien, J.**, *Dynamics of Structures*, McGraw-Hill, New York, 1975.

[17] **Cole, R. H.**, *Underwater Explosions*, Princeton University Press, Princeton, NJ, 1948.

[18] **Cook, R. D., Malkus, D. S., and Plesha, M. E.**, *Concepts and Applications of Finite Element Analysis*, 3rd ed., Wiley, New York, 1989.

[19] **Crandall, S. H., Dahl, N. C., and Lardner, T. J.**, *An Introduction to the Mechanics of Solids*, Wiley, New York, 1972.

[20] **Crighton, D. G.**, "The 1988 Rayleigh Medal Lecture: Fluid loading – The interaction between sound and vibration," *Journal of Sound and Vibration* **133**(1), 1–27, 1989.

[21] **Crisfield, M. A.**, *Nonlinear Finite Element Analysis of Solids and Structures*, Vol. 1: *Essentials*, Wiley, New York, 1991.

[22] **Crisfield, M. A.**, *Nonlinear Finite Element Analysis of Solids and Structures*, Vol. 2: *Advanced Topics*, Wiley, New York, 1997.

[23] **Crowley, F. B., Phillips, J. W., and Taylor, C. E.**, "Pulse propagation in straight and curved beams: Theory and experiment," *Journal of Applied Mechanics* **39**, 1–6, 1972.

[24] **Dally, J. W. and Riley, W. F.**, *Experimental Stress Analysis*, 3rd ed., McGraw-Hill, New York, 1991.

[25] **Davendralingam, N. and Doyle, J. F.**, "Nonlinear identification problems under large deflections," *Experimental Mechanics* **48**, 529–538, 2008.

[26] **Dhondt, G.**, *The Finite Element Method for Three-Dimensional Thermomechanical Applications*, Wiley, Chichester, England, 2004.

[27] **Doyle, J. F. and Sun, C.-T.**, *Theory of Elasticity: An Introduction to Fundamental Principles and Methods of Analysis*, A&AE 553 Class Notes, Purdue University, 2006.

[28] **Doyle, J. F.**, *Wave Propagation in Structures*, Springer-Verlag, New York, 1989, 2nd ed., 1997.

[29] **Doyle, J. F.**, *Static and Dynamic Analysis of Structures*, Kluwer, Dordrecht, The Netherlands, 1991.

[30] **Doyle, J. F.**, *Nonlinear Analysis of Thin-Walled Structures: Statics, Dynamics, and Stability*, Springer-Verlag, New York, 2001.

[31] **Doyle, J. F.**, *Modern Experimental Stress Analysis: Completing the Solution of Partially Specified Problems*, Wiley, Chichester, England, 2004.

[32] **Doyle, J. F.**, *Nonlinear Mechanics: A Modern Introduction*, A&AE 690D Class Notes, Purdue University, 2006.

[33] **Doyle, J. F.**, *Structural Dynamics and Stability: A Modern Introduction*, A&AE 546 Class Notes, Purdue University, 2006.

[34] **Doyle, J. F.**, *Mechanics of Structures: A Modern Course on Principles and Methods of Solution*, A&AE 453 Class Notes, Purdue University, 2007.

[35] **Doyle, J. F.**, *Experimental Methods in the Mechanics of Solids & Structures: A Laboratory Guidebook*, A&AE 204L, 352L, 547 Class Notes, Purdue University, 2009.

[36] **Elmore, W. C. and Heald, M. A.**, *Physics of Waves*, Dover, New York, 1985.

[37] **Eriksson, A., Pacoste, C. and Zdunek, A.**, "Numerical analysis of complex instability behavior using incremental–iterative strategies," *Computer Methods in Applied Mechanics and Engineering* **179**, 265–305, 1999.

[38] **Eringen, A. C.**, *Nonlinear Theory of Continuous Media*, McGraw-Hill, New York, 1962.

[39] **Ewing, W. M. and Jardetzky, W. S.**, *Elastic Waves in Layered Media*, McGraw-Hill, New York, 1957.

[40] **Ewins, D. J.**, *Modal Testing: Theory and Practice*, Wiley, New York, 1984.

[41] **Fahy, F. J.**, *Sound and Structural Vibration: Radiation, Transmission and Response*, Academic, New York, 1985.

[42] **Freund, L. B.**, *Dynamic Fracture Mechanics*, Cambridge University Press, Cambridge, 1990.

[43] **Friswell, M. I. and Mottershead, J. E.**, *Finite Element Model Updating in Structural Dynamics*, Kluwer, Dordrecht, The Netherlands, 1995.

[44] **Gere, J. M.**, *Mechanics of Materials*, 5th ed., Brooks/Cole, Pacific Grove, CA, 2000.

[45] **Glendinning, P.**, *Stability, Instability and Chaos*, Cambridge University Press, Cambridge, 1994.

[46] **Goldsmith, W.**, *Impact*, Edward Arnold, London, 1960.

[47] **Gordon, J. E.**, *Structures, or Why Things Don't Fall Down*, Penguin, London, 1978.

[48] **Graff, K. F.**, *Wave Motion in Elastic Solids*, Ohio State University Press, Columbus, 1975.

[49] **Grandt, A. F.**, *Fundamentals of Structural Integrity: Damage Tolerance Design and Nondestructive Evaluation*, Wiley-IEEE, New York, 2004.

[50] **Gupta, A. D.**, "Dynamic elasto-plastic response of a generic vehicle floor model to coupled transient loads," in *14th U.S. Army Symposium on Solid Mechanics*, K.R. Iyer and S.-C. Chou, editors, pp. 507–520, Batelle Press, Columbus, OH, 1996.

[51] **Haines, D. W.**, "Approximate theories for wave propagation and vibrations in elastic rings and helical coils of small pitch," *International Journal of Solids and Structures* **10**, 1405–1416, 1974.

[52] **Hamilton, W. R.**, *The Mathematical Papers of Sir W.R. Hamilton*, Cambridge University Press, Cambridge, 1940.

[53] **Hibbeler, R. C.**, *Structural Analysis*, 2nd ed., Macmillan, New York, 1975, 1990.

[54] **Hsieh, D. Y. and Lee, J. P.**, "Experimental study of pulse propagation in curved elastic rods," *The Journal of the Acoustical Society of America* **54**, 1052–1055, 1973.

[55] **James, M. L., Smith, G. M., Wolford, J. C., and Whaley, P. W.**, *Vibration of Mechanical and Structural Systems*, Harper & Row, New York, 1989.

[56] **Johnson, K. L.**, *Contact Mechanics*, Cambridge University Press, Cambridge, 1985.

[57] **Johnson, W. and Mellor, P. B.**, *Plasticity for Mechanical Engineers*, Van Nostrand, London, 1962.

[58] **Jordon, D. W. and Smith, P.**, *Nonlinear Ordinary Differentia Equations*, 2nd ed., Clarendon, Oxford, 1987.

[59] **Junger, M. C. and Feit, D.**, *Sound, Structures, and Their Interaction*, MIT Press, Cambridge, MA, 1986.

[60] **Kachanov, L. M.**, *Fundamentals of the Theory of Plasticity*, Mir Publishers, Moscow, 1974.

[61] **Kane, T. R.**, "Reflection of flexural waves at the edge of a plate," *Journal of Applied Mechanics* **21–22**, 213–220, 1954.

[62] **Kanninen, M. F.**, "An augmented double cantilever beam model for studying crack propagation and arrest," *International Journal of Fracture* **9**, 83–92, 1973.

[63] **Kanninen, M. F.**, "A dynamic analysis of unstable crack propagation and arrest in the DCB test specimen," *International Journal of Fracture* **10**, 415–430, 1974.

[64] **Kawata, K. and Hashimoto, S.**, "On some differences between dynamic and static stress distributions," *Experimental Mechanics* **7**, 91–96, 1967.

[65] **Kleiber, M., Kotula, W., and Saran, M.**, "Numerical analysis of dynamic quasi-bifurcation," *Engineering Computations* **4**, 48–52, 1987.

[66] **Kobayashi, A. S. (Ed.)**, *Experimental Techniques in Fracture Mechanics*, Iowa State University Press, Ames, 1973.

[67] **Kolsky, H.**, *Stress Waves in Solids*, Dover, New York, 1963.

[68] **Kounadis, A. N. and Raftoyiannis, J.**, "Dynamic stability criteria of nonlinear elastic damped/undamped systems under step loading," *AIAA Journal* **28**, 1217–1223, 1990.

[69] **Kuske, A.**, "Photoelastic research on dynamic stresses," *Experimental Mechanics* **6**, 105–112, 1966.

[70] **Kuske, A.**, "Photoelastic stress analysis of machines under dynamic load," *Experimental Mechanics* **17**, 88–96, 1977.

[71] **Lamb, H.**, "On the propagation of tremors over the surfaces of an elastic solid," *Philosophical Transactions of the Royal Society of London* **A 203**, 1–42, 1904.

[72] **Lanczos, C.**, *The Variational Principles of Mechanics*, University of Toronto Press, Toronto, 1966.

[73] **Langhaar, H. L.**, *Dimensional Analysis and Theory of Models*, Wiley, New York, 1951.

[74] **Langhaar, H. L.**, *Energy Methods in Applied Mechanics*, Wiley, New York, 1962.

[75] **Lau, J. H.**, "Large deflections of beams with combined loads," *Engineering Mechanics ASCE* **108**, 180–185, 1982.

[76] **Leet, K. M.**, *Fundamentals of Structural Analysis*, Macmillan, New York, 1988, 2nd ed., 1990.

[77] **Leipholz, H.**, *Stability Theory, An Introduction to the Stability of Dynamic Systems and Rigid Bodies*, 2nd ed., Wiley and B. G. Teubner, Stuttgart, 1987.

[78] **Leissa, A. W.**, *Vibration of Plates*, NASA SP-160, 1969.

[79] **Leissa, A. W.**, *Vibration of Shells*, NASA SP-288, 1973.

[80] **Ljunggren, S.**, "Generation of waves in an elastic plate by a vertical force and by a moment in the vertical plane," *Journal of Sound and Vibration* **90**, 559–584, 1983.

[81] **Main, I. G.**, *Vibrations and Waves in Physics*, 3rd ed., Cambridge University Press, Cambridge, 1993.

[82] **Malvern, L. E.**, *Introduction to the Mechanics of a Continuous Medium*, Prentice-Hall, Englewood Cliffs, NJ, 1969.

[83] **Markus, S.**, *Mechanics of Vibrations of Cylindrical Shells*, Elsevier, New York, 1988.

[84] **McCoy, J. J.**, "Effects of non-propagating plate waves on dynamical stress concentrations," *International Journal of Solids and Structures* **4**, 355–370, 1968.

[85] **McCoy, R. W.**, *Dynamic Analysis of a Thick-Section Composite Cylinder Subjected to Underwater Blast Loading*, Ph.D. Dissertation, Purdue University, August, 1996.

[86] **Medick, M. A.**, "On classical plate theory and wave propagation," *Journal of Applied Mechanics* **28**, 223–228, 1961.

[87] **Megson, T. H. G.**, *Aircraft Structures*, Halsted Press, New York, 1990.

[88] **Meirovitch, L.**, *Elements of Vibration Analysis*, McGraw-Hill, 1986.

[89] **Milne-Thomson, L. M.**, *Theoretical Hydrodynamics*, 4th ed., Macmillan, New York, 1960.

[90] **Mottershead, J. E. and Friswell, M. I.**, "Model updating in structural dynamics: A survey," *Journal of Sound and Vibration* **162**, 347–375, 1993.

[91] **Norton, M. P.**, *Fundamentals of Noise and Vibration Analysis for Engineers*, Cambridge University Press, Cambridge, 1989.

[92] **Novozhilov, V. V.**, *Foundations of the Nonlinear Theory of Elasticity*, Graylock Press, Rochester, NY, 1953.

[93] **Oden, J. T.**, *Mechanics of Elastic Structures*, McGraw-Hill, New York, 1967.

[94] **Pacoste, C. and Eriksson, A.**, "Beam elements in instability problems," *Computer Methods in Applied Mechanics and Engineering* **144**, 163–197, 1997.

[95] **Pacoste, C.**, "Co-rotational flat facet triangular elements for shell instability analyses," *Computer Methods in Applied Mechanics and Engineering* **156**, 75–110, 1998.

[96] **Panovko, Y. G. and Gubanova, I. I.**, *Stability and Oscillations of Elastic Systems: Paradoxes, Fallacies, and New Concepts*, Consultants Bureau, New York, 1965.

[97] **Pao, Y-H. and Mow, C. C.**, *The Diffraction of Elastic Waves and Dynamic Stress Concentrations*, Crane, Russak, New York, 1973.

[98] **Pao, Y-H.**, "Dynamical stress concentration in an elastic plate," *Journal of Applied Mechanics* **29**, 299–305, 1962.

[99] **Peery, S. J. and Azar, J. J.**, *Aircraft Structures*, 2nd ed., McGraw-Hill, New York, 1982.

[100] **Pilkey, W. D.**, *Peterson's Stress Concentration Factors*, 2nd ed., Wiley, New York, 1997.

[101] **Press, W. H., Flannery, B. P., Teukolsky, S. A., and Vetterling, W. T.**, *Numerical Recipes*, Cambridge University Press, Cambridge, 1986, 2nd ed., 1992.

[102] **Przemieniecki, J. S.**, *Theory of Matrix Structural Analysis*, Dover, New York, 1985.

[103] **Redwood, M.**, *Mechanical Waveguides*, Pergamon, New York, 1960.

[104] **Riks, E., Rankin, C. C., and Brogan, F. A.**, "On the solution of mode jumping phenomena in thin-walled shell structures," *Computational Methods in Applied Mechanics and Engineering* **136**, 59–92, 1996.

[105] **Ripperger, E. A. and Abramson, H. N.**, "Reflection and transmission of elastic pulses in a bar at a discontinuity in cross section," in *Proceedings of the Third Midwestern Conference on Solid Mechanics*, University of Michigan, pp. 29–39, 1957.

[106] **Rooke, D. P. and Cartwright, D. J.**, *Compendium of Stress Intensity Factors*, Her Majesty's Stationery Office, London, 1976.

[107] **Roorda, J.**, "Some thoughts on the Southwell plot," *ASCE Journal of Engineering Mechanics* **93**, 37–48, 1967.

[108] **Rybicki, E. F. and Kanninen, M. F.**, "A finite element calculation of stress intensity factors by a modified crack-closure integral," *Engineering Fracture Mechanics* **9**, 931–938, 1977.

[109] **Schueller, J. K. and Wall, T. M. P.**, "Impact of fruit on flexible beams to sense modulus of elasticity," *Experimental Mechanics* **31**, 118–121, 1991.

[110] **Schwieger, H.**, "Central deflection of a transversely struck beam," *Experimental Mechanics* **10**, 166–169, 1970.

[111] **Seelig, J. M. and Hoppmann, W. H. II**, "Impact on an elastically connected double-beam system," *Journal of Applied Mechanics* **31**, 621–626, 1964.

[112] **Shames, I. H. and Dym, C. L.**, *Energy and Finite Element Methods in Structural Analysis*, Hemisphere, Washington, 1985.

[113] **Simitses, G. J.**, "Suddenly loaded structural configurations," *Journal of Engineering Mechanics ASCE* **110**, 1320–1334, 1984.

[114] **Singer, J., Arbocz, J., and Weller, T.**, *Buckling Experiments: Experimental Methods in Buckling of Thin-Walled Structures*, Vol. 1, Wiley, New York, 1998.

[115] **Soedel, W.**, *Vibrations of Shells and Plates*, Marcel Dekker, New York, 1981.

[116] **Speigel, M. R.**, *Fourier Analysis with Applications to Boundary Value Problems*, McGraw-Hill, New York, 1974.

[117] **Stein, M. L.**, *The Phenomenon of Change in Buckle Patern in Elastic Structures*, NASA Technical Report R-39, 1959.

[118] **Stricklin, J. A, Haisler, E. E., Tisdale, P. R., and Gunderson, R.**, "A rapidly converging triangular plate element," *AIAA Journal* **7**, 180–181, 1969.

[119] **Tedesco, J. W., McDougal, W. G., and Allen Ross, C.**, *Structural Dynamics: Theory and Applications*, Addison-Wesley, Menlo Park, CA, 1999.

[120] **Thompson, J. M. T. and Hunt, G. W.**, *A General Theory for Elastic Stability*, Wiley, London, 1973.

[121] **Thompson, J. M. T. and Hunt, G. W.**, *Elastic Stability*, Wiley, London, 1993.

[122] **Thomson, W. T.**, *Theory of Vibrations with Applications*, Prentice-Hall, Englewood Cliffs, NJ, 1981.

[123] **Timoshenko, S. P. and Gere, J. M.**, *Theory of Elastic Stability*, McGraw-Hill, New York, 1963.

[124] **Timoshenko, S. P. and Goodier, J. N.**, *Theory of Elasticity*, McGraw-Hill, New York, 1970.

[125] **Ulrich, K.**, *State of the Art in Numerical Methods for Continuation and Bifurcation Problems With Applications in Continuum Mechanics – A Survey and Comparative Study*, Laboratorio Nacional de Comutacao, Brazil, 1988.

[126] **Viktorov, I. A.**, *Rayleigh and Lamb Waves Physical Theory and Applications*, Plenum, New York, 1967.

[127] **Virgin, L. N.**, *Introduction to Experimental Nonlinear Dynamics*, Cambridge University Press, Cambridge, 2000.

[128] **Warnock, F. V. and Benham, P. P.**, *Mechanics of Solids and Strength of Materials*, Pitman Paperbacks, London, 1970.

[129] **Weaver, W. and Gere, J. M.**, *Matrix Analysis of Framed Structures*, Van Nostrand, New York, 1980.

[130] **Weaver, W. and Johnston, P. R.**, *Finite Elements for Structural Analysis*, Prentice-Hall, Englewood Cliffs, NJ, 1984.

[131] **Wright, D. V. and Bannister, R. C.**, "Plastic models for structural analysis. Part II: Experimental design," *Shock and Vibration Digest* **2**(12), 3–10, 1970.

[132] **Wright, D. V. and Bannister, R. C.**, "Plastic models for structural analysis. Part I: Testing types," *Shock and Vibration Digest* **2**(11), 2–10, 1970.

[133] **Xie, W.-C.**, *Dynamic Stability of Structures*, Cambridge University Press, Cambridge, 2006.

[134] **Yang, T. Y.**, *Finite Element Structural Analysis*, Prentice-Hall, Englewood Cliffs, NJ, 1986.

[135] **Young, D. F.**, "Similitude, modeling, and dimensional analysis," in *Handbook on Experimental Mechanics*, A. S. Kobayashi, editor, pp. 601–634, Society for Experimental Mechanics, Bethel, CT, 1993.

[136] **Yourgrau, W. and Mandelstam, S.**, *Variational Principles in Dynamics and Quantum Theory*, Dover, New York, 1979.

[137] **Zachary, L. W. and Burger, C. P.**, "Dynamic wave propagation in a single lap joint," *Experimental Mechanics* **20**, 162–166, 1980.

[138] **Zaveri, K.**, *Modal Analysis of Large Structures – Multiple Exciter Systems*, Bruel and Kjaer, Denmark, 1985.

[139] **Ziegler, H.**, *Principles of Structural Stability*, Ginn, Massachusetts, 1968.

Index